# CLASSICAL MECHANICS



S. N. BISWAS

CELEBRATING 55 YEARS 1960-2015

Since 1960

[ *For the Postgraduate (M.Sc.) students majoring in Physics* ]

# CLASSICAL MECHANICS

**S. N. BISWAS,** F.N.A., F.A.Sc., F.N.A.Sc.

*Professor of Physics (Retd.), University of Delhi, Delhi.*
*Formerly, TIFR, Bombay; JNU, New Delhi;*
*Visva-Bharati; UGC National Lecturer, 74;*
*Senior Associate Member, ICTP, Trieste, Italy, 76.*

**BOOKS AND ALLIED (P) Ltd.**

No.1-E(1) 'Shubham Plaza' (1st Floor)
83/1, Beliaghata Main Road, Kolkata 700010 (India)
***Tel*** : (033)6535-3844 • (033)6548-5530
***e-mail*** : books1960@ymail.com • booksandallied1960@gmail.com

# CLASSICAL MECHANICS

First Published : 1998
Reprinted : 1999
Reprinted (with Additions) : 2004
Reprinted : 2008, 2010, 2011, 2013, 2015
**Reprinted : 2016**



***Publisher*** : Arunabha Sen
BOOKS AND ALLIED (P) LTD.
8/1 Chintamoni Das Lane, Kolkata 700009

***Printer*** : EUREKA
91/A, Baithakkhana Road, Kolkata 700009

***Project Team*** : Salil Das and Dinesh Bhunya

**ISBN** 81-87134-18-6

***Price* : ₹ 325.00**

# Preface

This book on CLASSICAL MECHANICS is an outcome of the course of lectures I gave to the students of Physics at the University of Delhi for many years. The book deals with the mechanics of particles and of rigid bodies based on Lagrangian dynamics. The elegant formalism of Classical Mechanics through *Hamiltonian* and *Poisson Bracket* has been extended up to Hamilton-Jacobi and action angle variable theory so as to appreciate the development of the newer mechanics known as *Quantum Mechanics.* I have also attempted to include some recent developments in the subject motivated by some contemporary ideas in *Quantum Theory*. The book also contains a number of problems with solutions and some for exercises. I am very happy to thank the members of my family for their active co-operation and constant encouragement.

*1st February*
*Saraswati Puja 1998*

B7 Vasundhara Enclave
Ila Apartments Flat 6
Delhi 110 096

**S N BISWAS**

# Contents

Chapter I

# D'Alembert's Principle and Lagrangian Equation

## 1.1 Introduction

It is well known that Newtonian mechanics is extremely successful in describing the dynamics of macro-physical system in the non-relativistic domain, i.e., for system moving with velocities much less than the velocity of light. In the major part of the book we will concern ourselves with non-relativistic classical mechanics. We will consider here the dynamics of a point particle as well as the dynamics of a rigid body. The particle dynamics is devoted to the study of the motion of bodies treated as point object characterised by a position vector $\mathbf{X} = (x, y, z)$ and a parameter, $m$ representing the mass of the body. In Newtonian theory, the equation of motion of the body is given by

$$m\ddot{\mathbf{X}} = \mathbf{F}, \tag{1.1}$$

where $\mathbf{F}$ is the total force acting on the body. If $\mathbf{F}$ is known then the solution of (1.1) will, given initial conditions represent the trajectory of the system from which it is possible to extract all physical information of the system. Mechanics developed from (1.1) is known as *Vectorial Formulation* of Newtonian mechanics.

In the present text, we would like to develop the subject as *Scalar-Formulation* of the mechanics. In vectorial formulation, the basic ingredients are the vector quantities, acceleration $\ddot{\mathbf{X}}$ and the force $\mathbf{F}$ acting on the system. Alternatively, in the scalar formulation one can consider kinetic and potential energies which are scalar quantities in non-relativistic mechanics. The appropriate function is a scalar quantity and is called a *Lagrangian Function.* We will see below that Newtonian mechanics assumes an elegant structure extremely rich in contents and applications. The present form of the mechanics based on Lagrangian, Hamiltonian functions leads almost to the formulation of quantum mechanics valid for the descriptions of the micro-physical system, where classical mechanics becomes inadequate.

In this monograph, we would include a few of the modern developments in classical mechanics motivated by the present interesting developments in quantum mechanics. As for example, while discussing canonical transformations we introduce gauge transformations as canonical transformations in classical mechanics. There has also been a startling discovery in the phase of the wave function in quantum mechanics which is purely of topological origin. Analogously in classical mechanics, we have an additional piece in the angle of action angle variable theory which is also of geometrical origin. Finally, in the relativistic mechanics in addition to the mechanics of the continuous media we have included briefly relativistic Lagrangian of an extended object like *strings* which has been dominating high energy physics for a long time.

## 1.2 Co-ordinate System: Inertial Frame

At the begining of the last section we have mentioned that a point particle is characterised by the three co-ordinates $(x, y, z)$ in a three-dimensional space, and for equation (1.1) we also need to know its velocity $\dot{\mathbf{X}}$ for the determination of acceleration. We, therefore, must have a *Reference Co-ordinate System* with respect to which the measurement of position and velocity etc. can be performed. This reference co-ordinate system is taken, in such a way that its origin is fixed at a point in space with fixed directions forming an orthogonal triad, irrespective of the time. We call such a frame as *Inertial Frame* in which Newton's equation (1.1) is valid. Of course, we can take any other frame which is moving with constant velocity in any given direction, with respect to the fixed axes system. Such moving frames are also *Inertial* and Newton's equation has the same form as given in (1.1). We obviously cannot choose the system of axes fixed on the surface or at the centre of the earth as there is diarnal rotation of the earth around its axis and also the annual rotation of the earth round the sun. Ideal choice should be the centre of the universe which is extremely difficult to find. A non-rotating frame (axes pointing always towards same fixed star) with origin fixed at the centre of the earth is more nearly inertial. Non-rotating axes with origin fixed at the centre of the sun constitutes an excellent (though not perfect) inertial frame. It must be emphasised that the above discussion does not imply that non-inertial frame cannot be used. Newton's equation can be written in a non-inertial co-ordinate system. In treating dynamical problems associated with some physical system, the first step is to find a suitable co-ordinate in terms of which the problem may become solvable. In ordinary configuration space three variables $x, y$ and $z$ are assigned to a given mass point situated at a point $P$, say. Sometimes, it is useful to represent these configuration space co-ordinates in terms of polar variables, namely, $x = r\sin\theta\cos\phi$, $y = r\sin\theta\sin\phi$ and $z = r\cos\theta$, $0 \le \theta \le \pi$ and $0 \le \phi \le 2\pi$. Cylindrical co-ordinates, where $x = \rho\cos\phi$, $y = \rho\sin\phi$ and

$z = z$ are found useful for the description of the dynamics of a mass point. There are a number of other system of variables which can be profitably used. We can mention parabolic cylindrical co-ordinates given by

$$x = (\xi\eta); \quad y = \frac{1}{2}\left(\xi^2 - \eta^2\right) \quad \text{and} \quad z = z.$$

This is the most commonly used co-ordinate system for the solution of dynamical equation. More complex co-ordinate system would have to be used depending on the complexity of problems.

## 1.3 Constraints

Complexity in the choice of co-ordinate system arises when there are constraints on the co-ordinates used to represent the problem.
Constraints are expressed in terms of some algebraic relation among some or all of these co-ordinates. There are generally two types of contraints: ($a$) Geometrical. ($b$) Kinematical. An example of geometrical constraint can be given as follows.
Consider three mass points in a plane designated by cartesian co-ordinates. $(x_1, y_1)$, $(x_2, y_2)$ and $(x_3, y_3)$. The system is moving in such a way that the distances

$$\begin{aligned} &(x_1 - x_2)^2 + (y_1 - y_2)^2, \\ &(x_2 - x_3)^2 + (y_2 - y_3)^2 \\ \text{and } &(x_3 - x_1)^2 + (y_3 - y_1)^2 \end{aligned}$$

are always constant. There are 6-co-ordinates related by three-algebraic relations. Thus, the number of independent set is three. For Kinematical constraint, consider motion of a rod in a plane whose centre of mass is moving along the direction of the rod. Let $(x_1, y_1)$ and $(x_2, y_2)$ are the co-ordinates of the two end points of the rod. The centre of mass co-ordinate is

$$\frac{(x_1 + x_2)}{2} \quad \text{and} \quad \frac{(y_1 + y_2)}{2}.$$

Then from the problem

$$\left.\begin{aligned} (1/2)(\dot{x}_1 + \dot{x}_2) = \rho(x_1 - x_2) \\ (1/2)(\dot{y}_1 + \dot{y}_2) = \rho(y_1 - y_2) \end{aligned}\right\}. \tag{1.2}$$

where $\rho$ is a constant.
Equations (1.2) is an example of a Kinematic constraint.

If a constraint can be expressed as an algebraic equality it is called *holonomic*. For example, consider a plane pendulum, the bob of which has the co-ordinate $(x, y)$ and it is suspended from the origin as in the Fig. 1.1. Then, as the length of the pendulum is fixed, we have

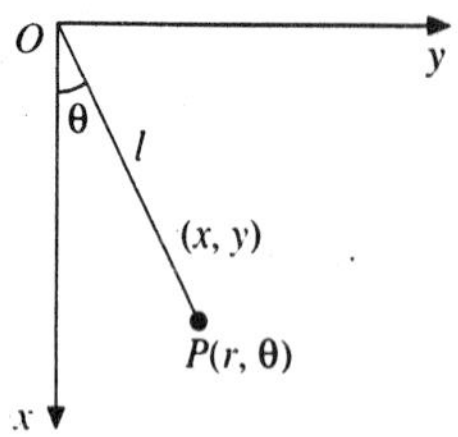

Fig 1.1

$$x^2 + y^2 = l^2.$$

This algebraic constraint is in the form of an equality, hence the constraint is a holonomic one. If a constraint relation obeys inequality then we call it non-holonomic. As an example consider a mass point which is constrained to move inside a sphere. The co-ordinate $(x, y, z)$ of the mass point obeys the constraint equation

$$\sqrt{(x^2 + y^2 + z^2)} < a, \tag{1.3}$$

when $a$ is the radius of the sphere. Equation (1.3) is a non-holonomic constraint. These equations need not be given only in terms of algebraic relations. They can also be given in terms of differential equation as in the example in equation (1.2). If the differential equation is non-integrable then the constraint is non-holonomic. Equation (1.2) is a non-holonomic one.

Sometimes we may have a time-dependent constraint: consider the motion of a particle in a $xy$-plane along a rod through origin which is rotating with constant angular velocity $\omega$. Then the polar angle $\theta = \omega t$; the corresponding co-ordinates are

$$\left.\begin{aligned} x &= r\cos\omega t \\ y &= r\sin\omega t \end{aligned}\right\}. \tag{1.4}$$

Time-dependent constraint in which the parameter time $t$, occurs explicitly is called *Rheonomous*. When the constraint is time-independent we call then *Seleronomous.*

Constraints in general produce two types of difficulties in solving dynamical problems. First, the co-ordinates $x, y, z$ are no longer independent, since there exists algebraic relation between them. As a consequence the resulting equation of motions are no longer independent. Second, the forces of constraint, introduce unknown element in the equations of motion. For example, when a bead is constrained to move on a wire of a given shape, the force exerted on the bead by wire is an unknown element in the equation and must be obtained from the solutions we seek.

Indeed imposing a constraint in a problem is simply another way of stating that there are other forces present in the problem that cannot be specified directly, but are known in terms of their effect on the motion of the system. In general, forces are required to maintain the constraint in a system. As mentioned, these forces are not known initially. A knowledge of their values is not often important and they are normally eliminated from the equation of

motion at an early stage, but methods can be devised for their determination if required. If the conditions of the constraints are timse-independent the associated forces of constraints, although not necessarily constant, do no work in the course of the motion. Such constraints are usually called *workless*. Time-dependent constraints are not *workless*.
Constraints are idealised concepts introduced to assist in the solution of mechanical problems. Generally, they represent highly simplified forms of complex systems. This is the case with the problem of the particle moving over a horizontal plane. In the real case, the plane would be the surface of an elastic solid and would be slightly deformed under the weight of the object. In the process of idealisation to a geometrical plane the continuous real forces of reaction are replaced by discontinuous constraint forces. Such discontinuous forces are not envisaged in Newtonian mechanics and necessitates a new postulate before they can be included in a conventional mathematical scheme.
The first difficultly, mentioned above can be solved by the generalised co-ordinates as shown by a few examples of a holonomic system.

## 1.4 Generalised Co-ordinates

Consider the case of the simple plane pendulum. The bob $P$ has the co-ordinate $(x, y)$. In terms of polar variable $(r, \theta)$ (Fig. 1.1).

$$\left.\begin{aligned} x = r\cos\theta = f_1(r,\theta) \\ y = r\sin\theta = f_2(r,\theta) \end{aligned}\right\}. \tag{1.5}$$

But the length of the pendulum is fixed and equal to $l$ a constant. Then

$$r = l = \text{ constant.}$$

Therefore,

$$x = f_1(\theta) \quad \text{and} \quad y = f_2(\theta).$$

After using the constraint $x^2 + y^2 = l^2$, the co-ordinates $x$ and $y$ are given by a single quantity $\theta$. Then $\theta$ is the generalised co-ordinate for the description of motion of the pendulum.
In a similar way, consider a particle which is constrained to move on the surface of an inverted cone (Fig. 1.2).
Here we have

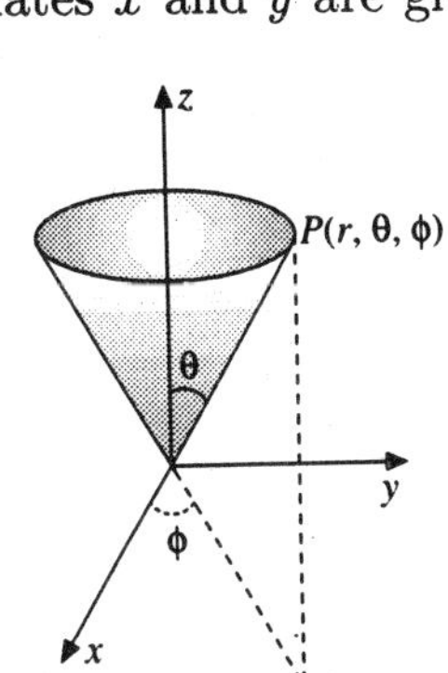

Fig 1.2

$$x = r\sin\theta\cos\phi = f_1(r,\theta,\phi)$$
$$y = r\sin\theta\sin\phi = f_2(r,\theta,\phi)$$
$$z = r\cos\theta = f_3(r,\theta,\phi).$$

Since the point moves on the surface of the cone the angle $\theta$ is fixed. Then.

$$\begin{aligned} x &= f_1(r,\phi), \\ y &= f_2(r,\phi) \\ \text{and} \quad z &= f_3(r,\phi). \end{aligned}$$

Given $r$ and $\phi$, then any point $P$ with $(x,y,z)$ is determined. Here $r$ and $\phi$ are the two generalised co-ordinates. Consider in general, a system of $N$ point masses each with position co-ordinates $\mathbf{r}_i$ when $\mathbf{r}_i$ stands for the $i$th particle. Suppose, there are $K$-relations among the $3N$ position co-ordinates. Therefore, $3N - K$ is the number of independent co-ordinates which we call as generalised co-ordinates. If we denote these by $q_1, q_2, \cdots, q_n$ where $n = 3N - K$, then,

$$\mathbf{r}_1 = \mathbf{f}_1\left(q_1, \cdots, q_n\right)$$
$$\cdots\cdots\cdots\cdots\cdots\cdots\cdots$$
$$\mathbf{r}_i = \mathbf{f}_i\left(q_1, q_2, \cdots, q_n\right).$$

Here $q_1, q_2$, etc. play the same role as $\theta$ and $r, \phi$ of the above two examples. If the constraint were time dependent then $\mathbf{f}_1, \mathbf{f}_2, \cdots$, etc. also contain $t$, the time parameter occurs explicitly. Writing the same as $\mathbf{r}_i$ for $\mathbf{f}_i$, etc. we have

$$\mathbf{r}_1 = \mathbf{r}_1\left(q_1, q_2, \cdots, q_n; t\right), \text{ etc.}$$

For the general $\mathbf{r}_i$, we have

$$\mathbf{r}_i = \mathbf{r}_i\left(q_1, q_2, \cdots, q_n; t\right), \quad i = 1, 2, \cdots, N \text{ and } n = 3N - K. \tag{1.6}$$

Equation (1.6) is the most general expression relating the configuration space co-ordinates $\mathbf{r}_i$ in terms of the generalised co-ordinates $q_1, q_2, \cdots, q_n$.

## 1.5 Generalised Velocity

We obtain an expression for the velocity of a particle in terms of generalised co-ordinates. We define velocity by

$$\lim_{\Delta t \to 0} \frac{\Delta \mathbf{s}}{\Delta t},$$

where $\Delta\mathbf{s}$ is an infinitesimal arc of a curve covered by the body in time $\Delta t$. We call

$$\lim_{\Delta t \to 0} \frac{\Delta \mathbf{s}}{\Delta t} = \dot{\mathbf{s}} \tag{1.7}$$

where the velocity $\dot{\mathbf{s}}$ is along the tangent to the curve at a particular point $P$ on the curve at any given instant. In configuration space $\dot{\mathbf{s}}$ is given by

$$\dot{\mathbf{s}} = \frac{ds}{dt} = \left(\frac{dX}{dt}, \frac{dY}{dt}, \frac{dZ}{dt}\right).$$

The magnitude of the velocity $\mathbf{v} = (d\mathbf{s}/dt)$ is

$$v^2 = \left(\frac{dX}{dt}\right)^2 + \left(\frac{dY}{dt}\right)^2 + \left(\frac{dZ}{dt}\right)^2. \tag{1.8}$$

If the parametric representation of a position is given by spherical polar variables.

$$\begin{aligned} x &= r\sin\theta\cos\phi, \\ y &= r\sin\theta\sin\phi, \\ z &= r\cos\theta, \end{aligned}$$

then,

$$d\mathbf{s} = (dr, r\,d\theta, r\sin\theta\,d\phi)$$

or the velocity

$$\left.\begin{aligned} \mathbf{v} &= \frac{d\mathbf{s}}{dt} = \left(\frac{dr}{dt}, r\frac{d\theta}{dt}, r\sin\theta\frac{d\phi}{dt}\right) \\ \text{and } v^2 &= (\dot{r})^2 + r^2\left(\dot{\theta}\right)^2 + r^2\sin^2\theta\left(\dot{\phi}\right)^2 \end{aligned}\right\}. \tag{1.9}$$

In case, the particle is constrained to more on the surface of a sphere of constant radius $a$, then

$$\dot{r} = \frac{dr}{dt} = 0, \quad \text{and} \quad \mathbf{v} = \left(0, a\dot{\theta}, a\sin\theta\dot{\phi}\right). \tag{1.10}$$

The quantities $\dot{\theta}$ and $\dot{\phi}$ are the generalised velocities and $\theta, \phi$ are the generalised co-ordinate. In the absence of any constraints the three co-ordinates $x, y$ and $z$ are given in terms of three parameters $r, \theta$ and $\phi$. The expression

$$\begin{aligned} x &= r\sin\theta\cos\phi, \\ y &- r\sin\theta\sin\phi \\ \text{and} \quad z &= r\cos\theta \end{aligned}$$

are called the *parametric representation of the configuration space co-ordinates*. The quantities $\dot{r}, \dot{\theta}$ and $\dot{\phi}$ are the velocities along $r, \theta$ and $\varphi$ directions. Consider $N$-particle system; the $i$th particle of which has the co-ordinate $\mathbf{r}_i$. Expressing $\mathbf{r}_i$ in terms of generalised co-ordinates $q_1, q_2, \cdots, q_n$ we have,

$$\mathbf{r}_i = \mathbf{r}_i(q_1, \cdots, q_n; t)$$

then

$$\delta\mathbf{r}_i = \sum_j \frac{\partial\mathbf{r}_i}{\partial q_j}\cdot\delta q_j + \frac{\partial\mathbf{r}_i}{\partial t}\cdot\delta t. \tag{1.11}$$

Hence,

$$\dot{\mathbf{r}}_i = \frac{d\mathbf{r}_i}{dt} = \sum_{j=1}^{n} \frac{\partial \mathbf{r}_i}{\partial q_j}\frac{dq_j}{dt} + \frac{\partial \mathbf{r}_i}{\partial t}$$

$$\text{or, } \dot{\mathbf{r}}_i = \text{velocity} = \sum_j \frac{\partial \mathbf{r}_i}{\partial q_j}\dot{q}_j + \frac{\partial \mathbf{r}_i}{\partial t}. \tag{1.12}$$

Thus,

$$\mathbf{v}_i \equiv \dot{\mathbf{r}}_i = \sum_{j=1}^{n} a_{ij}\dot{q}_j + \bar{\alpha}_i, \tag{1.13}$$

where

$$a_{ij} = a_{ij}(q_1, \cdots, q_n, t) \equiv \frac{\partial \mathbf{r}_i}{\partial q_j} \tag{1.14}$$

and $\bar{\alpha}_i = (\partial \mathbf{r}_i/\partial t)$.
If $t$ does not occur explicitly then

$$\bar{\alpha}_i = 0 \tag{1.15}$$

and velocity expression becomes

$$\mathbf{v}_i = \sum_j a_{ij}\dot{q}_j. \tag{1.16}$$

This is a matrix-equation which can be inverted if

$$\det|a_{ij}| \neq 0 \quad .$$

and we can express

$$\dot{q}_i = \sum_j A_{ij}\mathbf{v}_j. \tag{1.17}$$

$A_{ij} \equiv (a_{ij})^{-1}$; $\dot{q}_i$ are called *generalised velocity.* In the next section, we consider an expression for *Generalised Force.*

## 1.6 Generalised Force

Consider a $N$-particle system with $i$th particle having the position co-ordinate $\mathbf{r}_i$ there are $k$-number of holonomic constraint then $3N - k = n$ is the number of generalised co-ordinates $q_1, \cdots, q_n$. Then

$$\mathbf{r}_i = \mathbf{r}_i(q_1, \cdots, q_n). \tag{1.18}$$

We will be considering conservative system, so that the forces do not dependent on time explicitly. As a result, we have

$$\mathbf{r}_i = \mathbf{r}_i(q_1, \cdots, q_n).$$

Let $F^{(i)}(\mathbf{r}_i)$ be a force acting on the $i$th particle of the system of $N$-particles. Then, the work done by the force for an arbitrary displacement $\delta\mathbf{r}_i$ is

$$\mathbf{F}^{(i)}\cdot\delta\mathbf{r}_i. \tag{1.19}$$

Therefore, the total work done by all the force is

$$\sum_i \mathbf{F}^{(i)}(\mathbf{r}_i)\cdot\delta\mathbf{r}_i. \tag{1.20}$$

But

$$\mathbf{r}_i = \mathbf{r}_i(q_1,\cdots,q_n),$$

so $\delta\mathbf{r}_i$ is then

$$\delta\mathbf{r}_i = \sum_{j=1}^{n}\frac{\partial\mathbf{r}_i}{\partial q_j}\delta q_j. \tag{1.21}$$

Substituting (1.21) in (1.20) we get the work done $\delta W$ as

$$\delta W = \sum_{i=1}^{N}\mathbf{F}^{(i)}(\mathbf{r}_i)\sum_{j=1}^{n}\frac{\partial\mathbf{r}_i}{\partial q_j}\delta q_j = \sum_j\left[\delta q_j\left(\sum_{i=1}^{N}\mathbf{F}^{(i)}\frac{\partial\mathbf{r}_i}{\partial q_j}\right)\right].$$

We call

$$Q_j = \sum_{i=1}^{N}F^{(i)}\frac{\partial\mathbf{r}_i}{\partial q_j}. \tag{1.22}$$

Then

$$\delta W = \sum_j Q_j\delta q_j. \tag{1.23}$$

Thus, the quantity $Q_j$ given by (1.22) is called *Generalised Force*.
The expression $\sum_j Q_j\delta q_j$ denote the work done by all the generalised force $Q_1$'s in the arbitrary displacements $\delta q_j$'s.
We will make use of the generalised force to obtain Lagrange's equation of motion using D'Alembert's principle.

## 1.7 Expression for Kinetic Energy using Generalised Velocity

Consider $N$-particle. The $i$th particle has co-ordinate $\mathbf{r}_i$. Removing the holonomic constraints we introduce the generalised co-ordinate $q_1,\cdots,q_n$. Then,

$$\mathbf{r}_i = \mathbf{r}_i(q_1,\cdots,q_n;t). \tag{1.24}$$

We assume constraints are time-dependent. Then

$$\dot{\mathbf{r}}_i = \sum_{j=1}^{n} \frac{\partial \mathbf{r}_i}{\partial q_j}\dot{q}_j + \frac{\partial \mathbf{r}_i}{\partial t} \tag{1.25}$$

$$= \sum_{j=1}^{n} a_{ij}\dot{q}_j + \bar{\alpha}_i. \tag{1.26}$$

where $a_{ij} = (\partial \mathbf{r}_i/\partial q_j)$ and $\bar{\alpha}_i = (\partial \mathbf{r}_i/\partial t)$.

The kinetic energy $T$ is

$$T = \frac{1}{2}\sum_i m_i \dot{\mathbf{r}}_i^2 \tag{1.27}$$

for $N$-particle systems with $m_i$ being the mass of the $i$th particle. Using (1.26) in (1.27) we get for $T$.

$$T = \frac{1}{2}\sum_i m_i \left(\sum_j a_{ij}\dot{q}_j + \alpha_i\right)^2. \tag{1.28}$$

Expanding the right hand side of (1.28) we get

$$\frac{1}{2}\sum_i m_i \left[\sum_{j=1}^{n} a_{ij}\dot{q}_j \sum_{k=1}^{n} a_{ik}\dot{q}_k + \bar{\alpha}_i \left(\sum_{j=1}^{n} a_{ij}\dot{q}_j + \sum_{k=1}^{n} a_{ik}\dot{q}_k\right) + \alpha_i^2\right] \tag{1.29}$$

we call

$$C = \frac{1}{2}\sum_j m_j \alpha_j^2.$$

The first term of (1.29) can be put as

$$\sum_{j=1}^{n}\sum_{k=1}^{n}\left(\frac{1}{2}\sum_i m_i a_{ij} a_{ik}\right)\dot{q}_j\dot{q}_k = \sum_j\sum_k A_{jk}\dot{q}_j\dot{q}_k \tag{1.30}$$

we note

$$A_{jk} = A_{kj}. \tag{1.31}$$

The second term becomes

$$\sum_j \left(\frac{1}{2}\sum_i \alpha_i m_i a_{ij}\right)\dot{q}_j + \sum_k \left(\frac{1}{2}\sum_i \alpha_i m_i a_{ik}\right)\dot{q}_k$$

$$= \sum_j A_j\dot{q}_j + \sum_k A_k\dot{q}_k = 2\sum_j A_j\dot{q}_j. \tag{1.32}$$

Collecting all these results.

$$T = \sum_j\sum_k A_{jk}\dot{q}_j\dot{q}_k + 2\sum_j A_j\dot{q}_j + C. \tag{1.33}$$

This is a quadratic expression for $T$ in terms of $\dot{q}_k, \dot{q}_j$. If the constraints were time independent, then $a_j = 0$ and $T$ becomes a homogeneous function of second-degree in $\dot{q}_j$'s.
From (1.33) we find

$$T = \sum_j \sum_k A_{jk} \dot{q}_j \dot{q}_k. \tag{1.34}$$

where

$$A_{jk} = A_{kj}. \tag{1.35}$$

From (1.33) and (1.34),

$$\sum_r \dot{q}_r \frac{\partial T}{\partial \dot{q}_r} = 2T. \tag{1.36}$$

## 1.8 Virtual Displacement; Virtual Work; Forces of Constraint

Any real displacement $\delta \mathbf{r}$ is always associated with a change in time $\delta t$. But a displacement $\delta \mathbf{r}$ is said to be *virtual*, if the displacement takes place without any change in time, i.e., the displacement takes place *instantaneously*. Consider a system of $N$-particles with $\mathbf{r}_i$ as the position vector for the $i$th particle. Therefore, for holonomic time-dependent constraints

$$\mathbf{r}_i = \mathbf{r}_i (q_1, \cdots, q_n; t). \tag{1.37}$$

where $q_1, \cdots, q_n$ are the generalised co-ordinates. If $\delta \mathbf{r}_i$ is a virtual displacement then

$$\delta \mathbf{r}_i = \sum_j \frac{\partial \mathbf{r}_i}{\partial q_j} \delta q_j. \tag{1.38}$$

But for a *real* displacement

$$\delta \mathbf{r}_i = \sum_j \frac{\partial \mathbf{r}_i}{\partial q_j} \delta q_j + \frac{\partial \mathbf{r}_i}{\partial t} \delta t. \tag{1.39}$$

From (1.39) we also get

$$\dot{\mathbf{r}}_i = \frac{d\mathbf{r}_i}{dt} = \sum_j \frac{\partial \mathbf{r}_i}{\partial q_j} \dot{q}_j + \frac{\partial \mathbf{r}_i}{\partial t}. \tag{1.40}$$

From (1.40) we find an important relation

$$\frac{\partial \dot{\mathbf{r}}_i}{\partial \dot{q}_j} = \frac{\partial \mathbf{r}_i}{\partial q_j}. \tag{1.41}$$

Let $F^{(i)}(\mathbf{r}_i)$ be the total force acting on the $i$th particle at the position $\mathbf{r}_i$. The $i$th particle is given an arbitrary virtual displacement $\delta\mathbf{r}_i$ then the work done by $F^{(i)}$ on the $i$th particle is

$$\delta W = F^{(i)}(\mathbf{r}_i) \cdot \delta\mathbf{r}_i \tag{1.42}$$

$\delta W$ given by (1.42), where $\delta\mathbf{r}_i$ is a virtual displacement is called *virtual work*. In a physical system, when the system is constrained to move in a particular way, there are always one or more *forces of constraint.* It is an experience of facts that virtual work done by *forces of constraint* is always zero. Although no direct proof is available one can convince oneself by taking specific examples to show this.

## 1.9 Principle of Virtual Work

Consider a system of $N$-particles, the $i$th particle having the configuration-space co-ordinates $\mathbf{r}_i$. Let there be time-dependent holonomic constraint. If $q_1, \cdots, q_n$ are the generalised co-ordinates, then the virtual displacement $\delta\mathbf{r}_i$ is

$$\sum_j \frac{\partial\mathbf{r}_i}{\partial q_j} \cdot \delta q_j;$$

as was defined in the last section. Suppose that under the action of all external applied and forces of constraints, the system is in equilibrium. Let $F^{(i)}(\mathbf{r}_i)$ be the total force acting on the $i$th particle of the system. As the system is in equilibrium, then

$$F^{(i)}(\mathbf{r}_i) = 0.$$

Therefore,

$$F^{(i)} \cdot \delta\mathbf{r}_i = 0,$$

where $\delta\mathbf{r}_i$ is a virtual displacement. Since the whole system is in equilibrium,

$$\sum_i F^{(i)} \cdot \delta\mathbf{r}_i = 0. \tag{1.43}$$

But

$$F^{(i)} = F^{(i)\,app} + F^{(i)c};\ F^{(i)\,app}$$

is the external applied force and $F^{(i)c}$ is the forces of constraints. Then equation (1.43) reduces to

$$\sum_i F^{(i)\,app} \cdot \delta\mathbf{r}_i + \sum_i F^{(i)c}\delta\mathbf{r}_i = 0.$$

But forces of constraint do not perform any virtual work, so

$$\sum_i F^{(i)c}\delta\mathbf{r}_i = 0.$$

We then have

$$\sum_i \mathbf{F}^{(i)\,app}.\,\delta\mathbf{r}_i = 0. \tag{1.44}$$

Thus equation (1.44) implies that if a system of $N$-particles are in equilibrium under the action of both applied and forces of constraints. then the total virtual work done by the external applied forces on the system is zero. i.e..

$$\sum_i F^{(i)\,app}.\,\delta\mathbf{r}_i = 0$$

where $\delta\mathbf{r}_i$ is a virtual displacement and

$$\sum_i F^{(i)\,app}.\,\delta\mathbf{r}_i$$

is the total virtual work done by only the applied forces on the system. This is known as *principle of virtual work*. As such. the principle is applicable to problems in *statics*. We should note that althrough $\delta\mathbf{r}_i$ an arbitrary variation

$$\sum_i F^{(i)\,app}.\,\delta\mathbf{r}_i = 0$$

does not mean $F^{(i)\,app} = 0$ as the co-ordinate $\mathbf{r}_i$'s are not independent but releated by constraints.

## 1.10 D'Alembert's Principle

In the last section. we have introduced the principle of virtual work which applies to a system in equilibrium. i.e.. suitable to the problems in statics. We want to extend the principle valid for problems in dynamics.
Consider $N$-particle system. the $i$th particle has mass $m_i$ and co-ordinate vector $\mathbf{r}_i, i = 1, 2, \cdots, N$. Then $m_i\dot{\mathbf{r}}_i$ is the momentum of the $i$th particle. We call $\mathbf{p}_i = m_i\dot{\mathbf{r}}_i$. Let $\mathbf{F}^{(i)}(\mathbf{r}_i)$ be the total force acting on the $i$th particle. Then by Newton's equation of motion we have

$$\frac{d}{dt}\mathbf{p}_i = \mathbf{F}^{(i)} \tag{1.45}$$

$$\text{or.}\quad m_i\ddot{\mathbf{r}}_i = \mathbf{F}^{(i)}. \tag{1.46}$$

Due to the application of the total force $\mathbf{F}^{(i)}$ the $i$th particle develops a motion with acceleration $\ddot{\mathbf{r}}_i$ and as a result *an effective force* $= m_i\ddot{\mathbf{r}}_i$ is generated in the system. Equation (1.46) merely states that total external forces acting on $i$th particle is equal to the effective force generated on the $i$th particle. We rewrite (1.46) as

$$\left(-m_i\ddot{\mathbf{r}}_i + F^{(i)}\right) = 0. \tag{1.47}$$

Equation (1.47) means that the $i$th particle under the combined applications of the *reversed effective force* and the total external force $\mathbf{F}^{(i)}$ is in equilibrium. Looking this way the dynamical problem reduces to a problem in Statics. We now apply the principle of virtual work to the dynamical $N$-particle system. Since, the $N$-particle system is in equilibrium under the combined effect of reversed effective force and the external forces, the virtual work done by these forces must be zero. That is

$$\sum_i \left(-m_i\ddot{\mathbf{r}}_i + \mathbf{F}^{(i)}\right)\cdot\delta\mathbf{r}_i = 0. \tag{1.48}$$

Here

$$\mathbf{F}^{(i)} = \mathbf{F}^{(i)app} + f^c$$

when $\mathbf{F}^{(i)app}$ is the external applied force and $f^c$ is the force of constraint. Since forces of constraint do not perform any virtual work equation (1.48) reduces to

$$\sum_i \left(-\dot{\mathbf{p}}_i + \mathbf{F}^{(i)app}\right)\cdot\delta\mathbf{r}_i = 0. \tag{1.49}$$

where

$$\dot{\mathbf{p}}_i = m_i\dot{\mathbf{r}}'_i \tag{1.50}$$

and $\delta\mathbf{r}_i$ is a virtual displacement.

From equation (1.49) we see that forces of constraint disappear from the problem. Equation (1.49) is the content of the D'Alembert's principle.

We will use equation (1.49) to derive Lagrange's equation for a $N$-particle system.

## 1.11 Lagrange's Equation of Motion

We now use D'Alembert's principle to obtain Lagrange's equation of motion. Consider $N$-particle system. For $i$th particle of mass $m_i$ and co-ordinate $\mathbf{r}_i$ we write

$$\mathbf{r}_i = \mathbf{r}_i\left(q_1, \cdots, q_n; t\right) \tag{1.51}$$

where $q_1, \cdots, q_n$ are the generalised co-ordinates. For virtual displacement

$$\delta\mathbf{r}_i = \sum_{j=1}^{n}\frac{\partial\mathbf{r}_i}{\partial q_j}\delta q_j. \tag{1.52}$$

For real $\dot{\mathbf{r}}'_i$ we get

$$\dot{\mathbf{r}}_i = \sum_{j=1}^{n}\frac{\partial r_i}{\partial q_j}\dot{q}_j + \frac{\partial r_i}{\partial t} \tag{1.53}$$

and from (1.53) we also get

$$\frac{\partial\mathbf{r}_i}{\partial q_j} = \frac{\partial\dot{\mathbf{r}}_i}{\partial\dot{q}_j}. \tag{1.54}$$

Note that kinetic energy

$$T = \sum_i \frac{1}{2} m_i \dot{\mathbf{r}}_i^2.$$

so

$$\frac{\partial T}{\partial \dot{q}_j} = \sum_i m_i \dot{\mathbf{r}}_i \frac{\partial \dot{\mathbf{r}}_i}{\partial \dot{q}_j}. \tag{1.55}$$

From

$$\frac{d}{dt}(\mathbf{r}_i) = \sum_k \frac{\partial \mathbf{r}_i}{\partial q_k} \dot{q}_k + \frac{\partial r_i}{\partial t}$$

we get

$$\begin{aligned}
\frac{d}{dt}\left(\frac{\partial \mathbf{r}_i}{\partial q_j}\right) &= \sum_k \frac{\partial}{\partial q_k}\left(\frac{\partial \mathbf{r}_i}{\partial q_j}\right)\dot{q}_k + \frac{\partial}{\partial t}\left(\frac{\partial r_i}{\partial q_j}\right) \\
&= \sum_{k=1}^{n} \frac{\partial}{\partial q_j}\left(\frac{\partial \mathbf{r}_i}{\partial q_k}\dot{q}_k\right) + \frac{\partial}{\partial q_j}\left(\frac{\partial r_i}{\partial t}\right) \\
&= \frac{\partial}{\partial q_j}\left(\sum_{k=1}^{n} \frac{\partial \mathbf{r}_i}{\partial q_k}\dot{q}_k + \frac{\partial \mathbf{r}_i}{\partial t}\right) = \frac{\partial}{\partial q_j}\dot{\mathbf{r}}_i.
\end{aligned}$$

$$\text{or,} \quad \frac{d}{dt}\left(\frac{\partial r_i}{\partial q_j}\right) = \frac{\partial}{\partial q_j}\left(\frac{d}{dt}\mathbf{r}_i\right) = \frac{\partial}{\partial q_j}\dot{\mathbf{r}}_i. \tag{1.56}$$

Consider now the virtual work

$$\left(-m_i \ddot{\mathbf{r}}_i + F^{(i)app}\right) \cdot \delta \mathbf{r}_i = 0.$$

Since we have only external applied forces we will write

$$\mathbf{F}^{(i)} \text{ for } F^{(i)app}.$$

Hence $F^{(i)app}\delta r_i$ term becomes ($\delta \mathbf{r}_i$ is virtual)

$$\begin{aligned}
\sum_i F^{(i)} \delta \mathbf{r}_i &= \sum_i F^{(i)} \sum_j \left(\frac{\partial \mathbf{r}_i}{\partial q_j}\delta q_j\right) \\
&= \sum_j \delta q_j \left(\sum_i \mathbf{F}^{(i)} \cdot \frac{\partial \mathbf{r}_i}{\delta q_j}\right) = \sum_j \delta q_j Q_j.
\end{aligned} \tag{1.57}$$

where

$$Q_j = \text{generalised force} = \sum_i \mathbf{F}^{(i)} \cdot \frac{\partial \mathbf{r}_i}{\partial q_j}.$$

We now take up

$$\sum_j m_i \ddot{\mathbf{r}}_i \cdot \delta \mathbf{r}_i$$

term:

$$\sum_i m_i \ddot{\mathbf{r}}_i \delta \mathbf{r}_i = \sum_i m_i \frac{d}{dt}(\dot{\mathbf{r}}_i)\cdot \sum_{j=1}^{n} \frac{\partial \mathbf{r}_i}{\partial q_j}\delta q_j = \sum_j \delta q_j \left(\sum_i m_i \frac{d}{dt}\dot{\mathbf{r}}_i \frac{\partial \mathbf{r}_i}{\partial q_j}\right)$$

$$= \sum_j \delta q_j \left[\sum_i m_i \frac{d}{dt}\left(\dot{\mathbf{r}}_i \frac{\partial \mathbf{r}_i}{\partial q_j}\right) - \sum_j \delta q_j \sum_i m_i \dot{\mathbf{r}}_i \frac{d}{dt}\left(\frac{\partial \mathbf{r}_i}{\partial q_j}\right)\right]$$

$$= \sum_j \delta q_j \left[\sum_i m_i \frac{d}{dt}\left(\dot{\mathbf{r}}_i \frac{\partial \dot{\mathbf{r}}_i}{\partial \dot{q}_j}\right) - \sum_i m_i \dot{r}_i \frac{\partial}{\partial q_j}\dot{\mathbf{r}}_i\right]$$

$$= \sum_j \partial q_j \left[\frac{d}{dt}\left(\frac{\partial T}{\partial \dot{q}_j}\right)\right] - \sum_j \delta q_j \frac{\partial}{\partial q_j}(T).$$

where

$$T = \frac{1}{2}\sum_i m_i \dot{\mathbf{r}}_i^{\,2}.$$

Therefore, from

$$\sum_i \left(-m_i \ddot{\mathbf{r}}_i + \mathbf{F}^{(i)}\right)\delta \mathbf{r}_i = 0.$$

we get

$$\sum_j \delta q_j Q_j = \sum_j \delta q_j \left[\frac{d}{dt}\left(\frac{\partial T}{\partial \dot{q}_j}\right) - \frac{\partial T}{\partial q_j}\right]. \tag{1.58}$$

Since $q_j$'s are all independent, equating the coefficients of $\delta q_j$ we get

$$\frac{d}{dt}\left(\frac{\partial T}{\partial \dot{q}_j}\right) = Q_j + \frac{\partial T}{\partial q_j}. \tag{1.59}$$

In deriving (1.59) we used (1.52–1.56), writing

$$Q_j = -\frac{\partial V}{\partial q_j}.$$

where $V$ is now the generalised potential, we get,

$$\frac{d}{dt}\frac{\partial}{\partial \dot{q}_j}(T - V) = \frac{\partial}{\partial q_j}(T - V). \tag{1.60}$$

In writing down (1.60) we have used

$$\frac{\partial}{\partial \dot{q}_j}V = 0.$$

Since for a particular case

$$Q_j = \mathbf{F}\frac{\partial \mathbf{r}}{\partial q_j}$$

when the system consists of one particle so.

$$Q_j = F_x \frac{\partial x}{\partial q_j} + F_y \frac{\partial y}{\partial q_j} + F_z \frac{\partial z}{\partial q_j}$$
$$= -\left( \frac{\partial V}{\partial x}\frac{\partial x}{\partial q_j} + \frac{\partial V}{\partial y}\frac{\partial y}{\partial q_j} + \frac{\partial V}{\partial z}\frac{\partial z}{\partial q_j} \right) = -\frac{\partial V}{\partial q_j}.$$

Hence. for general case

$$Q_j = \sum_i F^{(i)} \frac{\partial \mathbf{r}_i}{\partial q_j} = -\sum_i \frac{\partial V^{(i)}}{\partial q_j} = -\frac{\partial V}{\partial q_j},$$

where

$$V = \sum_i V^{(i)}.$$

the generalised potential.

Put $L = T - V$ (kinetic energy—potential energy). then equation (1.60) becomes

$$\frac{d}{dt}\frac{\partial L}{\partial \dot{q}_j} = \frac{\partial L}{\partial q_j} \quad \text{for } j = 1, 2, \cdots, n. \tag{1.61}$$

$L = T - V$ is called the *Lagrangian function* and equation (1.61) is the Lagrangian equation of motion expressed in terms of generalised co-ordinates $q_1, \cdots, q_n$.

The usefulness of Lagrangian's form of equation of motion given by equation (1.61) can be seen from the following examples.

For instance. consider a one-dimensional harmonic oscillator problem for which the kinetic energy

$$T = \frac{1}{2}m\dot{x}^2,$$

where $\dot{x}$ is the velocity and potential energy

$$V = \frac{1}{2}m\omega^2 x^2,$$

$\omega$ being the frequency of oscillation and $m$ is the mass. Thus. the Lagrangian function

$$L = T - V = \frac{1}{2}m\dot{x}^2 - \frac{1}{2}m\omega^2 x^2.$$

Therefore.

$$\frac{\partial L}{\partial \dot{x}} = m\dot{x} \quad \text{and} \quad \frac{d}{dt}\left(\frac{\partial L}{\partial \dot{x}}\right) = m\ddot{x} \quad \text{and} \quad \frac{\partial L}{\partial x} = -m\omega^2 x.$$

Therefore. Lagrange equation for one co-ordinate $x$ becomes

$$m\ddot{x} = -m\omega^2 x, \quad \text{or,} \quad \ddot{x} + \omega^2 x = 0. \tag{1.62}$$

Equation (1.62) is the equation for one-dimensional harmonic oscillator problem. Quite generally we can put Newton's equation in Lagrange's form directly. For example, Newton's equation of motion is

$$m\ddot{\mathbf{x}} = \mathbf{F} = -\frac{\partial V}{\partial \mathbf{x}}. \tag{1.63}$$

Equation (1.63) is the Newton's equation for a conservative system. $V$ is the potential function.
If

$$T = \frac{1}{2}m\dot{\mathbf{x}}^2.$$

then

$$\frac{d}{dt}\frac{\partial T}{\partial \dot{\mathbf{x}}} = m\ddot{\mathbf{X}}.$$

then (1.63) is

$$\frac{d}{dt}\left(\frac{\partial T}{\partial \dot{\mathbf{x}}}\right) = -\frac{\partial V}{\partial \mathbf{x}}. \tag{1.64}$$

As

$$\frac{\partial T}{\partial \mathbf{x}} = 0 \quad \text{and} \quad \frac{\partial V}{\partial \dot{\mathbf{x}}} = 0.$$

as $T$ is a function of $\dot{\mathbf{X}}$ only and $V$ is a function of $x, y, z$ only we can rewrite (1.65) as follows

$$\frac{d}{dt}\frac{\partial}{\partial \dot{\mathbf{x}}}(T - V) = \frac{\partial}{\partial \mathbf{x}}(T - V), \quad \text{i.e.,} \quad \frac{d}{dt}\frac{\partial}{\partial \dot{\mathbf{X}}}L = \frac{\partial}{\partial \mathbf{X}}L \tag{1.65}$$

$$L = \frac{1}{2}m\dot{\mathbf{X}}^2 - V(x, y, z). \tag{1.66}$$

Thus, Lagrange's approach avoids the use of detailed Newtonian equations involving forces of constraints etc. For a simple plane pendulum we have only one generalised co-ordinate $\theta$, the angle made by the bob of the pendulum of length $l$ with the vertical. The velocity is therefore $l\dot{\theta}$; then kinetic energy

$$T = \frac{1}{2}ml^2\dot{\theta}^2.$$

The potential energy

$$V = mgl(l - \cos\theta).$$

Hence for small amplitude

$$1 - \cos\theta \simeq \frac{\theta^2}{2}.$$

Hence

$$L = \frac{1}{2}ml^2\dot{\theta}^2 - mgl\frac{1}{2}\theta^2.$$

The Lagrange's equation

$$\frac{d}{dt}\frac{\partial L}{\partial \dot{\theta}} = \frac{\partial L}{\partial \theta}$$

becomes

$$ml^2\ddot{\theta} = -mgl\theta, \quad \text{i.e.,} \quad l\ddot{\theta} + g\theta = 0. \tag{1.67}$$

which is the familiar equation for simple pendulum. In the next section, we will put the Lorentz force equation of electron in Lagrange's form.

## 1.12 Lorentz-force Equation in Lagrange's form

The equation of motion of a charged particle of mass $m$ and charge $e$ in presence of an electro-magnetic field is given by

$$m\ddot{\mathbf{X}} = e\left[\mathbf{E} + \frac{1}{c}(\mathbf{v}\times\mathbf{B})\right]. \tag{1.68}$$

where $\ddot{\mathbf{X}}$ is the acceleration, $\mathbf{E}$ is the electric field and $\mathbf{B}$ stands for the magnetic field. The electric field

$$\mathbf{E} = -\nabla\phi - \frac{1}{c}\frac{\partial \mathbf{A}}{\partial t} \tag{1.69}$$

$$\text{and} \quad \mathbf{B} = (\nabla\times\mathbf{A}). \tag{1.70}$$

The $\mathbf{A}$ and $\phi$ are vector and scalar potentials respectively. The right hand side of equation (1.68) is called the *Lorentz force*, $\mathbf{F}$. Writing the $x$-component of equation (1.68) we get

$$m\ddot{x} = F_x = e\left[E_x + \frac{1}{c}(\mathbf{v}\times\mathbf{B})_x\right] \tag{1.71}$$

$$F_x = e\left[-\frac{\partial\phi}{\partial x} - \frac{1}{c}\frac{\partial A_x}{\partial t} + \frac{1}{c}(\mathbf{v}\times\mathbf{B})_x\right]. \tag{1.72}$$

Since $\mathbf{B} = \bar{\nabla}\times\mathbf{A}$ we can find

$$\frac{1}{c}(\mathbf{v}\times\mathbf{B})_x = \frac{1}{c}\left[\left(v_y\frac{\partial}{\partial x}A_y + v_z\frac{\partial}{\partial x}A_z\right) - \left(v_y\frac{\partial}{\partial y}A_x + v_z\frac{\partial}{\partial z}A_x\right)\right].$$

We use this expression in (1.72) to get

$$F_x = e\left[-\frac{\partial\phi}{\partial x} - \frac{1}{c}\frac{\partial A_x}{\partial t} + \frac{1}{c}\left(v_y\frac{\partial}{\partial x}A_y + v_z\frac{\partial}{\partial x}A_z\right) - \frac{1}{c}\left(v_y\frac{\partial}{\partial y} + v_z\frac{\partial}{\partial z}\right)A_x\right]. \tag{1.73}$$

Since $A_x = A_x(x, y, z, t)$ we get

$$\frac{1}{c}\frac{dA_x}{dt} = \frac{1}{c}\frac{\partial A_x}{\partial t} + \frac{1}{c}\left(v_x\frac{\partial}{\partial x} + v_y\frac{\partial}{\partial y} + v_z\frac{\partial}{\partial z}\right)A_x. \tag{1.74}$$

We use (1.74) in (1.73), then

$$\begin{aligned} F_x &= e\left[-\frac{\partial\phi}{\partial x} - \frac{1}{c}\frac{dA_x}{dt} + \frac{1}{c}\left(v_x\frac{\partial}{\partial x}A_x + v_y\frac{\partial}{\partial y}A_y + v_z\frac{\partial}{\partial z}A_z\right)\right] \\ &= e\left[-\frac{\partial\phi}{\partial x} - \frac{1}{c}\frac{dA_x}{dt} + \frac{1}{c}\left(\mathbf{v}\frac{\partial}{\partial x}\mathbf{A}\right)\right] \\ &= e\left[-\frac{\partial}{\partial x}\left(\phi - \frac{1}{c}\mathbf{v}\cdot\mathbf{A}\right) + \frac{d}{dt}\frac{\partial}{\partial v_x}\left(\phi - \frac{1}{c}\mathbf{v}\cdot\mathbf{A}\right)\right]. \end{aligned} \tag{1.75}$$

In deriving (1.75) we have used $(\partial v/\partial x) = 0$ as velocity does not depend on $x$, explicitly and

$$\frac{\partial\phi}{\partial v_x} = 0, \quad \phi = \phi(x, y, z, t)$$

only and

$$\frac{d}{dt}\frac{\partial}{\partial v_x}(\mathbf{v}\cdot\mathbf{A}) = \frac{dA_x}{dt}.$$

Again

$$m\ddot{x} = \frac{d}{dt}\frac{\partial}{\partial v_x}T,$$

where

$$T = \frac{1}{2}m\left(v_x^2 + v_y^2 + v_z^2\right).$$

We obtain for the Lorentz force equation (1.68), the following

$$\frac{d}{dt}\frac{\partial}{\partial v_x}\left[T - e\left(\phi - \frac{1}{c}\mathbf{v}\cdot\mathbf{A}\right)\right] = -\frac{\partial}{\partial x}e\left(\phi - \frac{1}{c}\mathbf{v}\cdot\mathbf{A}\right), \tag{1.76}$$

since $(\partial T/\partial x) = 0$, we can rewrite (1.76) as

$$\frac{d}{dt}\frac{\partial}{\partial v_x}(T - M) = \frac{\partial}{\partial x}(T - M), \tag{1.77}$$

where

$$M = e\left[\phi - \frac{1}{c}\mathbf{v}\cdot\mathbf{A}\right].$$

Thus $M$ is a velocity and time dependent potential. We put $T - M = L$, the Lagrange function. Equation (1.77) reduces to

$$\frac{d}{dt}\frac{\partial L}{\partial v_x} = \frac{\partial L}{\partial x}$$

and similarly

$$\frac{d}{dt}\frac{\partial L}{\partial v_y} = \frac{\partial L}{\partial y} \quad \text{and} \quad \frac{d}{dt}\frac{\partial L}{\partial v_z} = \frac{\partial L}{\partial z}$$

when $L = T - M$, where $M$ is the generalised velocity dependent potential.

To conclude this section we give a simple example of a velocity dependent potential. The Newton's equation is given as

$$m\ddot{r} = F, \tag{1.78}$$

where

$$F = \frac{1}{r^2}\left[1 - \frac{\dot{r}^2 - 2r\ddot{r})}{c^2}\right].$$

To find the potential.
Let us try with the following potential

$$V = \left(1 + \frac{\dot{r}^2}{c^2}\right) \cdot \frac{1}{r}.$$

Then it is easy to see that

$$\frac{d}{dt}\frac{\partial V}{\partial \dot{r}} - \frac{\partial V}{\partial r} = F.$$

For

$$\frac{d}{dt}\left(\frac{\partial V}{\partial \dot{r}}\right) = -\frac{1}{r^2}\left(\frac{2\dot{r}^2}{c^2} - \frac{2r\ddot{r}}{c^2}\right) \quad \text{and} \quad \frac{\partial V}{\partial r} = -\frac{1}{r^2}\left(1 + \frac{\dot{r}^2}{c^2}\right).$$

Again

$$m\ddot{r} = \frac{d}{dt}\frac{\partial T}{\partial \dot{r}},$$

where $T = (1/2)m\dot{r}^2$.
Hence the Newton's equation (1.78) becomes

$$\frac{d}{dt}\frac{\partial}{\partial \dot{r}}(T - V) = \frac{\partial}{\partial r}(T - V), \quad \text{i.e.,} \quad \frac{d}{dt}\frac{\partial L}{\partial \dot{r}} = \frac{\partial L}{\partial r}$$

where $L = T - V$.

## Exercises

1. Four masses $m_1$, $m_2$, $m_3$ and $m_4$ are hanging and they can move vertically only as in the figure. How many independent co-ordinates are necessary? Identify the constraint equations. Write down the kinetic and potential energy of the system.

   [*Hints:* Let $x_1$, $x_2$, $x_3$ and $x_4$ are vertical distances of $m_1$, $m_2$, $m_3$ and $m_4$ respectively.

   But the constraints are

   $$x_1 + x_3 = l_1 \quad \text{(length of string over } A\text{)}$$
   $$\text{and} \quad x_2 + x_4 = l_2 \quad \text{(length of string over } B\text{)}.$$

   Hence only two independent co-ordinates.]

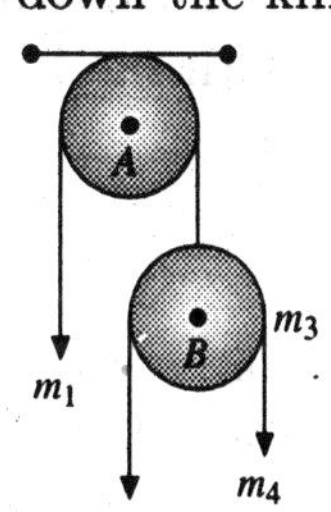


Fig 1.3

2. Consider the arrangement of springs and masses as shown in the following diagram. Find the degree of freedom where: (*a*) when

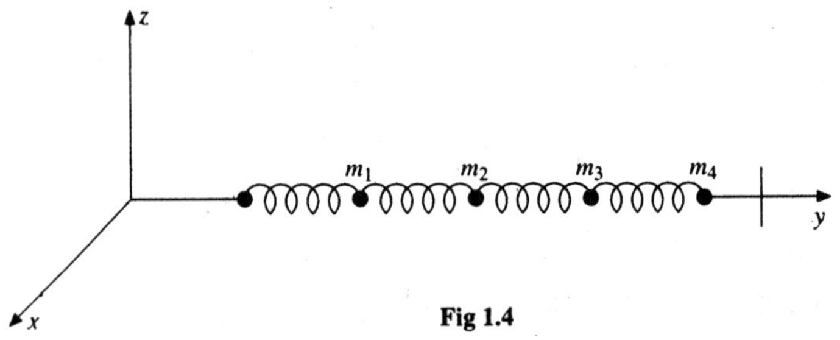


Fig 1.4

the motion of all of them is along $y$-direction (*b*) when their motion is restricted in $xy$-plane (*c*) when $m_1$ and $m_2$ have their motion along $y$ but $m_3$ and $m_4$ have their motion restricted to the $xy$-plane.

3. Are the following holonomic constraints?

   (*a*) $x\,dy - y\,dx = cx^2$;

   (*b*) $x\,dx + y\,dy = k$;

   (*c*) $x\,dy + y\,dx = c\,dx - f\,dy$, $c, f$ and $k$ constants;

   (*d*) $(x + y)\,dx + (x - y)\,dy = 0$.

4. Express the kinetic energy of a particle in terms of oblique co-ordinates as shown in the diagram.

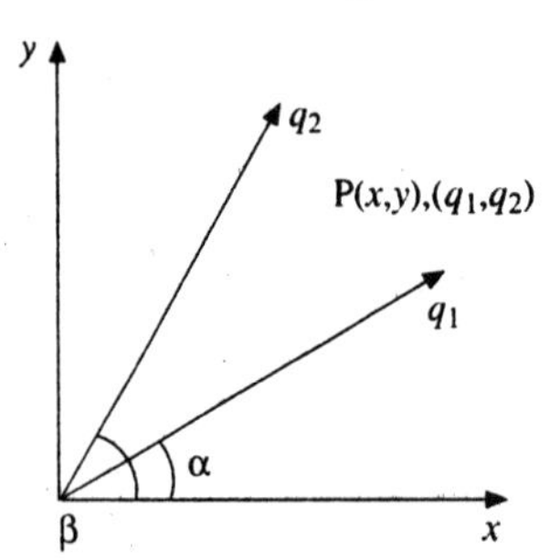


Fig 1.5

[*Hints:* $(x, y)$ = orthogonal co-ordinate system.

$(q_1, q_2)$ oblique co-ordinate system.

$q_1$ axis makes an angle $\alpha$ with $x$-axis.

$q_2$ axis makes an angle $\beta$ with $x$-axis.

$$T = \frac{1}{2}m[\dot{q}_1^2 + \dot{q}_2^2 + 2\dot{q}_1\dot{q}_2\cos(\alpha - \beta)].]$$

5. A particle of mass $m$ is tied to one end of a spring of spring constant $k$ and unstretched length $r_0$. The other end is fixed to a point $P$ on a horizontal plane. Write down the Lagrangian equation.

   [*Hints:*

$$T = \frac{1}{2}m(\dot{r}^2 + r^2\dot{\theta}^2),$$

where $(r, \theta)$ is the polar co-ordinates of the mass point on the table. Force $F = -k(r - r_0)$. Hence

$$F = -\frac{\partial V}{\partial r}, \text{ gives } V = +\frac{1}{2}k(r - r_0)^2.$$

Then Lagrangian

$$L = \frac{1}{2}m(\dot{r}^2 + r^2\dot{\theta}^2) - \frac{1}{2}k(r - r_0)^2.$$

$$\frac{d}{dt}\left(\frac{\partial L}{\partial \dot{r}}\right) = \frac{\partial L}{\partial r}, \text{ gives } m\ddot{r} = mr\dot{\theta}^2 - k(r - r_0)$$

$$\text{and } \frac{d}{dt}\left(\frac{\partial L}{\partial \dot{\theta}}\right) = \frac{\partial L}{\partial \theta}, \text{ gives } \frac{d}{dt}(mr^2\dot{\theta}) = 0$$

which gives $mr^2\dot{\theta}$ = constant $= l$ say, then $\dot{\theta} = \frac{l}{mr^2}$. Substituting this in differential equation of $r$ we get the final equation.]

6. A particle is constrainted to move on the plane curve $xy = c$ (where $c$ is a constant) under gravity ($y$-axis vertical). Obtain Lagrangian equation.

[*Hints:*

$$T = \frac{1}{2}m(\dot{x}^2 + \dot{y}^2), \quad V = mgy.$$

But

$$y = \frac{c}{x} \text{ so } \dot{y} = -\frac{c}{x^2}\dot{x}.$$

Hence

$$L = T - V = \frac{1}{2}m\left(\dot{x}^2 + \frac{c^2}{x^4}\dot{x}^2\right) - (mgc)/x$$

$$L = \frac{1}{2}m\left(1 + \frac{c^2}{x^4}\right)\dot{x}^2 - \frac{mgc}{x}.$$

Now find

$$\frac{d}{dt}\left(\frac{\partial L}{\partial \dot{x}}\right) = \frac{\partial L}{\partial x}.]$$

7. Write down the Lagrangian equation from the Lagrangian

$$L = \frac{1}{2}q^2\dot{q}^2 - q^3.$$

Find the solution of the resulting equation.

8. A particle is constrained to move on the surface of a smooth sphere of radius $a$ under the action of gravity. Obtain the Lagrangian equation of motion.

[*Hints:* Here co-ordinates are $(r, \theta, \phi)$ of which $r$ is constant as the particle is constrained to move on the surface of a sphere of given radius $r = a$.

In spherical co-ordinates

$$T = \frac{1}{2}m(\dot{r}^2 + r^2\sin^2\theta\dot{\phi}^2 + r^2\dot{\theta}^2) = \frac{1}{2}m(a^2\dot{\theta}^2 + a^2\sin^2\theta\dot{\phi}^2).$$

$$V = mgz = mga\cos\theta.$$

Lagrangian

$$L = \frac{1}{2}m(a^2\dot{\theta}^2 + a^2\sin^2\theta\cdot\dot{\phi}^2) - mga\cos\theta.$$

$$\frac{\partial L}{\partial\dot{\theta}} = ma^2\dot{\theta}. \quad \frac{\partial L}{\partial\theta} = ma^2\sin\theta\cos\theta\cdot\dot{\phi}^2 + m\cdot ag\sin\theta.$$

$$\frac{d}{dt}\left(\frac{\partial L}{\partial\dot{\theta}}\right) = \frac{\partial L}{\partial\theta}. \text{ gives } ma^2\ddot{\theta} = ma^2\sin\theta\cos\theta\cdot\dot{\phi}^2 + amg\sin\theta.$$

$$\frac{\partial L}{\partial\dot{\phi}} = ma^2\sin^2\theta\cdot\dot{\phi}. \quad \frac{\partial L}{\partial\varphi} = 0.$$

So

$$\frac{d}{dt}\left(\frac{\partial L}{\partial\dot{\phi}}\right) = 0.$$

hence

$$\frac{\partial L}{\partial\dot{\phi}} = \text{constant} = l.$$

then

$$ma^2\sin^2\theta\dot{\phi} = l. \quad \text{or.} \quad \dot{\phi} = l/ma^2\sin^2\theta.$$

therefore the $\theta$-equation becomes

$$\ddot{\theta} = \frac{1\sin\theta\cos\theta\cdot l^2}{(m^2a^4\cdot\sin^4\theta)} + \left(\frac{g}{a^2}\right)a\sin\theta$$

$$\ddot{\theta} = \frac{1}{(m^2a^4)}\cdot\operatorname{cosec}^2\theta\cot\theta + \left(\frac{g}{a^2}\right)a\sin\theta.$$

$$\text{or.} \quad \dot{\theta}\ddot{\theta} = \frac{1}{(m^2a^4)}\operatorname{cosec}^2\theta\cot\theta\cdot\dot{\theta} + \left(\frac{g}{a^2}\right)a\sin\theta\cdot\dot{\theta}.$$

$$\text{or.} \quad \frac{1}{2}\frac{d}{dt}(\dot{\theta}^2) = -(m^2a^4)_2^{-1}\frac{d}{dt}(\cot\theta)^2 + \frac{g}{a^2}\cdot\frac{d}{dt}(-a\cos\theta).$$

Hence the equation is

$$\dot{\theta}^2 = -(m^2a^4)^{-1}\cot^2\theta - \frac{g}{a^2}\cos\theta\cdot a\cdot 2 + \text{constant}.$$

The solution is

$$t = c + \int d\theta\left(k - \frac{g}{a}2\cos\theta - \frac{\cot^2\theta}{m^2a^2}\right)^{-1}.$$

where $k$ and $c$ are constant.]

9. The same problem as in Example 8 except. the particle is constrained to move on the surface of a cylinder of fixed radius.

[*Hints:* Co-ordinates on the surface of a cylinder is $x = r\cos\theta$. $y = r\sin\theta$. $z = z$.

$$T = \frac{1}{2}m(\dot{r}^2 + r^2\dot{\theta}^2) + \frac{1}{2}m\dot{z}^2. \quad V = mgz$$

$$L = \frac{1}{2}m(\dot{r}^2 + r^2\dot{\theta}^2) + \frac{1}{2}m\dot{z}^2 - mgz.$$

But $r$ = constant for the particle is constraint to move in the surface of a cylinder. $r$ = constant $= a$.

$$L = \frac{1}{2}m(a^2\dot{\theta}^2) + \frac{1}{2}m\dot{z}^2 - mgz. \quad \frac{\partial L}{\partial \theta} = 0. \quad \frac{\partial L}{\partial \dot{\theta}} = ma^2\dot{\theta}.$$

So

$$\frac{d}{dt}\left(\frac{\partial L}{\partial \dot{\theta}}\right) = \frac{\partial L}{\partial \theta} = 0$$

gives $ma^2\dot{\theta} = l$ ($a$ constant), i.e., $\theta = At + B$. $A = (ma^2/l)^{-1}$. $B$ = constant of integration.

And

$$\frac{\partial L}{\partial z} = -mg. \quad \frac{\partial L}{\partial \dot{z}} = m\dot{z}.$$

Here $m\ddot{z} = -mg$. or. $\ddot{z} = -g$; $z = ut - \frac{1}{2}gt^2 +$ constant.

Let at $t = 0$. $z = 0$ and $\theta = \theta_0$ then

$$z = ut - \frac{1}{2}gt^2. \quad \theta = \frac{l}{ma^2}t + \theta_0.]$$

10. A mass $m$ is tied to one end of a string and is moving on a horizontal table. The other end of the string passes through a hole in the table and a mass $M$ is hanging vertically from this end. Write down the Lagrangian equation of motion and find its solution.

[*Hints:* Let $(r, \theta)$ be the polar co-ordinate of $m$ on the table. the origin is $O$. the hole in the table.

$l$ = length of string. Hence co-ordinate of hanging mass $M$ is $(l - r)$.

$$\text{The K.E.} = \frac{1}{2}M\left[\left(\dot{\overline{l-r}}\right)\right]^2 + \frac{1}{2}m\left(\dot{r}^2 + r^2\dot{\theta}^2\right).$$

The potential energy $V = -Mg(l - r)$.

$$L = T - V = \frac{1}{2}(M+m)\dot{r}^2 + \frac{1}{2}mr^2\dot{\theta}^2 + Mg(l-r)$$

$$\frac{\partial L}{\partial r} - mr\dot{\theta}^2 - Mg = \left(\frac{\partial}{\partial t}\cdot\frac{\partial L}{\partial \dot{r}}\right) = (M+m)\ddot{r}$$

$$\frac{\partial L}{\partial \theta} = 0 = \frac{d}{dt}\left(\frac{\partial L}{\partial \dot{\theta}}\right). \text{ gives } mr^2\dot{\theta} = l = \text{ constant}.$$

Hence equation is

$$(M+m)\ddot{r} = -Mg + mr\frac{l^2}{(m^2r^4)} = -Mg + \frac{1}{m}l^2\cdot\frac{1}{r^3}.]$$

11. The point of support of a plane simple pendulum of mass $m$ and length $l$ is moving along horizontal axis with constant velocity $V$. Write down the Lagrangian equation.

[*Hints:* The diagram is given below.

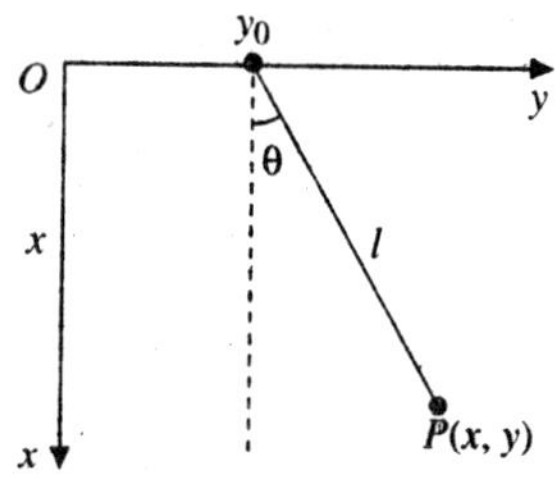


**Fig 1.6**

Use Co-ordinate $x = l\cos\theta$, $y = y_0 + l\sin\theta$; $y_0 = Vt$.

Hence

$$\begin{aligned} T &= \frac{1}{2}m\left(\dot{x}^2 + \dot{y}^2\right) \\ &= \frac{1}{2}m\left[l^2\sin^2\theta\dot{\theta}^2 + V^2 + l^2\cos^2\dot{\theta}^2 + 2Vl\cos\theta\cdot\dot{\theta}\right] \\ V &= mgl(1-\cos\theta) \\ L &= T - V = \frac{1}{2}m\left(l^2\dot{\theta}^2 + V^2 + 2Vl\cos\theta\cdot\dot{\theta}\right) - mgl(1-\cos\theta). \end{aligned}$$

Hence

$$\begin{aligned} \frac{\partial L}{\partial\dot{\theta}} &= ml^2\dot{\theta} + 2Vl\cos\theta\frac{m}{2}. \\ \frac{d}{dt}\left(\frac{\partial L}{\partial\dot{\theta}}\right) &= ml^2\ddot{\theta} - mVl\sin\theta\cdot\dot{\theta}; \\ \frac{\partial L}{\partial\theta} &= -mVl\sin\theta\cdot\dot{\theta} - mgl\sin\theta. \end{aligned}$$

Therefore

$$\frac{d}{dt}\left(\frac{\partial L}{\partial\dot{\theta}}\right) = \frac{\partial L}{\partial\theta},$$

gives $ml^2\ddot{\theta} + mgl\sin\theta = 0$.

For small $\theta$, $l\ddot{\theta} + g\theta = 0$. Time period $= 2\pi\sqrt{\frac{l}{g}}$.

This is the same one result as finds in simple pendulum. The term $mVl\cos\theta\cdot\dot{\theta}$ did not contribute because it is equal to

$$\frac{d}{dt}(Vl\sin\theta) = \frac{d}{dt}F(t).$$

One can always add a term which is a total time derivative to $L$. (See next section).]

Chapter II

# Variational Principle and Lagrangian Dynamics

## 2.1 Introduction

The variational principles have been proposed in many branches of Physics. In some cases they have been surrounded with philosophical mysticism, which has delayed an appreciation of their values. The importance of such principles lies in their extreme economy of expression. Here we will consider the application of Hamilton's least action principle to derive the Lagrange's equation of motion. We will see below how simply the equations can be derived as compared to the method adopted while using D'Alembert's principle. In the last section, we have introduced the Lagrangian function. $L$ is equal to $T - V$, where $T$ and $V$ are the kinetic and potential energy respectively. Since $T$ depends on velocity and potential for a conservative system depends on co-ordinate we will consider quite generally Lagrangian $L$ as a function of generalised co-ordinates and velocity. And time parameter $t$ may also occur explicitly. Then, for a single particle system

$$L = L[q(t), \dot{q}(t); t].$$

Here, $q(t)$ is the trajectory of a single particle system characterised by a single generalised co-ordinate. Consider a path of the particle starting at $t = t_0$ from $A$ and ending the journey at $B$ when $t = t_1$.
In the $[t, q(t)]$ diagram Fig 2.1 we draw several nearby trajectories beginning at $A$ and ending at $B$ via $C, C'$, etc.
Let $ACB$ denote the trajectory $q(t)$. $AC'B$ denotes the nearby path

$$q'(t) = q(t) + \delta q(t), \text{etc.}$$

Let $S, S', \cdots$, are the actions corresponding to the path described by

$$q(t), q'(t), \cdots, \text{ etc.}$$

Hamilton's least action principle states that of the various trajectories given by

$$q(t), q'(t), \cdots, \text{ etc.}$$

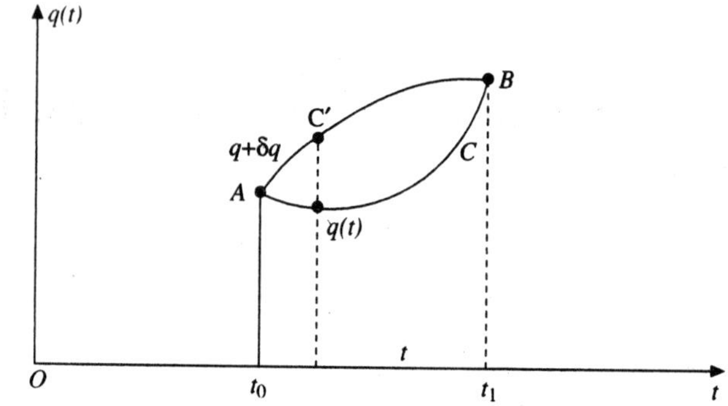


**Fig 2.1**

the path followed by the system is the one for which the corresponding action is minimum. We define the action $S$ by

$$S = \int_{t_0}^{t_1} L \cdot dt. \tag{2.1}$$

Corresponding to $q(t)$, the $L$ is

$$L\left(q(t), \dot{q}(t); t\right).$$

Similarly for $q'(t)$ the Lagrangian is

$$L(\dot{q}'(t), \dot{q}'(t); t), \text{ etc.}$$

so that the corresponding action $S'$ is

$$S' = \int_{t_0}^{t_1} L\left(q'(t), \dot{q}'(t); t\right) dt \tag{2.2}$$

$$\text{and } S = \int_{t_0}^{t_1} L\left(q(t), \dot{q}(t); t\right) dt. \tag{2.3}$$

We have to find out $q(t)$ for which $S$ is least. For $S$ to be least we must have $\delta S = 0$, where $\delta S$ stands for the difference of $S$ and $S'$ corresponding to two nearby paths. Then,

$$\begin{aligned} \delta S = S' - S &= \int_{t_0}^{t_1} \left[L(q', \dot{q}'; t) - L(q, \dot{q}; t)\right] dt \\ &= \int_{t_0}^{t_1} \left[L(q + \delta q; \dot{q} + \delta\dot{q}; t) - L(q, \dot{q}; t)\right] dt, \end{aligned} \tag{2.4}$$

where $q' = q(t) + \delta q(t)$ is used. Hence, using Taylor's expansion,

$$\delta S = \int_{t_0}^{t_1} \left[L\left(q, \dot{q}, t\right) + \frac{\partial L}{\partial q}\delta q + \frac{\partial L}{\partial \dot{q}}\delta\dot{q} - L\left(q, \dot{q}; t\right)\right] dt. \tag{2.5}$$

In (2.5) we have kept up to first-order terms in the Taylor's expansion of

$$L((q+\delta q),\ (\dot{q}+\delta\dot{q});t).$$

Hence

$$\delta S = \int_{t_0}^{t_1} dt \left(\frac{\partial L}{\partial q}\delta q + \frac{\partial L}{\partial \dot{q}}\delta\dot{q}\right). \tag{2.6}$$

Consider the term

$$\int_{t_0}^{t_1} \frac{\partial L}{\partial \dot{q}}\delta\dot{q}dt.$$

which is equal to

$$\int_{t_0}^{t_1} dt\frac{\partial L}{\partial \dot{q}}\frac{d}{dt}(\delta q) = \int_{t_0}^{t_1} dt\frac{d}{dt}\left(\frac{\partial L}{\partial \dot{q}}\delta q\right) - \int_{t_0}^{t_1} dt\left[\frac{d}{dt}\left(\frac{\partial L}{\partial \dot{q}}\right)\right]\delta q.$$

Since $\delta q = 0$ at $t = t_0$ and $t = t_1$ [see figure (2.11)] we find

$$\int_{t_0}^{t_1} dt\frac{\partial L}{\partial \dot{q}}\frac{d}{dt}\delta q = -\int_{t_0}^{t_1} dt\frac{d}{dt}\left(\frac{\partial L}{\partial \dot{q}}\right)\cdot\delta q. \tag{2.7}$$

We use equation (2.7) in equation (2.6), then

$$\delta S = \int_{t_0}^{t_1} dt\left[\frac{\partial L}{\partial q} - \frac{d}{dt}\left(\frac{\partial L}{\partial \dot{q}}\right)\right]\delta q. \tag{2.8}$$

Least action requires $\delta S = 0$. Since $\delta q$ is any arbitrary variation, [because the trajectory $q'(t)$ is an arbitrary path near $q(t)$], we then find from equation (2.8) that the integrand of equation (2.8) must vanish. That is,

$$\frac{\partial L}{\partial q} - \frac{d}{dt}\left(\frac{\partial L}{\partial \dot{q}}\right) = 0. \tag{2.9}$$

Equation (2.9) is the Lagrange's equation of motion as we derived in the last section.

We now generalize equation (2.9) for the $N$-particle system. Consider a $N$-particle system characterised by a Lagrangian function $L$.

$$L = L(q_1, q_2, \cdots, q_n; \dot{q}_1, \dot{q}_2, \cdots, \dot{q}_n; t).$$

The generalised co-ordinate $q_1, \cdots, q_n$ are obtained by removing the holonomic constraints in the configuration space co-ordinates of the $N$-particle system. The action for this Lagrangian is

$$S = \int_{t_0}^{t_1} L(q_1, \cdots, q_r, \cdots; \dot{q}_1, \cdots, \dot{q}_r, \cdots; t)\, dt. \tag{2.10}$$

Therefore,

$$\delta S = \int_{t_0}^{t_1} \delta L(q_1, \cdots, q_r, \cdots; \dot{q}_1, \cdots, \dot{q}_r, \cdots; t)\, dt. \tag{2.11}$$

Keeping only first-order terms in the variation of $L$ due to arbitrary variation of $q$'s we get for equation (2.11), the following

$$\delta S = \int_{t_0}^{t_1} \left[\frac{\partial L}{\partial q_1}\delta q_1 + \cdots + \frac{\partial L}{\partial q_r}\delta q_r + \cdots + \frac{\partial L}{\partial \dot{q}_1}\delta \dot{q}_1 + \cdots + \frac{\partial L}{\partial \dot{q}_r}\delta \dot{q}_r + \cdots\right] dt. \tag{2.12}$$

We rewrite equation (2.12) as

$$\delta S = \sum_r \int_{t_0}^{t_1} dt \left[\frac{\partial L}{\partial q_r}\delta q_r + \frac{\partial L}{\partial \dot{q}_r}\delta \dot{q}_r\right];$$

the summation over $r$ is from 1 to $n$ the numebr of independent co-ordinates. Again, using the fact that

$$\int dt \frac{\partial L}{\partial \dot{q}_r}\delta \dot{q}_r = \int dt \frac{\partial L}{\partial \dot{q}_r}\frac{d}{dt}\delta q_r,$$

we obtain, following equation (2.7),

$$\int dt \frac{\partial L}{\partial \dot{q}_r}\delta \dot{q}_r = -\int dt \frac{d}{dt}\left(\frac{\partial L}{\partial \dot{q}_r}\right)\delta q_r.$$

Hence equation (2.12) becomes

$$\delta S = \sum_{r=1}^{n} \int dt \left[\frac{\partial L}{\partial q_r} - \frac{d}{dt}\left(\frac{\partial L}{\partial \dot{q}_r}\right)\right]\delta q_r. \tag{2.13}$$

Least action demand $\delta S = 0$. Since, $\delta q_r$'s are arbitrary variation so we choose $\delta q_r$ in such a way that all $\delta q_1, \delta q_2, \cdots$ are zero except $\delta q_r$th one. Then equation (2.13) yields

$$\delta S = \int dt \left[\frac{\partial L}{\partial q_r} - \frac{d}{dt}\left(\frac{\partial L}{\partial \dot{q}_r}\right)\right]\delta q_r. \tag{2.14}$$

Therefore vanishing of $\delta S$ implies

$$\int dt \left[\frac{\partial L}{\partial q_r} - \frac{d}{dt}\left(\frac{\partial L}{\partial \dot{q}_r}\right)\right]\delta q_r = 0. \tag{2.15}$$

But $\delta q_r$ itself is an arbitrary variation. So the vanishing of equation (2.15) further implies

$$\frac{\partial L}{\partial q_r} - \frac{d}{dt}\left(\frac{\partial L}{\partial \dot{q}_r}\right) = 0. \tag{2.16}$$

In deriving above we have assumed that end-point variations of $q_r = 0$. Since $r$ can take up any value from 1 to $n$, we have the Lagrange's equation for each of the $n$-co-ordinates as follows,

$$\frac{\partial L}{\partial q_r} - \frac{d}{dt}\left(\frac{\partial L}{\partial \dot{q}_r}\right) = 0; \quad r = 1, 2, \cdots, n, \cdots. \tag{2.17}$$

We must mention here the condition we required in the derivation of equation (2.17) is only an *extremum* condition on the action $S$. We must also note that $S$ has the dimension of energy $\times$ time. Since $L$ by definition has the *dimension of energy*.

## 2.2 Non-uniqueness in the choice of Lagrangian

We have noticed that in the variational approach, the equation of motion of a particle follows from the vanishing of $\delta S$, where $S$ is the action given by

$$\int L(q, \dot{q}, t)\, dt.$$

From this we easily find that to the Lagrangian $L(q, \dot{q}, t)$ we can add a term which is a total time-derivative of any function $F(t)$. That is

$$S = \int_{t_0}^{t_1} \left[ L(q, \dot{q}, t) + \frac{dF}{dt} \right] dt. \tag{2.18}$$

The equation of motion is unchanged.
This is because, from (2.18), we get

$$\begin{aligned} S &= \int_{t_0}^{t_1} L(q, \dot{q}, t)\, dt + \int_{t_0}^{t_1} \frac{dF}{dt}(t)\, dt \\ &= \int_{t_0}^{t_1} L(q, \dot{q}, t)\, dt + [F(t_1) - F(t_0)]. \end{aligned} \tag{2.19}$$

The last term in the third bracket appearing in the right hand side of equation (2.19) is a constant. Now taking the variation of $S$, we get

$$\delta S = \delta \int L(q, \dot{q}, t)\, dt + \delta[F(t_1) - F(t_0)] = \delta \int L(q, \dot{q}; t)\, dt$$

as $\delta[F(t_1) - F(t_0)] = 0$. Hence $\delta S = 0$ implies

$$\int \delta L dt = 0$$

which immediately gives the same equation of motion. Therefore,

$$\int L(q, \dot{q}, t)\, dt \quad \text{and} \quad \int \left[ L(q, \dot{q}, t) + \frac{dF}{dt} \right] dt.$$

both give the same equation of motion. Thus, we have the following theorem.
If two Lagrangians differ by a term which is a total time-derivative of a function then both give rise to the same equation of motion.
As a simple example, consider the motion of a particle moving freely in a given inertial frame. Let $x_i$ denotes the position of the particle

$$x_i = (x_1, x_2, x_3) \equiv (x, y, z).$$

Then the Lagrangian $L_I$ is (summation over $i$ implied)

$$L_I = \frac{1}{2} m \dot{x}_i^2. \tag{2.20}$$

Consider now another Lagrangian in another inertial frame II such that the frame is moving with a constant velocity **v** with respect to the frame I for which the Lagrangian is $L_{II}$. Hence, if $v_i$ be the velocity of frame II then

$$x_i' = x_i + v_i t. \tag{2.21}$$

Hence

$$L_{II} = \frac{1}{2}m\dot{x}_i'^2$$

becomes

$$\begin{aligned} L_{II} &= \frac{1}{2}m(\dot{x}_i + v_i)^2 = \frac{1}{2}m\dot{x}_i^2 + \frac{1}{2}mv_i^2 + mv_i \cdot \dot{x}_i \\ &= L_I + \frac{1}{2}mv^2 \cdot \frac{d}{dt} \cdot t + mv_i \cdot \frac{d}{dt}x_i \\ &= L_I + \frac{d}{dt}\left(\frac{1}{2}mv^2 t + mv_i x_i\right) = L_I + \frac{d}{dt}F(t). \end{aligned} \tag{2.22}$$

where

$$F(t) = \frac{1}{2}mv^2 t + m\mathbf{v} \cdot \mathbf{x}.$$

So, $L_I$ and $L_{II}$ differ by total time-derivative. The equation of motion remains of same in frame I and II. Thus if two inertial frames are such that one is moving with constant velocity with respect to the other, the Newton's equation remains of same form.

To mention another choice of non-uniqueness of the Lagrangian, we may mention that we chose the $L$ as only a function of $q$ and $\dot{q}$ only. Suppose $L$ is a function of

$$L = L(q, \dot{q}, \ddot{q}; t). \tag{2.23}$$

In this case

$$S = \int L(q, \dot{q}, \ddot{q}; t)\, dt$$

$$\text{and} \quad \delta S = \int \left(\frac{\partial L}{\partial q}\delta q + \frac{\partial L}{\partial \dot{q}}\delta \dot{q} + \frac{\partial L}{\partial \ddot{q}}\delta \ddot{q}\right) dt. \tag{2.24}$$

We have for the first two terms

$$\delta S = \int \left\{\left[\frac{\partial L}{\partial q} - \frac{d}{dt}\left(\frac{\partial L}{\partial \dot{q}}\right)\right]\delta q + \frac{\partial L}{\partial \ddot{q}}\delta \ddot{q}\right\} dt. \tag{2.25}$$

For the last term of (2.24) we have

$$\int_{t_0}^{t_1} \frac{\partial L}{\partial \ddot{q}}\frac{d}{dt}(\delta \dot{q})\, dt = \int_{t_0}^{t_1} \frac{d}{dt}\left(\frac{\partial L}{\partial \ddot{q}}\delta \dot{q}\right) dt - \int_{t_0}^{t_1} \frac{d}{dt}\left(\frac{\partial L}{\partial \ddot{q}}\right)\delta \dot{q}dt. \tag{2.26}$$

Assuming $\delta \dot{q} = 0$ at the end points, (2.26) becomes

$$\int \frac{\partial L}{\partial \ddot{q}}\frac{d}{dt}\delta \dot{q}dt = -\int \frac{d}{dt}\left(\frac{\partial L}{\partial \ddot{q}}\right)\delta \dot{q}dt$$

for the term

$$\left(\frac{\partial L}{\partial \ddot{q}}\delta\dot{q}\right)_{t_0}^{t_1} = 0$$

as $\delta\dot{q} = 0$ at $t = t_0$ and $t = t_1$. Again

$$\begin{aligned}\int \frac{d}{dt}\left(\frac{\partial L}{\partial \ddot{q}}\right)\delta\dot{q}dt &= \int \frac{d}{dt}\left(\frac{\partial L}{\partial \ddot{q}}\right)\frac{d}{dt}\delta q dt \\ &= \int \frac{d}{dt}\left[\frac{d}{dt}\left(\frac{\partial L}{\partial \ddot{q}}\right)\delta q\right]dt - \int \frac{d^2}{dt^2}\left(\frac{\partial L}{\partial \ddot{q}}\right)\delta q dt \\ &= \left[\frac{d}{dt}\left(\frac{\partial L}{\partial \ddot{q}}\right)\delta q\right]_{t_0}^{t_1} - \int \frac{d^2}{dt^2}\left(\frac{\partial L}{\partial \ddot{q}}\right)\delta q dt.\end{aligned}$$

As $\delta q = 0$ at $t = t_0$ and $t = t_1$, so

$$\int \frac{d}{dt}\left(\frac{\partial L}{\partial \ddot{q}}\right)\delta\dot{q}dt = -\int \frac{d^2}{dt^2}\left(\frac{\partial L}{\partial \ddot{q}}\right)\delta q dt.$$

Using this result we get from (2.26)

$$\int \frac{\partial L}{\partial \ddot{q}}\delta\ddot{q}dt = +\int \frac{d^2}{dt^2}\left(\frac{\partial L}{\partial \ddot{q}}\right)\delta q dt.$$

Hence

$$\delta S = \int \left[\frac{\partial L}{\partial q} - \frac{d}{dt}\left(\frac{\partial L}{\partial \dot{q}}\right) + \frac{d^2}{dt^2}\left(\frac{\partial L}{\partial \ddot{q}}\right)\right]\delta q dt.$$

From $\delta S = 0$, the equation of motion is

$$\frac{\partial L}{\partial q} - \frac{d}{dt}\left(\frac{\partial L}{\partial \dot{q}}\right) + \frac{d^2}{dt^2}\left(\frac{\partial L}{\partial \ddot{q}}\right) = 0. \tag{2.27}$$

This equation will give a third-order differential equation for Newton's equation. Hence, keeping the $n$th time derivative of $q(t)$, the resulting equation will become of a much higher order differential equation for the equation of motion.

The Lagrangian equation of motion

$$\frac{\partial L}{\partial q} = \frac{d}{dt}\left(\frac{\partial L}{\partial \dot{q}}\right),$$

which we have obtained from the vanishing of $\delta S$, $S$ being the action is known in the literature as *Euler-Lagrangian Equation*. It is extremely interesting to note its application to problems in Calculus of Variation apart from its main application to physics. Below we will solve a few problems of interest to us. Before concluding this section, we must emphasize the importance of the theorem on the non-uniqueness of the choice of Lagrangians made at the beginning of this section. This theorem will play a fundamental role in the theory of Canonical Transformation which we will take up later.

## 2.3 Some Simple Application to problems in Calculus of Variation

We have found in section (2.1) that if

$$S = \int L(q, \dot{q}, t)\, dt; \dot{q} = \frac{dq}{dt}.$$

then for the extremum of $S$, we must have the Euler Lagrange equation for $L$ to be satisfied, i.e.,

$$\frac{\partial L}{\partial q} = \frac{d}{dt}\left(\frac{\partial L}{\partial \dot{q}}\right).$$

Consider an elementary example. For example, to find that the shortest distance between two points in a plane is a straight line. The elementary distance $ds$ in two-dimension is

$$ds = \sqrt{dx^2 + dy^2} = \sqrt{(1 + y'^2)} \cdot dx \tag{2.28}$$

where $y' = dy/dx$. Hence the distance $S$ is

$$S = \int \sqrt{1 + y'^2}\, dx.$$

We write the last integral as

$$S = \int L(y, y'; x)\, dx. \tag{2.29}$$

Then

$$L = (1 + y'^2)^{1/2}.$$

Therefore, $S$ to be minimum, the Euler-Lagrange equation is

$$\frac{d}{dx}\left(\frac{\partial L}{\partial y'}\right) = \frac{\partial L}{\partial y}. \tag{2.30}$$

But $\partial L/\partial y = 0$, so from (2.30)

$$\frac{\partial L}{\partial y'} = \text{constant}.$$

Hence,

$$\frac{y'}{\sqrt{1 + y'^2}} = k \text{ (a constant)}. \tag{2.31}$$

(2.31) gives $dy/dx = m$ (another constant). That is

$$y = mx + c$$

which is a straight line.

In a similar way we find the shortest distance between two points on the surface of three-dimensional sphere as great circles. Since the element of distance $ds$ on the surface of a three-dimensional sphere of constant radius $r = a$, is

$$ds = \sqrt{\left[a^2(d\theta)^2 + a^2\sin^2\theta(d\phi)^2\right]}.$$

we get

$$ds = \left\{\sqrt{\left[a^2 + a^2\sin^2\theta\left(\frac{d\phi}{d\theta}\right)^2\right]}\right\}d\theta \equiv L(\phi, \phi'; \theta)d\theta.$$

where $\phi' = d\phi/d\theta$. Then

$$S = \int L(\phi, \phi'; \theta)\, d\theta.$$

Therefore, for $S$ to be minimum we must have

$$\frac{\partial L}{\partial \phi} = \frac{d}{d\theta}\left(\frac{\partial L}{\partial \phi'}\right).$$

But $L$ does not depend on $\phi$, so

$$\frac{\partial L}{\partial \phi'} = \text{constant}.$$

This gives

$$\frac{(a^2\sin^2\theta\cdot\phi')}{\sqrt{(a^2 + a^2\sin^2\theta\phi'^2)}} = \text{constant} = c \text{ say}.$$

Then we get

$$d\phi = c\, d\theta\left(a^2\sin^4\theta - c^2\sin^2\theta\right)^{-1/2}.$$

Hence, integrating ($k$ = constant of integration)

$$(\phi + k) = c\int d\theta\left(a^2\sin^4\theta - c^2\sin^2\theta\right)^{-1/2}.$$

This can be easily integrated by substituting

$$t = c\cot\theta,$$

we then find

$$\phi + k = \sin^{-1}\frac{t}{b},$$

where $b^2 = a^2 - c^2$. That is

$$b\sin(\phi + k) = c\cot\theta.$$

Therefore,

$$c\cos\theta = b\sin\theta\sin\phi\cdot\cos k + b\sin\theta\cos\phi\cdot\sin k.$$

Multiplying $r$ on both sides this gives $z = Ax + By$. ($A$ and $B$ are suitable constant), which is an equation of a plane passing through origin. This cuts the sphere in great circles (called *Geodesics*). An important application of these varieties of results lies in the determination of the equation of the curve representing the shortest distance between two points on the surface of a right circular cylinder. The element of distance on the surface of a cylinder is given by

$$ds = \sqrt{[(dz)^2 + (dr)^2 + r^2(d\theta)^2]}.$$

But $r = a =$ constant radius of the base so

$$ds = \sqrt{[(dz)^2 + a^2(d\theta)^2]} = \left\{\sqrt{\left[\left(\frac{dz}{d\theta}\right)^2 + a^2\right]}\right\} d\theta = L(z, z'; \theta)\, d\theta$$

with $z' = dz/d\theta$. Hence for minimum $S$,

$$\frac{\partial L}{\partial z'} = \text{constant} = k.$$

Therefore,

$$z'\left[\sqrt{(z'^2 + a^2)}\right]^{-1/2} = k$$

which gives $z = a\theta + b$ ($a$ and $b$ are constants).

This is an equation for a helix.

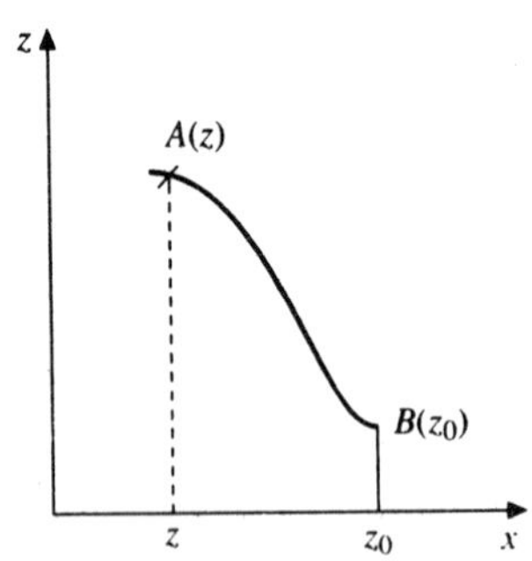

Fig 2.2

As another application of Euler-Lagrange-equation we take up the following problem. We want to find a plane curve along which a particle falls under gravity from a higher to a lower point on the curve in a shortest possible time.

Let $AB$ be the plane curve; Fig 2.2 a particle of mass $m$ at an instant of time $t$ be at $A$ at a height $z$ and $v$ be the velocity along the curve. Let $B$ at a height $z_0$ be the final position and $v_0$ be the velocity of the particle.

Total energy

$$\text{at}\quad A = (1/2)mv^2 + mgz$$

$$\text{and at}\quad B = (m/2)v_0^2 + mgz_0.$$

By conservation,

$$\frac{1}{2}mv^2 + mgz = \frac{1}{2}mv_0^2 + mgz_0.$$

Then from this we get

$$v^2 = v_0^2 + 2g(z_0 - z). \tag{2.32}$$

If $ds$ denotes an element of arc of $AB$ covered by the particle in time $dt$, then

$$v^2 = \left(\frac{ds}{dt}\right)^2.$$

Hence, from (2.32)

$$dt = ds\left[v_0^2 + 2g(z_0 - z)\right]^{-1/2}. \tag{2.33}$$

Therefore time taken $t$ is

$$t = \int ds\left[v_0^2 + 2g(z_0 - z)\right]^{-1/2}. \tag{2.34}$$

But

$$ds = \left[(dx)^2 + (dz)^2\right]^{1/2};$$

we then have

$$ds = dz(1 + x'^2)^{1/2},$$

where $x' = dx/dz$. Therefore,

$$t = \int (1 + x'^2)^{1/2} dz\left[v_0^2 + 2g(z_0 - z)\right]^{-1/2} = \int L(x, x', z)\, dz. \tag{2.35}$$

For minimum $t$,

$$\frac{\partial L}{\partial x} = \frac{d}{dz}\frac{\partial L}{\partial x'}$$

must hold. As $L$ does not depend on $x$, we get

$$\frac{\partial L}{\partial x'} = \text{constant} = c_1 \text{ say}.$$

Then from $\partial L/\partial x'$, we find

$$x'^2(1 + x'^2)^{-1}\left[v_0^2 + 2g(z_0 - z)\right]^{-1} = c^2;$$

we can simplify to obtain

$$x'^2 = \left(\frac{dx}{dz}\right)^2 = \frac{(b - z)}{(a + z)}, \tag{2.36}$$

where

$$a = \frac{(1 - c^2 v_0^2 - 2gc^2 z_0)}{2gc^2} \quad \text{and} \quad b = \frac{(c^2 v_0^2 + 2gc^2 z_0)}{2gc^2}.$$

We integrate (2.36), by substituting

$$z = -\frac{a-b}{2} - \frac{a+b}{2}\cos\theta \tag{2.37}$$

then

$$x = \frac{a+b}{2}(\theta - \sin\theta) + \text{constant}. \tag{2.38}$$

This pair of equations (2.37) and (2.38) yield a parametric representation of the plane curve known as *epicycloid.*

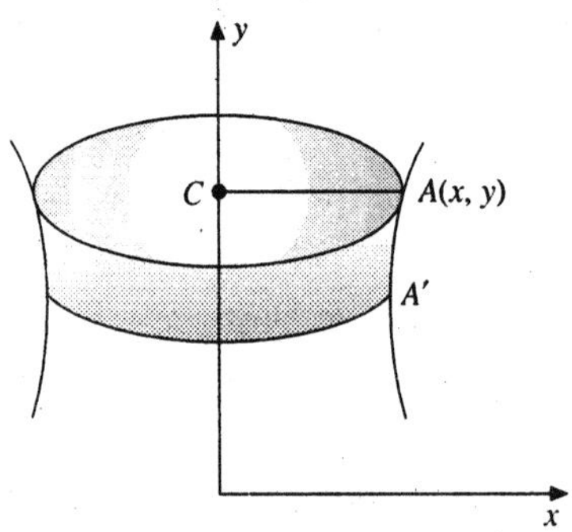


Fig 2.3

As a final example we consider the application to a geometrical problem: To find the equation of a plane curve whose surface of revolution has a minimum area Fig 2.3.

$A$ and $A'$ are nearby points on a plane curve in $xy$ plane. The curve is revolving around $y$-axis. $C$ is the centre of the circle with radius $CA$. A has the co-ordinate $(x, y)$. Then surface area of revolution covered by $AA'$ is then $2\pi\cdot CA\cdot AA'$. Then $AA' = ds$, element of arc of the curve. $CA = x$. Hence, $dA$, element of surface of revolution is

$$2\pi x\, ds, \quad \text{or,} \quad dA = 2\pi x\left[(dx)^2 + (dy)^2\right]^{1/2} = 2\pi x\left[1 + \left(\frac{dy}{dx}\right)^2\right]^{1/2} dx$$

$$= 2\pi L(y, y'; x)dx.$$

For minimum area, we have

$$\frac{\partial L}{\partial y} = \frac{d}{dx}\frac{\partial L}{\partial y'}$$

as $L$ does not depend on $y$, we get

$$\frac{\partial L}{\partial y'} = \text{constant} = k, \text{ say.}$$

Then from

$$L = 2\pi x(1 + y'^2)^{1/2}$$

we get

$$\frac{2\pi x\cdot y'}{(1 + y'^2)^{1/2}} = k.$$

Therefore,

$$\frac{dy}{dx} = y' = \frac{c}{\sqrt{(x^2 - c^2)}}; \quad c^2 = \frac{k^2}{4\pi^2}.$$

Hence,

$$y = c\int dx(x^2 - c^2)^{-1/2}.$$

This can easily be integrated by substituting

$$x = c\cosh\theta.$$

Then, $y = c\theta$, we have put constant of integration $= 0$, for simplicity. Therefore,

$$\theta = \frac{y}{c}, \quad \text{i.e.,} \quad \frac{y}{c} = \cosh^{-1}\frac{x}{c}, \quad \text{or,} \quad x = c\cosh\frac{y}{c}. \tag{2.39}$$

This is the equation for catenary.

## 2.4 Symmetry and Conservation Laws

We believe the space around us is homogeneous and isotropic. In fact, if we close ourselves in a room in a particular place it is extremely difficult to know whether we are in Bombay or any other place. *We really do not know where we are.* A particular physical observation is made at one place will not be different at another place. That is physical result is unchanged by *space translation.* Let us therefore consider a Lagrangian of $N$-particle system with co-ordinates in configuration space given by

$$\mathbf{r}_i, \quad i = 1, \cdots, N$$

and these have masses

$$m_i, \quad i = 1, \cdots, N, \text{ respectively.}$$

We are assuming they are free from any constraint. For the present discussion, the Lagrangian is therefore a function of $\mathbf{r}_i$, $\dot{\mathbf{r}}_i$. Thus

$$L = L(\mathbf{r}_1, \cdots, \mathbf{r}_i \cdots; \dot{\mathbf{r}}_1, \cdots, \dot{\mathbf{r}}_i \cdots) \tag{2.40}$$

Consider an infinitesimal translation of the co-ordinate,

$$\mathbf{r}_i \to \mathbf{r}'_i, \quad \text{where } \mathbf{r}'_i = \mathbf{r}_i + \varepsilon \tag{2.41}$$

Let, under the translation (2.41) the Lagrangian $L \to L + \delta L$.
The change $\delta L$ in $L$ due to the infinitesimal change $\varepsilon$ in co-ordinate is given by

$$\delta L = \sum \frac{\partial L}{\partial \mathbf{r}_i}\varepsilon. \tag{2.42}$$

If under the translation Lagrangian has to remain invariant so that physical results remain unchanged, then $\delta L = 0$. Thus (2.42) implies

$$\sum_i \frac{\partial L}{\partial \mathbf{r}_i} \cdot \varepsilon = 0 \tag{2.43}$$

As $\varepsilon$ is an arbitrary non-vanishing infinitesimal displacement, then (2.43) implies

$$\sum_i \frac{\partial L}{\partial \mathbf{r}_i} = 0. \tag{2.44}$$

By Lagrange's equation of motion

$$\frac{\partial L}{\partial \mathbf{r}_i} = \frac{d}{dt}\left(\frac{\partial L}{\partial \dot{\mathbf{r}}_i}\right).$$

So equation (2.44) reduces to

$$\frac{d}{dt}\sum_i \frac{\partial L}{\partial \dot{\mathbf{r}}_i} = 0. \tag{2.45}$$

But by our choice,

$$L = T - V = \sum_i \frac{1}{2} m_i \dot{\mathbf{r}}_i^2 - V(\mathbf{r}_1, \cdots, \mathbf{r}_N).$$

Hence

$$\frac{\partial L}{\partial \dot{\mathbf{r}}_i} = m_i \dot{\mathbf{r}}_i. \tag{2.46}$$

Therefore from (2.45) and (2.46) we have

$$\frac{d}{dt}\sum_i m_i \dot{r}_i = 0; \quad \text{i.e.,} \quad \sum_i m_i \dot{\mathbf{r}}_i = \text{constant}.$$

i.e., total linear momentum is a constant of motion.

We thus obtain the following theorem: If a Lagrangian for $N$-particle system given by $L(\mathbf{r}_i, \cdots, \dot{\mathbf{r}}_i)$ be invariant under infinitesimal translation

$$\mathbf{r}_i \to \mathbf{r}_i' = \mathbf{r}_i + \varepsilon,$$

then total linear momentum is conserved.

If in anology with space translation we assume the Lagrangian to be invariant under time-translation another fundamental conservation law can be derived, i.e., the conservation of energy. Consider the time translation $t \to t' = t + \delta t$ and assume that $L \to L + \delta L$ under $t \to t + \delta t$. Then

$$\delta L = \frac{\partial L}{\partial t} \cdot \delta t. \tag{2.47}$$

If $L$ remains invariant under $t \to t+\delta t$, then $\delta L = 0$, this means from (2.47) that

$$\frac{\partial L}{\partial t} = 0 \tag{2.48}$$

as $\delta t$ is arbitrary non-vanishing infinitesimal change in time. From $dL/dt$, we find

$$\begin{aligned}\frac{dL}{dt} &= \sum_i \frac{\partial L}{\partial \mathbf{r}_i}\cdot\frac{d\mathbf{r}_i}{dt} + \sum_i \frac{\partial L}{\partial \dot{\mathbf{r}}_i}\frac{d\dot{\mathbf{r}}}{dt} + \frac{\partial L}{\partial t} \\ &= \sum_i \frac{d}{dt}\left(\frac{\partial L}{\partial \dot{\mathbf{r}}_i}\right)\dot{\mathbf{r}}_i + \sum_i \frac{\partial L}{\partial \dot{\mathbf{r}}_i}\frac{d\dot{\mathbf{r}}_i}{dt} + \frac{\partial L}{\partial t} = \frac{d}{dt}\sum_i\left(\frac{\partial L}{\partial \dot{\mathbf{r}}_i}\dot{\mathbf{r}}_i\right) + \frac{\partial L}{\partial t}.\end{aligned} \tag{2.49}$$

But for $L$ to be invariant, $\partial L/\partial t = 0$, so (2.49) reduces to

$$\frac{d}{dt}\left[L - \sum_i\left(\frac{\partial L}{\partial \dot{\mathbf{r}}_i}\dot{\mathbf{r}}_i\right)\right] = 0. \tag{2.50}$$

(2.50) implies

$$\sum_i\left(\frac{\partial L}{\partial \dot{\mathbf{r}}_i}\right)\dot{\mathbf{r}}_i - L \text{ is a constant of motion.} \tag{2.51}$$

Since

$$\begin{aligned}L &= L(\mathbf{r}_i,\cdots,\dot{\mathbf{r}}_i,\cdots) = T - V \\ &= \sum_i \frac{1}{2}m_i\dot{\mathbf{r}}_i^2 - V(\mathbf{r}_i\cdots).\end{aligned}$$

therefore,

$$\frac{\partial L}{\partial \dot{\mathbf{r}}_i} = m_i\dot{\mathbf{r}}_i = \text{momentum of } i\text{th particle}$$

and

$$\sum_i(m_i\dot{\mathbf{r}}_i)\dot{\mathbf{r}}_i = \sum_i m\dot{\mathbf{r}}_i^2.$$

Therefore,

$$\begin{aligned}\sum_i\left(\frac{\partial L}{\partial \dot{\mathbf{r}}_i}\right)\dot{\mathbf{r}}_i - L &= \sum_i m_i\dot{\mathbf{r}}_i^2 - \frac{1}{2}\sum_i m_i\dot{\mathbf{r}}_i^2 + V \\ &= \sum_i \frac{1}{2}m_i\dot{\mathbf{r}}_i^2 + V = T + V.\end{aligned}$$

Therefore, from (2.51) it implies total energy $T+V$ is a constant of motion. We could have proved the result with generalised co-ordinate also. For example, let

$$L = L(q_1,\cdots,q_r;\dot{q}_1,\cdots,\dot{q}_r,\cdots;t)$$

under the time-translation. $t \to t + \delta t$, changes to $L + \delta L$, where

$$\delta L = \frac{\partial L}{\partial t}\delta t.$$

Hence invariance of $L$ under time-translation requires $\partial L/\partial t = 0$. That is, $L$ does not contain $t$, explicitly. Now, let us calculate

$$\begin{aligned}
\frac{dL}{dt} &= \frac{\partial L}{\partial t} + \sum_r \frac{\partial L}{\partial q_r}\dot{q}_r + \sum_r \frac{\partial L}{\partial \dot{q}_r}\frac{d}{dt}\dot{q}_r \\
&= \frac{\partial L}{\partial t} + \sum_r \left[\frac{d}{dt}\left(\frac{\partial L}{\partial \dot{q}_r}\right)\dot{q}_r + \frac{\partial L}{\partial \dot{q}_r}\frac{d}{dt}\dot{q}_r\right] \\
&= \frac{\partial L}{\partial t} + \sum_r \frac{d}{dt}\left(\frac{\partial L}{\partial \dot{q}_r}\dot{q}_r\right).
\end{aligned}$$

As $\partial L/\partial t = 0$, so

$$\frac{d}{dt}\left(L - \sum_r \frac{\partial L}{\partial \dot{q}_r}\dot{q}_r\right) = 0.$$

From $L = T - V$, and since $V$ does not contain $\dot{q}_r$, we get

$$\frac{d}{dt}\left(L - \sum_r \frac{\partial T}{\partial \dot{q}_r}\dot{q}_r\right) = 0.$$

Since $t$ does not occur explicitly,

$$\sum_r \dot{q}_r \frac{\partial T}{\partial \dot{q}_r} = 2T.$$

Therefore,

$$\frac{d}{dt}\left(L - \sum_r \frac{\partial T}{\partial \dot{q}_r}\dot{q}_r\right) = \frac{d}{dt}(L - 2T) = \frac{d}{dt}(T - V - 2T) = 0.$$

$$\text{or, } \frac{d}{dt}(T + V) = 0,$$

i.e., $T + V = E$ is conserved.

If there are many particle systems then the total Lagrangian is obtained by adding them so time-translation invariance implies

$$\sum_a \frac{d}{dt}\left(L^{(a)} - \sum_r \frac{\partial L^{(a)}}{\partial \dot{q}_r}\dot{q}_r\right) = 0,$$

i.e., $\sum_\alpha E^\alpha$ is conserved. Hence energy is additively conserved.

## 2.5 Rotational Symmetry and Conservation of Angular Momentum

Let the vector $\overline{OA}$ represent $\mathbf{r}$ which makes an angle $\theta$ with $z$-axis Fig 2.4. The vector $\mathbf{r}$ rotates around $z$-axis and takes the position $OA'$, which we denote by $\mathbf{r}'$. But

$$\overline{OA'} = \overline{OA} + \overline{AA'}.$$

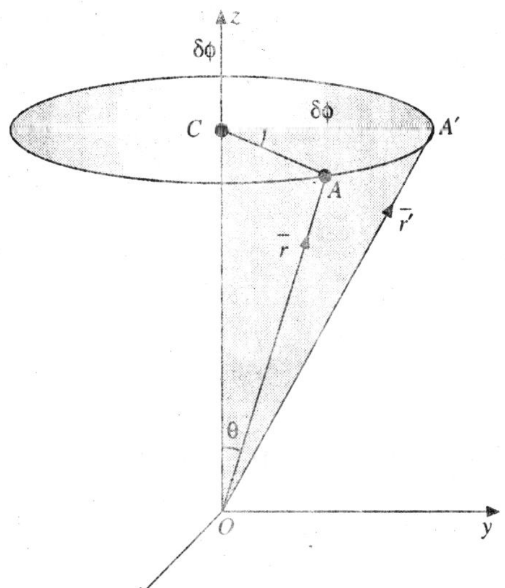


Fig 2.4

$AA'$ is equal to $r \sin\theta \cdot \delta\phi$ in magnitude. By three-dimensional rotation,

$$\mathbf{r} \to \mathbf{r}' = \mathbf{r} + \delta\mathbf{r}$$

where $\delta\mathbf{r} = \overline{AA'}$. Since

$$|\overline{AA'}| = r \sin\theta \cdot \delta\phi,$$

so

$$\delta\mathbf{r} - \overline{AA'} = (\delta\bar{\phi} \times \mathbf{r}),$$

where now the $\delta\bar{\phi}$ is along the positive $z$-axis.

Then if a vector $\mathbf{r}$ is changed to $\mathbf{r}'$ by rotation then $\mathbf{r}' = \mathbf{r} + \delta\mathbf{r}$ where $\delta\mathbf{r} = (\delta\bar{\phi} \times \mathbf{r})$. Hence if $\mathbf{v}$, is a velocity vector, then by rotation

$$\mathbf{v} \to \mathbf{v}' = \mathbf{v} + \delta\mathbf{v},$$

where $\delta\mathbf{v} = (\delta\bar{\phi} \times \mathbf{v})$.

Consider now the Lagrangian

$$L = L(\mathbf{r}_1, \cdots, \mathbf{r}_i; \dot{\mathbf{r}}_1, \cdots, \dot{\mathbf{r}}_i).$$

Suppose under rotation $L$ changes to $L'$ which is $L + \delta L$, then

$$\delta L = \sum_i \frac{\partial L}{\partial \mathbf{r}} \delta\mathbf{r} + \sum_i \frac{\partial L}{\partial \dot{\mathbf{r}}_i} \delta\dot{\mathbf{r}}_i. \qquad (2.52)$$

We use

$$\dot{\mathbf{r}}_i = \mathbf{v} \quad \text{and} \quad \frac{\partial L}{\partial \mathbf{r}_i} = \frac{d}{dt}\left(\frac{\partial L}{\partial \dot{\mathbf{r}}_i}\right)$$

by Euler-Lagrange equation. therefore. (2.52) reduces to

$$\delta L = -\sum_i \frac{d}{dt}\left(\frac{\partial L}{\partial \dot{\mathbf{r}}_i}\right)(\mathbf{r} \times \delta\bar{\phi}) - \sum_i \frac{\partial L}{\partial \dot{\mathbf{r}}_i}(v \times \delta\phi). \tag{2.53}$$

where we used

$$\begin{aligned} \delta\mathbf{r} &= +\delta\bar{\phi} \times \mathbf{r} \\ \text{and} \quad \delta\dot{\mathbf{r}} &= \delta\mathbf{v} = +\delta\bar{\phi} \times \mathbf{v}. \end{aligned}$$

Since

$$L = \sum_i \frac{1}{2} m_i \dot{\mathbf{r}}_i^2 - V(r_1, \cdots, r_i).$$

we have

$$\frac{\partial L}{\partial \dot{\mathbf{r}}_i} = m_i \dot{\mathbf{r}}_i = \mathbf{p}_i = \text{momentum}.$$

Then (2.52) reduces to

$$\delta L = -\sum_i \left[\dot{\mathbf{p}}_i(\mathbf{r}_i \times \delta\bar{\phi}) + \mathbf{p}_i(\dot{r}_i \times \delta\phi)\right] \tag{2.54}$$

$$= +\delta\bar{\phi} \cdot \sum_i \frac{d}{dt}(\mathbf{r}_i \times p_i). \tag{2.55}$$

If $L$ is invariant under rotation then $\delta L = 0$. and from (2.55).

$$\frac{d}{dt}\sum_i (\mathbf{r}_i \times \mathbf{p}_i) = 0. \tag{2.56}$$

But

$$\mathbf{r}_i \times \mathbf{p}_i = \mathbf{L}_i = \text{ angular momentum.}$$

Thus from (2.56)

$$\frac{d}{dt}\sum_i \mathbf{L}_i = 0.$$

i.e.. total angular momentum is conserved.

In obtaining the conservation of angular momentum we had to assume the rotational invariance of the Lagrangian. It is indeed justified. because the sapce around is completely *isotopic*. Any direction in which we look appears same. In other words. we do not know which direction we are looking.

## 2.6 Application of Variational Principle to Non-conservative System

To obtain Lagrange type equation for a system with non-conservative forces, we choose the Lagrangian $L$, as

$$L = T + W \tag{2.57}$$

where $T$ is the kinetic energy and $W$ is the *virtual-work* done by the non-conservative force $\mathbf{F}^{(i)nc}$. That is

$$\delta W = \sum_{i=1}^{N} \mathbf{F}^{(i)nc} \cdot \delta \mathbf{r}_i. \tag{2.58}$$

where $\delta \mathbf{r}_i$ is a virtual displacement. The co-ordinates $\mathbf{r}_i$ is given in terms of generalised co-ordinates namely,

$$\mathbf{r}_i = \mathbf{r}_i(q_1, \cdots, q_r; t). \tag{2.59}$$

Hence for virtual $\delta \mathbf{r}_i$,

$$\delta \mathbf{r}_i = \sum_j \frac{\partial \mathbf{r}_i}{\partial q_j} \delta q_j.$$

Therefore,

$$\delta W = \sum_j \left[ \sum_i \mathbf{F}^{(i)nc} \frac{\partial \mathbf{r}_i}{\partial q_j} \right] \delta q_j = \sum_j Q_j \delta q_j \tag{2.60}$$

$$\text{and} \quad Q_j = \sum_i F^{(i)nc} \frac{\delta \mathbf{r}_i}{\partial \mathbf{q}_j}. \tag{2.61}$$

The kinetic energy $T$, is in term of generalised variables is

$$T = T(q_j, \dot{q}_j, \cdots). \tag{2.62}$$

The action

$$S = \int L dt = \int (T + W)\, dt. \tag{2.63}$$

Variation of $S$ gives

$$\begin{aligned}
\delta S &= \int \delta T dt + \int \delta W dt \\
&= \int \sum_j \left( \frac{\partial T}{\partial q_j} \partial q_j + \frac{\partial T}{\delta \dot{q}_j} \delta \dot{q}_j \right) dt + \sum_j \int Q_j \cdot \delta q_j dt \\
&= \sum_j \int dt \left[ \frac{\partial T}{\partial q_j} - \frac{d}{dt} \left( \frac{\partial T}{\partial \dot{q}_j} \right) + Q_j \right] \delta q_j.
\end{aligned} \tag{2.64}$$

Hence, from $\delta S = 0$, we find as $\delta q_j$ is arbitrary,

$$\frac{\partial T}{\partial q_j} - \frac{d}{dt}\left(\frac{\partial T}{\partial \dot{q}_j}\right) = -Q_j \tag{2.65}$$

when $Q_j$ is the generalised force given by (2.61).
Equation (2.65) is the desired Lagrange's equation for the non-conservative system. As an example, consider a dissipative system viz., the systems with frictional forces being present. The presence of friction induces a loss of energy in the system. An example of a dissipative force,

$$F_i^d = -k_i \dot{\mathbf{x}}_i. \tag{2.66}$$

i.e., a force proportional to velocity ($k_i$ = constant). This is descriped in terms of another function called *Rayleigh-Dissipative function*,

$$R = \frac{1}{2}\sum_i k_i \dot{x}_i^2. \tag{2.67}$$

Then

$$F_i^{(d)} = -\frac{\partial R}{\partial \dot{x}_i}. \tag{2.68}$$

Assuming that the system has otherwise a well defined Lagrangian

$$L = T - V,$$

the equation of motion in cartesion co-ordinates are

$$\frac{d}{dt}\left(\frac{\partial L}{\partial \dot{x}_i}\right) - \frac{\partial L}{\partial x_i} + \frac{\partial R}{\partial \dot{x}_i} = 0. \tag{2.69}$$

Transforming to generalised co-ordinate system, the generalised dissipative force $Q_j^d$ can be written as,

$$Q_i^d = \sum_j F_j^d \frac{\delta x_j}{\delta q_i} = -\sum_j k_j \dot{x}_j \frac{\partial \dot{x}_j}{\partial \dot{q}_i} = \frac{\partial}{\partial \dot{q}_i}\left(-\frac{1}{2}\sum_j k_j \dot{x}_j^2\right) = \frac{\partial R}{\partial \dot{q}_i}. \tag{2.70}$$

Combining this result with the result for the conservative part given interms of generalised co-ordinate we get

$$\frac{d}{dt}\left(\frac{\partial L}{\partial \dot{q}_j}\right) - \frac{\partial L}{\partial q_j} = Q_j^{(d)} = -\frac{\partial R}{\partial \dot{q}_j}.$$

$$\text{or,}\quad \frac{d}{dt}\left(\frac{\partial L}{\partial \dot{q}_j}\right) - \frac{\partial L}{\partial q_j} + \frac{\partial R}{\partial \dot{q}_j} = 0. \tag{2.71}$$

Then, for a particular choice of dissipative force the equations of motion can be put to the Lagrangian-equation form.

## 2.7 Extension to Non-Holonomic System

So far we have discussed holonomic system with $q_1, q_2, \cdots, q_n$ as $n$-independent generalised co-ordinates. Suppose that all these $n$-generalised co-ordinates are not independent. There are $m$ number of non-holonomic constraints of the following form with differential quantities, namely.

$$\sum_{i=1}^{n} a_{ri} dq_i + b_{rt}\, dt = 0; \quad r = 1, 2, \cdots, m. \tag{2.72}$$

$a_{ri}$ and $b_{rt}$ are in general functions of $q$'s and $t$.
The $q_i$'s are $n$ in number and there are $r$-non-holonomic constraints given by equation (2.72). Hence there are $(n-m)$ independent $q$'s present.
It is assumed that there is a Lagrangian $L$, involving time $t$ and all the $q_i$'s $(i = 1, 2, \cdots, n)$ and their time-derivatives are such that equations of motion follow from

$$\delta \int L dt = 0 \tag{2.73}$$

which immediately gives

$$\sum_j \int dt \left[\frac{\partial L}{\partial q_j} - \frac{d}{dt}\left(\frac{\partial L}{\partial \dot{q}_j}\right)\right] \delta q_j = 0. \tag{2.74}$$

We put

$$\delta q_j = (\delta\alpha) \cdot \eta_j.$$

Then $\eta_i$'s are not all independent as $q_i$'s are not, because of constraints given by (2.72). So far we have not used relations (2.72). Since $\delta q_j$'s are not all independent we cannot obtain the Lagrange's equation by equating the integrand to be zero as was done before. We must use (2.72) to reduce $q_i$'s from $n$ to $(n-m)$ in numbers. Then calculating

$$\delta \int L(q_i, \cdots, \dot{q}_i, \ldots t)\, dt = 0.$$

we may recall the parameter $t$, that occurs in $L$ under the integral sign is not varied. That is $\delta$-variation is at constant time $t$. Then the constraints (2.72) becomes

$$\sum_{i=1}^{n} a_{ri} dq_i = 0; \quad r = 1, 2, \cdots, m. \tag{2.75}$$

These constraints given in (2.75) can be incorporated in the $\delta$-variation equation by using Lagrange's method of undetermined multipliers.
Then equation (2.74) becomes

$$\delta\alpha \int dt \sum_i \left[\frac{\partial L}{\partial q_i} - \frac{d}{dt}\left(\frac{\partial L}{\partial \dot{q}_i}\right) + \sum_{r=1}^{m} \lambda_r a_{ri}\right] \eta_i = 0. \tag{2.76}$$

The quantities $\lambda_r, r = 1, \cdots, m$ are the unknown multipliers. We may choose the $\lambda_r$'s by equating

$$\frac{dL}{dq_i} - \frac{d}{dt}\left(\frac{\partial L}{\partial \dot{q}_i}\right) + \sum_{r=1}^{m} \lambda_r a_{ri} = 0 \text{ for } i = 1, 2, \cdots, m. \tag{2.77}$$

The choice makes (2.76) to the following

$$\delta\alpha \int dt \sum_{i=m+1}^{n} \left[\frac{\partial L}{\partial q_i} - \frac{d}{dt}\left(\frac{\partial L}{\partial \dot{q}_i}\right) + \sum_{r=1}^{m} \lambda_r a_{ri}\right] \eta_i = 0. \tag{2.78}$$

Since now $\eta_i$ are from $i = m + 1$ to $n$, so the $q_i$'s all independent. Hence for arbitrary $\delta\alpha$, we get from (2.78)

$$\frac{\partial L}{\partial q_i} - \frac{d}{dt}\left(\frac{\partial L}{\partial \dot{q}_i}\right) + \sum_{r=1}^{m} \lambda_r a_{ri} = 0 \text{ for } i = m + 1, m + 2, \cdots, n. \tag{2.79}$$

Combining (2.77) and (2.79) we derive

$$\frac{\partial L}{\partial q_i} - \frac{d}{dt}\left(\frac{\partial L}{\partial \dot{q}_i}\right) + \sum_{r=1}^{m} \lambda_r a_{ri} = 0; \quad \text{for } i = 1, \cdots, n. \tag{2.80}$$

Equation (2.77) determine $m$, $\lambda_r$'s, while (2.79) gives the $n$-equations of motion. We put (2.79) as

$$\frac{\partial L}{\partial q_i} - \frac{d}{dt}\left(\frac{\partial L}{\partial \dot{q}_i}\right) = Q_i \tag{2.81}$$

where

$$Q_i = -\sum_{r=1}^{m} \lambda_r a_{ri},$$

as generalised forces, where $i = 1, \cdots, n$.
Equation (2.81) is the Lagrange's equation with non-holomic constraints.

## Exercises

1. A particle is constrained to move on the arc of a parabola $z = 4ay^2$, ($z$-axis vertical) under the action of gravity. The parabola is rotating around the $z$-axis with constant angular velocity, $\omega$. Find at what condition the particle will be at rest on the parabola.

[*Hints:* The point $P$ on the parabola

$$z = 4ay^2$$

is rotating in a circle of radius $r$,

$$x^2 + y^2 = r^2.$$

Hence the equation for the paraboloid is

$$z = 4a(x^2 + y^2) = 4ar^2.$$

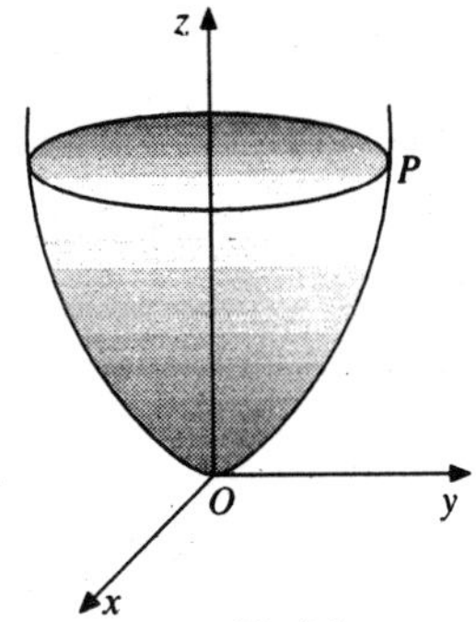

Fig 2.5

Hence the

$$\text{K.E.} = T = \frac{1}{2}m(\dot{x}^2 + \dot{y}^2 + \dot{z}^2) = \frac{1}{2}m(\dot{r}^2 + r^2\dot{\theta}^2 + \dot{z}^2).$$

$$\text{P.E.} = V = mgz = mg4ar^2; \ \dot{z} = 8ar\dot{r} \text{ and } \dot{\theta} = \omega.$$

Hence

$$L = T - V = \frac{1}{2}m(\dot{r}^2 + r^2\omega^2 + 64a^2r^2\dot{r}^2) - 4mgar^2.$$

The Lagrangian equation

$$\frac{d}{dt}\left(\frac{\partial L}{\partial \dot{r}}\right) = \frac{\partial L}{\partial r}$$

reduces to

$$\frac{d}{dt}(m\dot{r} + m64a^2r^2\dot{r}) = mr\omega^2 + m64a^2r\dot{r}^2 - 8mgar.$$

If the particle is at rest, $\dot{r} = 0$ which gives the condition

$$mr\omega^2 = 8mgar, \quad \text{i.e.,} \quad a = \omega^2/8g.]$$

2. A particle is constrained to move under gravity on the surface of a smooth sphere which is rotating with a constant angular velocity around the $z$-axis. Find the condition so that the particle may not fly off the surface.

3. A particle of mass $m$ is falling under gravity the potential $V = -mgz$ ($z$ = height) and describing the path ($a$) $z = (1/2)gt^2$, ($b$) $z = ct$ calculate the action for the paths and show that minimum occurs for the path ($a$). The constant $c$ in ($b$) is determined from the condition that the end points of the path agree at $t = t_1$.

[*Hints:* At $t = 0$, both the paths have same end points namely origin. At $t = t_1$ we have

$$\frac{1}{2}gt_1^2 = ct_1, \quad \text{i.e.,} \quad c = \frac{g}{2}t_1.$$

Hence for $(b)$

$$z = \frac{1}{2} g t_1 t$$

and $(a)$

$$z = \frac{1}{2} g t^2.$$

The $L$ for $(a)$ is

$$L_a = \frac{1}{2} m \dot{z}^2 + mg \frac{1}{2} g t^2 = \frac{1}{2} m g^2 t^2 + \frac{1}{2} m g^2 t^2 = g^2 t^2 m.$$

Action

$$S_a = \int_0^{t_1} dt\, m g^2\, t^2 = \frac{1}{3} m g^2\, t_1^3. \tag{1}$$

For path $(b)$

$$L_b = \frac{1}{2} m \left[\frac{1}{2} g t_1\right]^2 + mg \frac{1}{2}\, g t\, t_1.$$

Action

$$S_b = \int_0^{t_1} L_b dt = \frac{3}{8} m g^2 \cdot t_1^3. \tag{2}$$

Hence $S_a < S_b$. Action is minimum for path $(a)$.]

4. A particle of mass $m$ is connected by a spring of unstretched length $l_0$ and of spring constant $k$. The other end of the spring is connected to a point in the roof and the particle is swinging like a plane simple pendulum. Show that the Lagrangian equation are

$$m\ddot{r} = mr\dot{\theta}^2 - k(r - l_0) + mg\cos\theta$$

$$\text{and } \frac{d}{dt}(mr^2\dot{\theta}) = -mgr\sin\theta,$$

$\theta$ = angle made by spring with vertical.

[*Hints:* Here

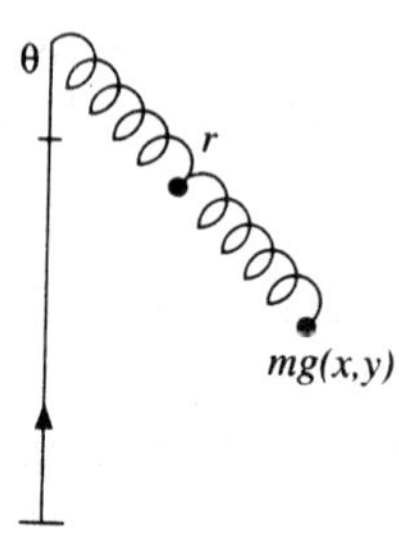


**Fig 2.6**

$$F = -k(r - r_0).$$

The potential

$$V = \frac{1}{2} k (r - r_0)^2$$

which satisfies

$$F = -\frac{\partial V}{\partial r}.$$

In addition, there is potential energy

$$V = -mgr\cos\theta.$$

$$\text{K.E.} = T = \frac{1}{2}m(\dot{x}^2 + \dot{y}^2) = \frac{1}{2}m(\dot{r}^2 + r^2\dot{\theta}^2)$$

in polar co-ordinates.

$$L = T - V = \frac{1}{2}m(\dot{r}^2 + r^2\dot{\theta}^2) - \frac{1}{2}k(r - r_0)^2 + mgr\cos\theta.$$

The equations:

$$\frac{\partial L}{\partial r} = \frac{d}{dt}\left(\frac{\partial L}{\partial \dot{r}}\right)$$

gives the first equation and

$$\frac{\partial L}{\partial \theta} = \frac{d}{dt}\left(\frac{\partial L}{\partial \dot{\theta}}\right)$$

gives the second equation.]

5. A Lagrangian for a system with one degree of freedom is

$$L = \frac{1}{2}q^2\dot{q}^2 - q^3.$$

Find Lagrangian equation and solve.

[*Hints:*

$$L = \frac{1}{2}q^2\dot{q}^2 - q^3.$$

Equation:

$$\frac{d}{dt}\left(\frac{\partial L}{\partial \dot{q}}\right) = \frac{\partial L}{\partial q} \Rightarrow \frac{d}{dt}(q^2\dot{q}) = q\dot{q}^2 - 3q^2, \quad \text{or,} \quad 2q\dot{q}^2 + q^2\ddot{q} = q\dot{q}^2 - 3q^2.$$

i.e., $q^2\ddot{q} + q\dot{q}^2 = -3q^2$, or, $q^2\ddot{q}\dot{q} + q\dot{q}^2\dot{q} = -3q^2\dot{q}$

or, $\dfrac{d}{dt}\left(\dfrac{1}{2}q^2\dot{q}^2\right) = -\dfrac{d}{dt}(q^3)$, or, $\dfrac{1}{2}q^2\dot{q}^2 = k - q^3$; $k = \text{constant}$.

This gives

$$\left[\frac{2(k - q^3)}{q^2}\right]^{1/2} = \frac{dq}{dt}. \quad \therefore t = \int dq\left(\frac{q}{\sqrt{k - q^3}}\right) + \text{constant.}]$$

6. A Lagrangian for a system with two degrees of freedom is

$$L = \dot{q}_1^2 + \dot{q}_2^2 + \dot{q}_1\dot{q}_2 - q_1^2 - q_2^2.$$

Obtain Lagrangian equation of motion.

[*Hints:*

$$L = \dot{q}_1^2 + \dot{q}_2^2 - \dot{q}_1\dot{q}_2 - q_1^2 - q_2^2$$

$$\frac{d}{dt}\left(\frac{\partial L}{\partial \dot{q}_1}\right) = \frac{\partial L}{\partial q_1}, \quad \frac{d}{dt}\left(\frac{\partial L}{\dot{q}_2}\right) = \frac{\partial L}{\partial q_2}$$

easily give:

$$2\ddot{q}_1 - \ddot{q}_2 = -2q_1$$
$$\text{and}\quad 2\ddot{q}_2 - \ddot{q}_1 = -2q_2.$$

Adding and subtracting,

$$\ddot{q}_1 + \ddot{q}_2 = -2(q_1 + q_2)$$
$$3(\ddot{q}_1 - \ddot{q}_2) = -2(q_1 - q_2).$$

Put, $X = q_1 + q_2$, $Y = q_1 - q_2$

$$\ddot{X} + 2X = 0, \quad \ddot{Y} + \frac{2}{3}Y = 0$$

both have simple harmonic oscillater solutions.]

7. A simple plane pendulum has its point of support is moving with a constant velocity, **v** along a horizontal line in the same vertical plane, where the pendulum is in motion. Find the effective Lagrangian.

8. Find the Lagrangian for a double pendulum (See figure).

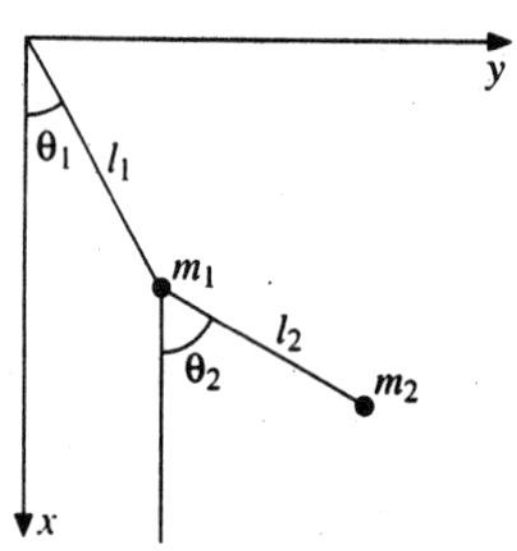

**Fig 2.7**

[*Hints:* Co-ordinate $m_1(x_1, y_1)$

$$x_1 = l_1 \cos\theta_1$$
$$y_1 = l_2 \sin\theta_1.$$

Mass $m_2(x_2, y_2)$;

$$x_2 = l_1 \cos\theta_1 + l_2 \cos\theta_2.$$
$$y_2 = (l_1 \sin\theta_1 + l_2 \sin\theta_2).$$

$$T = \frac{1}{2}m_1(\dot{x}_1^2 + \dot{y}_1^2) + \frac{1}{2}m\left(\dot{x}_2^2 + \dot{y}_2^2\right)$$
$$V = -m_1 g l_1 \cos\theta_1 - m_2 g(l_1 \cos\theta_1 + l_2 \cos\theta_2)$$
$$T = \frac{1}{2}m_1(l_1^2\dot{\theta}_1^2) + \frac{1}{2}m_2\left[l_1^2\dot{\theta}_1^2 + l_2^2\dot{\theta}_2^2 + 2l_1 l_2\dot{\theta}_1\dot{\theta}_2\cos(\theta_1 - \theta_2)\right]$$

$L = T - V$ can now be easily obtained.]

9. A particle is constrained to more on the surface of an inverted cone with semi-vertical angle $\theta$, under a force $V_1(r) + V_2(\phi)$. Find the Lagrangian equations and indicate a method of solution.

[*Hints:*

$$\text{K.E.} = T = \frac{1}{2}m(\dot{r}^2 + r^2 a^2 \dot{\phi}^2),$$

where $\theta$ is fixed. $a = \sin\theta$.

$$L = \frac{1}{2}m(\dot{r}^2 + r^2a^2\dot{\phi}^2) - V_1(r) - V_2(\phi).$$

Hence

$$\frac{d}{dt}\left(\frac{\partial L}{\partial \dot{r}}\right) = \frac{\partial L}{\partial r} \Rightarrow m\ddot{r} = -\frac{\partial V}{\partial r} + mra^2\dot{\phi}^2;$$

$$mr^2a^2\ddot{\phi} = -\frac{\partial V}{\partial \phi}.]$$

10. For the Liouville's problem the Lagrangian is given by

$$L = \sum_i f_i(q_i)\dot{q}_i^2 - \sum_i V_i(q_i).$$

Find Lagrangian equation.

[*Hints:*

$$\frac{\partial L}{\partial q_i} = -\frac{\partial V_i}{\partial q_i} + \dot{q}_i^2\frac{\partial f_i}{\partial q_i} \quad \text{and} \quad \frac{\partial L}{\partial \dot{q}_i} = 2f_i(q_i)\dot{q}_i.$$

Hence

$$2\frac{d}{dt}\{f_i(q_i)\dot{q}_i\} = -\frac{\partial V}{\partial q_i} + \dot{q}^2\frac{\partial f_i}{\partial q_i}.$$

$$\text{i.e., } 2f_i\ddot{q}_i + 2\frac{d}{dt}f_i(q_i)\dot{q} = -\frac{\partial V_i}{\partial q_i} + \frac{df_i}{dt}\dot{q}_i.$$

$$\text{i.e., } 2f_i\ddot{q}_i + \frac{df_i}{dt}\dot{q}_i = -\frac{\partial V_i}{\partial q_i}.$$

$$\text{or. } 2f_i(q_i)\ddot{q}_i\cdot\dot{q}_i + \frac{df_i(q_i)}{dt}\dot{q}_i^2 = -\frac{\partial V_i}{\partial q_i}\dot{q}_i = -\frac{d}{dt}V_i(q_i).$$

$$\text{i.e., } \frac{d}{dt}\{f_i(q_i)\dot{q}^2\} = -\frac{d}{dt}V_i(q_i).$$

Therefore,

$$f_i(q_i)\dot{q}_i^2 + V_i(q_i) = k \text{ (constant)}. \quad \dot{q}_i^2 = \frac{k - V_i(q_i)}{f_i(q_i)}.$$

Hence

$$t = \int f_i(q_i)\,dq_i\cdot[k - V_i(q_i)]^{-1/2} + \text{constant.}]$$

11. Find the differential equation of the *Geodesic* on the surface of an inverted cone with semi-vertical angle $\theta$. (The vertex is at the origin).

[*Hints:* The elementary distance $ds$ on the surface of the cone (where $\theta$ = fixed) is.

$$ds^2 = dr^2 + r^2d\theta^2 + r^2\sin^2\theta d\phi^2.$$

$d\theta = 0$, and put $\sin\theta = a$. Then

$$\begin{aligned} ds^2 &= dr^2 + a^2r^2d\phi^2; \\ ds &= \sqrt{(dr^2 + a^2r^2d\phi^2)}. \\ s &= \int \sqrt{a^2r^2\left(\frac{d\phi}{dr}\right)^2 + 1}\, dr = \int L(\phi, \phi', r)\, dr; \quad \phi' = \frac{d\phi}{dr} \end{aligned}$$

$\delta s = 0$ implies.

$$\frac{\partial L}{\partial \phi} = \frac{\partial}{\partial r}\left(\frac{\partial L}{\partial \phi'}\right) \text{ as } \frac{\partial L}{\partial \phi} = 0.$$

so we get

$$\frac{\partial L}{\partial \phi'} = \text{constant, i.e.,} \frac{a^2r^2\phi'}{\sqrt{a^2r^2\phi'^2 + 1}} = c \text{ (a constant)}.$$

$$\text{i.e.,} \quad a^4r^4\phi'^2 = c^2(a^2r^2\phi'^2 + 1), \text{ i.e.,} \quad \frac{d\phi}{dr} = \frac{c}{(a^4r^4 - c^2a^2r^2)^{1/2}}$$

is the required differential equation.]

12. A particle is moving on a cycloid $S = 4a\sin\theta$ under the action of gravity. Find the time-period of small oscillation.

[*Hints:* For the cycloid $S = 4a\sin\theta$, we can use the parametric representation.

$$\begin{aligned} x &= a(2\theta + \sin 2\theta) \\ y &= a(1 - \cos 2\theta). \end{aligned}$$

Then

$$\begin{aligned} dx &= 2a(1 + \cos 2\theta)\, d\theta, \\ dy &= +2a\sin 2\theta d\theta. \end{aligned}$$

Hence

$$ds^2 = 4a^2(2 + 2\cos 2\theta) = 16a^2\cos^2\theta.$$

This is consistent with

$$S = 4a\sin\theta.$$

The K.E. is

$$\begin{aligned} T &= \frac{1}{2}m\left(\frac{ds}{dt}\right)^2 = \frac{1}{2}m \times 16a^2\cos^2\theta\cdot\dot{\theta}^2; \\ V &= mgy = mga(1 - \cos 2\theta) \\ L &= T - V = \frac{1}{2}m \times 16a^2\cos^2\theta\cdot\dot{\theta}^2 - mg(1 - \cos 2\theta)a \\ \frac{\partial L}{\partial\dot{\theta}} &= \frac{1}{2}m \times 16a^2 \cdot \cos^2\theta\cdot\dot{\theta}. \\ \frac{\partial L}{\partial\theta} &= -\frac{1}{2}m \times 16a^2\cdot 2\cos\theta\sin\theta\cdot\dot{\theta}^2 - 2mga\sin 2\theta \\ \frac{d}{dt}\left(\frac{\partial L}{\partial\dot{\theta}}\right) &= \frac{\partial L}{\partial\theta} \end{aligned}$$

becomes

$$\frac{1}{2}m \times 16a^2\cos^2\theta\cdot\ddot{\theta} - \frac{1}{2}m \times 16a^2\cdot 2\cos\theta\sin\theta\cdot\dot{\theta}^2$$
$$= -\frac{1}{2}m \times 16a^2\cdot 2\cos\theta\sin\theta\cdot\dot{\theta}^2 - 2mga\sin 2\theta$$

or, we get

$$4 \times 2a\cdot\ddot{\theta} + 2g\frac{\sin 2\theta}{\cos^2\theta}.$$

For small amplitude $4a\ddot{\theta} = -2g\theta$.

The time period $T = 2\pi\sqrt{\frac{4a}{2g}}$.]

13. Two uniform spheres of masses $m_1$ and $m_2$ are free to move in space under the action of their gravitational attraction. Treating the spheres as point particles find their Lagrangian and equation of motion.

[*Hints:* Let $\mathbf{r}_1$ and $\mathbf{r}_2$ are the position vectors of masses $m_1$ and $m_2$. Gravitational attraction produces the potential as a function of distances between $m_1$ and $m_2$. Therefore, we introduce centre of mass co-ordinate and relative co-ordinate $\mathbf{R}$ and $\mathbf{r}$, given by

$$\mathbf{R} = \frac{(m_1\mathbf{r}_1 + m_1\mathbf{r}_2)}{(m_1 + m_2)}; \quad \mathbf{r} = \mathbf{r}_1 - \mathbf{r}_2.$$

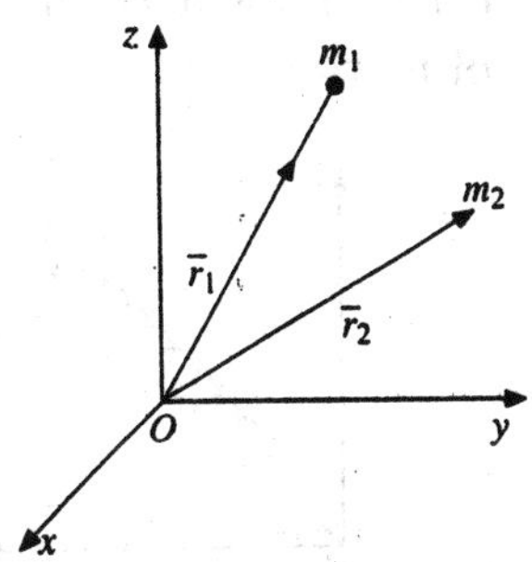

Fig 2.8

From these two equation we have

$$\mathbf{r}_1 = \mathbf{R} + \frac{m_2}{m_1 + m_2}\mathbf{r}; \quad \mathbf{r}_2 - \mathbf{R} - \frac{m_1}{m_1 + m_2}\mathbf{r}.$$

$$K.E. = T = \frac{1}{2}m_1\dot{\mathbf{r}}_1^2 + \frac{1}{2}m_2\dot{\mathbf{r}}_2^2$$
$$= \frac{1}{2}m_1\left(\dot{\mathbf{R}} + \frac{m_2}{m_1 + m_2}\dot{\mathbf{r}}\right)^2 + \frac{1}{2}m_2\left(\dot{\mathbf{R}} - \frac{m_1}{m_1 + m_2}\dot{\mathbf{r}}\right)^2$$
$$= \frac{1}{2}M\dot{\mathbf{R}}^2 + \frac{m_1 m_2}{2(m_1 + m_2)}\dot{\mathbf{r}}^2; \quad M = m_1 + m_2$$

and $$\frac{m_1 m_2}{(m_1 + m_2)} = \mu \text{ (reduced mass)}.$$

$$T = \frac{1}{2}M\dot{\mathbf{R}}^2 + \frac{1}{2}\mu\dot{\mathbf{r}}^2. \quad V = V(|\mathbf{r}_1 - \mathbf{r}_2|) = V(r) = -G\frac{m_1 m_2}{r}$$
$$L = \frac{1}{2}M\dot{\mathbf{R}}^2 + \frac{1}{2}\mu(\dot{\mathbf{r}})^2 - V(r)$$
$$= \frac{1}{2}M\dot{\mathbf{R}}^2 + \frac{1}{2}\mu(\dot{r}^2 + r^2\dot{\theta}^2 + r^2\sin^2\theta\cdot\dot{\phi}^2) - V(r)$$
$$\frac{d}{dt}\left(\frac{\partial L}{\partial \mathbf{R}}\right) = \frac{\partial L}{\partial \mathbf{R}} = 0$$

gives $M\ddot{\mathbf{R}} = 0$, C.M. motion is of free particle.

From

$$\frac{d}{dt}\frac{\partial L}{\partial \dot{r}} = \frac{\partial L}{\partial r}; \quad \frac{d}{dt}\left(\frac{\partial L}{\partial \dot{\theta}}\right) = \frac{\partial L}{\partial \theta}; \quad \frac{d}{dt}\left(\frac{\partial L}{\partial \dot{\phi}}\right) = \frac{\partial L}{\partial \phi}.$$

We get from $L$,

$$\mu\ddot{r} = \mu r\dot{\theta}^2 + \mu r \sin^2\theta\dot{\phi}^2 - G\frac{m_1 m_2}{r^2} \tag{1}$$

$$\frac{d}{dt}(\mu r^2\dot{\theta}) = \mu r^2 \sin 2\theta \cdot \dot{\phi}^2 \tag{2}$$

$$\frac{d}{dt}(\mu r^2 \sin^2\theta \cdot \dot{\phi}) = 0.]$$

14. A particle is tied to one end of a spring of spring constant $k$ and the other end is fixed on a point on a horizontal table. Write the equation of motion.

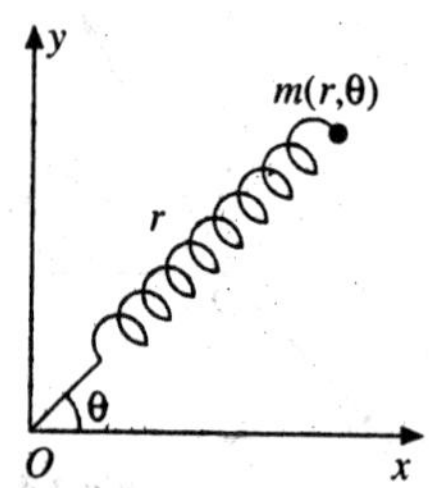


Fig 2.9

[*Hints:*

$$V = \frac{1}{2}kr^2; \quad T = \frac{1}{2}m(\dot{r}^2 + r^2\dot{\theta}^2)$$

as particle in two-dimensional.

Here

$$L = \frac{1}{2}(\dot{r}^2 + r^2\dot{\theta}^2) - \frac{1}{2}kr^2.$$

We get,

$$\frac{d}{dt}(mr^2\dot{\theta}) = 0; \quad m\ddot{r} = mr\dot{\theta}^2 - kr.]$$

15. From the Lagrangians given below find the conserved quantities.

(*a*) $L = \frac{1}{2}m(\dot{r}^2 + r^2\dot{\theta}^2) - \frac{1}{2}k(r - r_0)^2$;

(*b*) $L = \frac{1}{2}m(\dot{r}^2 + r^2\dot{\theta}^2 + r^2\sin^2\theta\dot{\phi}^2) - \mu r \sin\theta \sin\phi$;

(*c*) $L = \frac{1}{2}m(\dot{r}^2 + r^2\dot{\theta}^2 + r^2\sin^2\theta\dot{\phi}^2) - mgr\cos\theta$.

[*Hints:* (*a*) As $\partial L/\partial t = 0$, $E$ is conserved. Also angular momentum $mr^2\dot{\theta}$ = constant.

(*b*) Writing $L$ as $\frac{1}{2}m(\dot{x}^2+\dot{y}^2+\dot{z}^2)-\mu y$, we find $E$, $p_x$, $p_z$ and $L_y$ are conserved.

(*c*) $E$ and $L_z$ are conserved.

16. For the Lagrangian

$$L = \dot{\mathbf{r}}^2 + \mathbf{r}\cdot\dot{\mathbf{r}} + \mathbf{r}^2,$$

determine whether angular momentum is conserved. Is it invariant under translation?

[*Hints:* If a Lagrangian is invariant under rotation, then angular momentum is conserved.

Under rotation

$$\mathbf{r} \to \mathbf{r} + d\bar{\Omega} \times \mathbf{r}$$
$$\dot{\mathbf{r}} \to \dot{\mathbf{r}} + d\Omega \times \dot{\mathbf{r}}$$

and let $L \to L + \delta L$, then

$$\delta L = \frac{\partial L}{\partial \mathbf{r}}\cdot \delta\mathbf{r} + \frac{\delta L}{\delta \dot{\mathbf{r}}}\delta\dot{\mathbf{r}},$$
$$\text{or,}\ \delta L = (2\mathbf{r} + \dot{\mathbf{r}})\cdot(d\bar{\Omega} \times \mathbf{r}) + (\mathbf{r} + 2\dot{\mathbf{r}})(d\bar{\Omega} \times \dot{\mathbf{r}}) = 0.$$

Hence $L$ is invariant, so angular momentum is conserved.

If $L$ is to be invariant under translation then

$$\frac{\partial L}{\partial \mathbf{r}} = 0.$$

But

$$\frac{\partial L}{\partial \mathbf{r}} \neq 0$$

for the given Lagrangian. Here it is not invariant under translation.]

17. The point of support of a plane simple pendulum is moving on an ellipse

$$\frac{x^2}{a^2} + \frac{y^2}{b^2} = 1$$

fixed in the same plane with parametric equation

$$x = a\cos\omega t,$$
$$y = b\sin\omega t,$$
$$\omega = \text{constant}.$$

Find the Lagrangian.

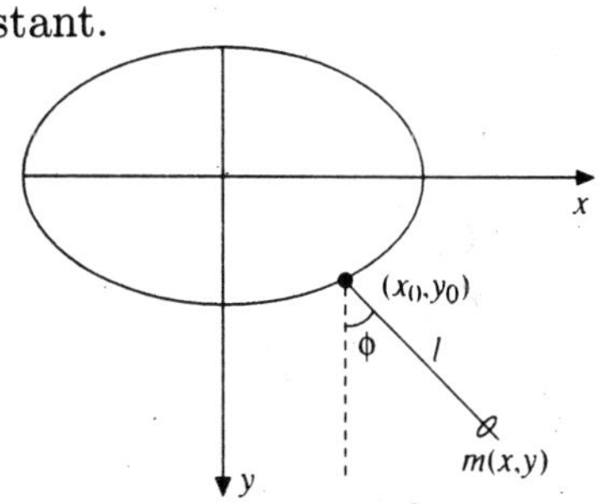

Fig 2.10

[*Hints:*

$$x_0 = a\cos\phi_0,$$
$$y_0 = b\sin\phi_0,$$

where

$$\phi_0 = \omega t; \quad \frac{x_0^2}{a^2} + \frac{y_0^2}{b^2} = 1.$$

The co-ordinate $(x, y)$ of the bob of the pendulum is

$$x = x_0 + l\sin\phi = a\cos\phi_0 + l\sin\phi = a\cos\omega t + l\sin\phi$$

$$y = y_0 + l\cos\phi = b\sin\phi_0 + l\cos\phi = +b\sin\omega t + l\cos\phi$$

$$L = \frac{1}{2}m(\dot{x}^2 + \dot{y}^2) - V.$$

where $V = -mgy = -mg(+b\sin\omega t + l\cos\phi)$.]

Chapter III

# Hamiltonian Dynamics and Poisson Bracket

## 3.1 Introduction

It is obvious from the last two chapters, that the study of classical mechanics based on Lagrangian dynamics is not only simpler but also advantageous in solving complex problems. Indeed, it has a very rich content. The study of classical theory of fields is extremely suitable via, the Lagrangian approach. In the following, we are going to develop further the structure of classical mechanics based on the Hamiltonian approach. This gives an elegant and most useful method in the theory of classical mechanics. To discuss this we first introduce a new concept called *conjugate momentum*. Consider a Lagrangian

$$L(q_1, q_2, \cdots, \dot{q}_1, \dot{q}_2, \cdots; t),$$

where $q_r(r = 1, 2, \cdots, n)$ are the generalised co-ordinates obtained after removing the holonomic constraints. We define a *conjugate momentum*, $p_r$ by

$$p_r = \frac{\partial L}{\partial \dot{q}_r} \tag{3.1}$$

$p_r$ is called the *conjugate momentum* corresponding to the generalised co-ordinate $q_r$. The pair $(q_r, p_r)$ is called the *canonical conjugate pair of variables*. In a situation, where there is no constraint over the configuration co-ordinate then we may write

$$L = T - V = \frac{1}{2}m\dot{x}^2 - V(x),$$

where $x$ is in one-dimension, then corresponding to the co-ordinate $x$, the canonical conjugate momentum $p_x$ is

$$p_x = \frac{\partial L}{\partial \dot{x}} = m\dot{x} = \text{linear momentum}.$$

Hence, for this case canonical momentum is ordinary linear momentum. Consider the situation of plane simple pendulum: the Lagrangian is

$$L = T - V = \frac{1}{2}ml^2\dot{\theta}^2 - V(\theta).$$

where $\theta$ is the generalised co-ordinate, the conjugate momentum $p_\theta$ corresponding to $\theta$ is

$$p_\theta = \frac{\partial L}{\partial \dot{\theta}} = ml^2\dot{\theta} = l(ml\dot{\theta}) = \text{angular momentum.} \tag{3.2}$$

Hence, in this case angular momentum is the canonical conjugate momentum. For a much more general situation, where the constraints do not depend on time,

$$T = \frac{1}{2}\sum_{i,j} a_{ij}\dot{q}_i\dot{q}_j; (a_{ij} = a_{ji}).$$

The potential

$$V = V(q_1, \cdots, q_n).$$

then from $L = T - V$, we find $p_r$, the conjugate momentum corresponding to $q_r$ is

$$p_r = \frac{\partial L}{\partial \dot{q}_r} = a_{rj}\dot{q}_j. \tag{3.3}$$

Hence from (3.3) we see that for $\dot{q}_r$ to have solution we must have $\det a_{rj} \neq 0$. In this connection, we introduce another definition. In a situation, when a Lagrangian $L$ does not depend on a particular co-ordinate $q_r$, say, then $q_r$ is called *cyclic*. That is for a *cyclic variable*, $q_r$,

$$\frac{\partial L}{\partial q_r} = 0. \tag{3.4}$$

Hence, by Lagrange's equation,

$$\frac{d}{dt}\left(\frac{\partial L}{\partial \dot{q}_r}\right) = 0.$$

Therefore,

$$\frac{d}{dt}(p_r) = 0. \tag{3.5}$$

That is $p_r$ is a constant of motion. So, if a Lagrangian is cyclic in one of its co-ordinate variables, then its corresponding conjugate momentum is a constant of motion (see equation 3.5).

The co-ordinate $q_r$ is also called *ignorable* co-ordinate.

## 3.2 Legendre Transformation

The main idea behind the construction of the Hamiltonian function, $H$ is to find the function

$$H(q_1, \cdots, q_n; p_1, \cdots, p_n; t)$$

from the Lagrangian $L$ which is a function of

$$q_1, \cdots, q_n; \dot{q}_1, \cdots, \dot{q}_n,$$

$$\text{i.e., } L = L(q_1, \cdots, q_r; \dot{q}_1, \cdots, \dot{q}_n; t)$$

by eliminating $\dot{q}_r$'s $(r = 1, \cdots, n)$ from $L$ in favour of $p_r$'s through the equation,

$$p_r = \frac{\partial L}{\partial \dot{q}_r}; \quad r = 1, 2, \cdots n.$$

This is achieved by means of *Legendre Transformation*. Let us consider for simplicity a function of two variables, $f(x, y)$. It is such that say,

$$u = \frac{\partial f}{\partial x}.$$

We want to construct another function $g$ of two variables $u$ and $y$ from the function $f(x, y)$. The answer to this problem is given by

$$g(u, y) = ux - f(x, y). \tag{3.6}$$

The correctness of this expression can easily be checked by calculating the differential of $g$ from the right hand side of (3.6). We have

$$\begin{aligned} dg &= du \cdot x + u\,dx - \frac{\partial f}{\partial x}\,dx - \frac{\partial f}{\partial y}\,dy \\ &= du \cdot x + u\,dx - u\,dx - \frac{\partial f}{\partial y}\,dy. \end{aligned}$$

$$\text{i.e., } dg = x\,du - \frac{\partial f}{\partial y}\,dy. \tag{3.7}$$

The right-hand side is of the form $Adu + Bdy$. If $g = g(u, y)$ then also $dg$ yields the same form, namely,

$$dg = \frac{\partial g}{\partial u}\,du + \frac{\partial g}{\partial y}\,dy.$$

where now

$$\frac{\partial g}{\partial u} = x \quad \text{and} \quad \frac{\partial g}{\partial y} = -\frac{\partial f}{\partial y}.$$

Suppose we have a relation like

$$v = \frac{\partial f}{\partial y}$$

along with

$$u = \frac{\partial f}{\partial x}.$$

We want to construct a function $g(u, v)$ from $f(x, y)$. Then

$$g(u, v) = ux + vy - f(x, y); \tag{3.8}$$

Again, calculating the differential $dg$ we find from (3.8)

$$\begin{aligned} dg &= u\,dx + x\,du + v\,dy + y\,dv - \frac{\partial f}{\partial x}\,dx - \frac{\partial f}{\partial y}\,dy \\ &= u\,dx + x\,du + v\,dy + ydv - u\,dx - v\,dy = x\,du + y\,dv; \end{aligned}$$

which shows $g$ is a function of $u$ and $v$ only. Then consider

$$L = L(\dot{q}, q) \quad \text{and} \quad p = \frac{\partial L}{\partial \dot{q}}.$$

Identify $L$ with $f$ and $\dot{q} \to x$; $q \to y$ then

$$p = \frac{\partial L}{\partial \dot{q}} = \frac{\partial f}{\partial x} = u.$$

Then, the function

$$g(u, y) \equiv g(p, q).$$

Therefore,

$$g(p, q) = g(u, y) = ux - f$$

gives

$$g(p, q) = p\dot{q} - L(\dot{q}, q).$$

We call $g(p, q)$ = Hamiltonian function and denote it by $H$. Therefore,

$$H(q, p) = p\dot{q} - L(q, \dot{q}). \tag{3.9}$$

If

$$L = L(q_1, q_2, \cdots, \dot{q}_1, \dot{q}_2, \cdots, t)$$

and

$$p_1 = \frac{\partial L}{\partial \dot{q}_1}, \quad p_2 = \frac{\partial L}{\partial \dot{q}_2} \text{ etc.}$$

then

$$\begin{aligned} &H(q_1, q_2, \cdots; p_1, p_2, \cdots; t) \\ &= p_1\dot{q}_1 + p_2\dot{q}_2 + \cdots - L(q_1, q_2, \cdots; \dot{q}_1, \dot{q}_2, \cdots; t). \end{aligned} \tag{3.10}$$

That is

$$
\begin{aligned}
&H(q_1, \cdots, q_r, \cdots; p_1, \cdots, p_r, \cdots; t) \\
&= \sum_{r=1}^{n} p_r \dot{q}_r - L(q_1, \cdots, q_r; \dot{q}_1, \cdots, \dot{q}_r; t).
\end{aligned} \tag{3.11}
$$

Equation (3.11) gives the definition of Hamiltonian $H(q_r, p_r)$. The quantities $q_r$ and $p_r$ form a set of independent variables in Hamiltonian formalism, whereas $q_r$ and $\dot{q}_r$ form the independent set in Lagrangian formalism. We now use (3.11) to obtain Hamilton's equation of motion.

## 3.3 Hamilton's Canonical Equations of Motion

We consider the Hamiltonian

$$H(q_1, \cdots; p_1, \cdots; t)$$

given by (3.11), namely

$$
\begin{aligned}
&H(q_1, \cdots, q_r, \cdots; p_1, \cdots, p_r, \cdots; t) \\
&= \sum_r p_r \dot{q}_r - L(q_1, \cdots, q_r, \cdots; \dot{q}_1, \cdots, \dot{q}_r, \cdots; t).
\end{aligned} \tag{3.12}
$$

Let us calculate the total differential $dH$ from both side: The left-hand side yields

$$dH = \sum_r \frac{\partial H}{\partial q_r} dq_r + \sum_r \frac{\partial H}{\partial p_r} dp_r + \frac{\partial H}{\partial t} dt. \tag{3.13}$$

Considering the right hand side of (3.12), we obtain,

$$dH = \sum_r dp_r \cdot \dot{q}_r + \sum_r p_r d\dot{q}_r - \sum_r \left( \frac{\partial L}{\partial q_r} dq_r + \frac{\partial L}{\partial \dot{q}_r} d\dot{q}_r \right) + \frac{\partial L}{\partial t} \cdot dt. \tag{3.14}$$

We use

$$\frac{\partial L}{\partial q_r} = \frac{d}{dt} \frac{\partial L}{\partial \dot{q}_r} = \frac{dp_r}{dt} = \dot{p}_r \quad \text{and} \quad \frac{\partial L}{\partial \dot{q}_r} = p_r.$$

Then (3.14) becomes

$$
\begin{aligned}
dH &= \sum_r dp_r \dot{q}_r + \sum_r p_r d\dot{q}_r - \sum_r (\dot{p}_r dq_r + p_r d\dot{q}_r) + \frac{\partial L}{\partial t} dt \\
&= \sum_r \dot{q}_r dp_r - \sum_r \dot{p}_r dq_r - \frac{\partial L}{\partial t} dt.
\end{aligned} \tag{3.15}
$$

We now equate (3.13) and (3.15). Then,

$$\sum_r \frac{\partial H}{\partial q_r} dq_r + \sum_r \frac{\partial H}{\partial p_r} dp_r + \frac{\partial H}{\partial t} dt = \sum_r \dot{q}_r dp_r - \sum_r \dot{p}_r dq_r - \frac{\partial L}{\partial t} dt.$$

Equating the coefficient of $dq_r$, $dp_r$ and $dt$ we get,

$$\left.\begin{aligned}\frac{\partial H}{\partial q_r} &= -\dot{p}_r\\[2ex] \frac{\partial H}{\partial p_r} &= \dot{q}_r\end{aligned}\right\} \tag{3.16}$$

and
$$\frac{\partial H}{\partial t} = -\frac{\partial L}{\partial t}. \tag{3.17}$$

Equation (3.16) are called *Hamilton's canonical equation of motion.* We note here, that equation (3.16) are also obtained from variational principle. Because from (3.12) we can write

$$L(q_r, \dot{q}_r, t) = \sum_r p_r\dot{q}_r - H(q_r, p_r).$$

Therefore,

$$\delta\int L\,dt = \delta\int\left\{\sum_r p_r\dot{q}_r - H\,(q_r, p_r, t)\right\} = 0 \text{ yields}$$

$$\int\left(\sum_r \delta p_r\dot{q}_r + \sum_r p_r\delta\dot{q}_r - \sum_r\frac{\partial H}{\partial q_r}\delta q_r - \sum_r\frac{\partial H}{\partial p_r}\delta p_r\right)dt = 0$$

$$\text{or,}\quad \sum_r\int dt\left(\dot{q}_r\delta p - \dot{p}_r\delta q_r - \frac{\partial H}{\partial q_r}\delta q_r - \frac{\partial H}{\partial p_r}\delta p_r\right) = 0.$$

Hence, as $\delta q_r$ and $\delta p_r$ are independent arbitrary variations, we get

$$\dot{q}_r - \frac{\partial H}{\partial p_r} = 0 \quad \text{and} \quad \dot{p}_r + \frac{\partial H}{\partial q_r} = 0.$$

These are equations (3.16).
We also note that

$$\frac{dH}{dt} = \frac{\partial H}{\partial t} + \sum_r\frac{\partial H}{\partial p_r}\dot{p}_r + \sum_r\frac{\partial H}{\partial q_r}\dot{q}_r. \tag{3.18}$$

Using (3.16), this reduces to

$$\frac{dH}{dt} = \frac{\partial H}{\partial t}.$$

Then,

$$-\frac{\partial L}{\partial t} = \frac{\partial H}{\partial t} = \frac{dH}{dt}.$$

Then if

$$\frac{\partial L}{\partial t} = 0$$

it implies that $H$ is a constant of motion.

We may recall our discussion on the symmetry of Lagrangian under time-translation $t \to t + \delta t$. If $L$ is invariant under time-translation, then

$$\frac{\partial L}{\partial t} = 0.$$

Again, we have seen that if

$$\frac{\partial L}{\partial t} = 0,$$

we must have

$$\frac{d}{dt}\left[\sum_i \frac{\partial L}{\partial \dot{\mathbf{r}}_i}\dot{r}_i - L(\mathbf{r}_i, \dot{\mathbf{r}}_i, t)\right] = 0.$$

The expression within the square bracket is the Hamiltanian if we note that

$$\frac{\partial L}{\partial \dot{\mathbf{r}}_i} = \mathbf{p}_i = \text{the momentum variable.}$$

That is we find

$$\frac{d}{dt}H = 0$$

when $H = H(\mathbf{r}_i; \mathbf{p}_i)$. For this case $H = T+V$, the total energy of the system. Thus, quite generally for the generalised co-ordinates and its canonical conjugate the Hamiltonian function $H(q_r, p_r)$ may be interpreted as the energy of the system. This gives a physical meaning of the Hamiltonian function.

## 3.4 Symmetry, Conservation Laws in Hamiltonian Picture

In the last chapter, we have discussed conservation laws based on the Lagrangian theory. We briefly mention how these laws follow in Hamiltonian dynamics. The motivation for discussing them in Hamiltonian theory is that this gives a direct passage for its discussion in quantum theory.
If a Hamiltonian $H(q, p; t)$ is invariant under time-translation, $t \to t + \delta t$, then, total energy is conserved. This can be proved in the following way. Let, under time-translation, the Hamiltonian $H$,

$$H \to H + \delta H. \tag{3.19}$$

then

$$\delta H = \frac{\partial H}{\partial t}\delta t. \tag{3.20}$$

For $H$ to remain invariant $\delta H = 0$. Therefore, from (3.20)

$$\frac{\partial H}{\partial t} = 0$$

as $\delta t$ is arbitrary variation. From (3.17) then

$$\frac{dH}{dt} = \frac{\partial H}{\partial t} = -\frac{\partial L}{\partial t} = 0.$$

Thus, $H$ is conserved. But we have already noticed $H = p\dot{q} - L$ and for $q = \mathbf{r}$, $p = m\dot{\mathbf{r}}$, and

$$L = \frac{1}{2}m\dot{\mathbf{r}}^2 - V(r),$$

the quantity, $p\dot{q} - L$ reduces to $T + V$, i.e., total energy. We interpret $H$ as energy of the system. Therefore, if $H$ is invariant under time-translation, then energy is conserved. Likewise, if $H$ is invariant under infinitesimal space-translation, $q_r \rightarrow q_r + \varepsilon$, then total linear momentum is conserved. Let us suppose that under

$$q_r \rightarrow q_r + \varepsilon, \quad H(q_r, p_r) \rightarrow H + \delta H.$$

Then,

$$\delta H = \sum_r \frac{\partial H}{\partial q_r}\delta q_r. \tag{3.21}$$

Using (3.16), we get for (3.21),

$$\delta H = -\sum_r \dot{p}_r \delta q_r = -\sum_r \dot{p}_r \varepsilon. \tag{3.22}$$

As $H$ is invariant, therefore, $\delta H = 0$. Hence, (3.22) yields

$$\sum_r \dot{p}_r = 0, \quad \text{i.e.,} \quad \sum_r p_r$$

is a constant of motion. Lastly, let us show that if $H$ is invariant under 3-dimensional Euclidean rotation, then angular momentum is conserved. We consider a Hamiltonian $H(\mathbf{q}, \mathbf{p})$ containing only one $q$-vector and a $p$-vector. We know, under 3-dimension Euclidean rotation

$$\mathbf{q} \rightarrow \mathbf{q} + \delta\mathbf{q} \tag{3.23}$$

$$\text{and} \quad \mathbf{p} \rightarrow \mathbf{p} + \delta\mathbf{p},$$

where $\delta\mathbf{q} = \delta\bar{\phi} \times \mathbf{q}$; $\delta\mathbf{p} = \delta\bar{\phi} \times \mathbf{p}$ and $\delta\bar{\phi}$ is the infinitesimal rotation vector along the axis of rotation. If $H \rightarrow H + \delta H$ under rotation, then

$$\delta H = \frac{\partial H}{\partial \mathbf{p}}\delta\mathbf{p} + \frac{\partial H}{\partial \mathbf{q}}\delta\mathbf{q}.$$

By (3.16) this reduces to

$$\begin{aligned}\delta H &= -(\dot{\mathbf{q}}\cdot\mathbf{p}\times\delta\bar{\phi} + \dot{\mathbf{p}}\cdot\mathbf{q}\times\delta\bar{\phi}) = -\left[(\dot{\mathbf{q}}\times\mathbf{p})\cdot\delta\bar{\phi} + (\mathbf{q}\times\dot{\mathbf{p}})\cdot\delta\bar{\phi}\right] \\ &= -\left\{\frac{d}{dt}(\mathbf{q}\times\mathbf{p})\right\}\delta\bar{\phi}.\end{aligned} \tag{3.24}$$

but $\mathbf{L} = \mathbf{q} \times \mathbf{p}$. is the angular momentum. Thus if $H$ is unchanged then $\delta H = 0$ and this gives us from (3.24). the following.

$$\frac{d}{dt}\mathbf{L} = 0. \quad \text{i.e.. } \mathbf{L} \text{ is a constant of motion.} \tag{3.25}$$

Generalisation to many particle system is obvious. We have represented $\mathbf{L}$ in terms of generalised co-ordinates and conjugate momentum. In spherical polar variable.

$$\begin{aligned} x &= r \sin\theta \cos\phi. \\ y &= r \sin\theta \sin\phi. \\ z &= r\cos\theta. \end{aligned}$$

The $L_z$ component is $(\mathbf{r} \times \mathbf{p})_z$.

$$L_z = xp_y - yp_x = xm\dot{y} - ym\dot{x} = m(x\dot{y} - y\dot{x}). \tag{3.26}$$

We have

$$\begin{aligned} \dot{x} &= \dot{r}\sin\theta\cos\phi + r\cos\theta\cdot\dot{\theta}\cos\phi - r\sin\theta\sin\phi\cdot\dot{\phi} \\ \dot{y} &= \dot{r}\sin\theta\sin\phi + r\cos\theta\dot{\theta}\sin\phi + r\sin\theta\cos\phi\dot{\phi} \\ \dot{z} &= \dot{r}\cos\theta - r\sin\theta\cdot\dot{\theta}. \end{aligned} \tag{3.27}$$

Then

$$\begin{aligned} L_z &= mr^2\sin^2\theta\cdot\dot{\phi} \\ L_y &= m(z\dot{x} - x\dot{z}) = mr^2(\dot{\theta}\cos\phi - \dot{\phi}\sin\theta\cos\theta\sin\phi) \\ L_x &= mr^2(y\dot{z} - z\dot{y}) = mr^2(\dot{\theta}\sin\phi + \dot{\phi}\sin\theta\cos\theta\cos\phi). \end{aligned} \tag{3.28}$$

As a simple application of conservation and symmetry. let us consider the motion of a particle in a two-dimensional infinite homogeneous plane. Suppose the plane is the $xy$-plane itself. Since. it is infinite and homogeneous the motion is translationally invariant along $x$ and $y$ direction so the $x$ and $y$ component of the linear momentum must be conserved. Again the plane is invariant under 3-dimension. Euclidean rotation around $z$-axis. So $L_z$ must be conserved. Consider the motion of the particle in an infinite homogeneous cylinder. In this case the infinite cylinder is rotationally invariant around $z$-axis (assuming the axis of the cylinder to be along $z$-axis) therefore. $L_z$ is conserved. As. the infinite cylinder has a finite base. but is infinitely long. and homogeneous. it is translationally invariant along $z$-axis. so $p_z$ is conserved.

As another application of symmetry consider a particle moving in the upper half of a $xy$-plane in the potential field $V_1$. with velocity $v_1$ and travels to the lower half-plane with velocity $v_2$ and in the field of potential $V_2$ Fig 3.1. If $\theta_1$

and $\theta_2$ are the directions made by the particle with $z$-axis in the upper and lower half-plane , we can find the change in the direction from symmetry properties. First as potential changes from $z > 0$ to $z < 0$ space the momentum components in $z > 0$ and $z < 0$ are different. But in the direction perpendicular to $z$-axis along $xy$-plane, is however conserved.

We can easily find,

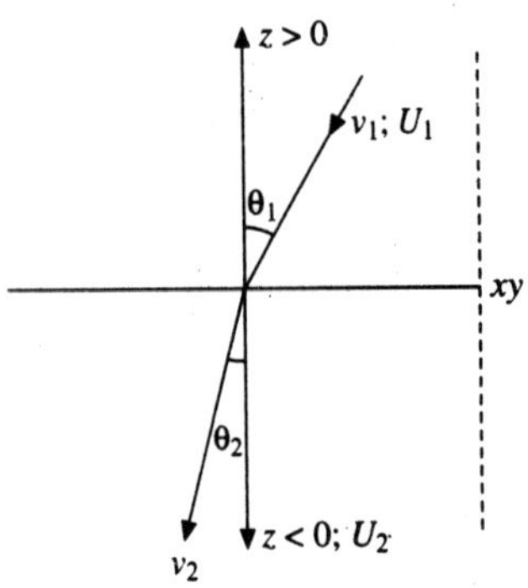


Fig 3.1

$$v_1 \sin\theta_1 = v_2 \sin\theta_2$$
$$\text{and } \frac{1}{2}mv_1^2 + V_1 = \frac{1}{2}mv_2^2 + V_2.$$

From these two relation we have

$$\frac{\sin\theta_1}{\sin\theta_2} = \frac{v_2}{v_1} = \sqrt{1 + \frac{2}{mv_1^2}(V_1 - V_2)}. \qquad (3.29)$$

We complete our discussion on symmetry by making one comment on the time-translational invariance of Hamiltonian. We have found that $H$ is conserved if the Hamiltonian does not have explicit time dependence. The energy-conservation follows if, $E = H$ and $(\partial H/\partial t) = 0$. Thus, for energy to be conserved it is necessary that $E = H$. Consider, the motion of a symmetrical top which is spinning around its vertical axis which is taken as the $z$-axis.

Therefore, the Lagrangian $L = T - V$, where

$$T = \frac{1}{2}c\dot{\phi}^2 \quad \text{and} \quad V = mgh,$$

$c$ is a constant and $\phi$ is the angle of rotation about $z$-axis. We now transform the Lagrangian with a new variable, $\alpha$

$$\alpha = \phi - \omega_0 t,$$

where now the axis system is rotating with a constant angular velocity $\omega_0$. Then $\dot{\alpha} = \dot{\phi} - \omega_0$, and $L$ is transformed to

$$L \to \frac{1}{2}c(\dot{\alpha} + \omega_0)^2 - mgh.$$

The conjugate

$$p_\alpha = \frac{\partial L}{\partial \dot{\alpha}} = c(\dot{\alpha} + \omega_0).$$

Hence, the Hamiltonian $H$ can be written as,

$$H = p_\alpha \cdot \dot{\alpha} - L = \frac{p_\alpha^2}{2c} + mgh - \omega_0 p_\alpha.$$

But

$$E = T + V = \frac{p_\alpha^2}{2c} + mgh \neq H.$$

This example shows although $t$ does not occur explicitly in $H$, i.e., $H$ is conserved but $E \neq H$. That is energy-conservation does not follow even though

$$\frac{dH}{dt} = \frac{\partial H}{dt} = 0.$$

The unusual character of $H$ arises here because of the co-ordinate system used. The co-ordinate system is rotating with respect to the frame of reference and it cannot be identified with the energy with respect to either the stationary or the moving system.

## 3.5 Construction of Hamiltonian

As an example of construction of Hamiltonian from the Lagrangian, we consider the Lagrangian for the motion of an electron in presence of electromagnetic field. The Lagrangian is given by

$$L = T - e\left(\phi - \frac{1}{c}\mathbf{v}\cdot \mathbf{A}\right) \tag{3.30}$$

with

$$T = \frac{1}{2}m\dot{\mathbf{x}}^2 = \sum_i \frac{1}{2}m\dot{x}_i^2.$$

The conjugate momentum $p_i$ is $p_i = (\partial L/\partial \dot{x}_i)$ and can be found from (3.30) to be

$$p_i = m\dot{x}_i + \frac{e}{c}A_i.$$

The Hamiltonian $H$ is given by

$$\begin{aligned} H = \sum_i p_i\dot{x}_i - L &= \sum_i \left(m\dot{x}_i + \frac{e}{c}A_i\right)\dot{x}_i - \frac{1}{2}m\dot{x}_i^2 + e\left(\phi - \frac{1}{c}\mathbf{v}.A\right) \\ &= \frac{1}{2}m\dot{x}_i^2 + e\phi = T + e\phi. \end{aligned} \tag{3.31}$$

Equation (3.31) can also be written as

$$H = \frac{1}{2m}\left(p_i - \frac{e}{c}A_i\right)^2 + e\phi. \tag{3.32}$$

As a second example we consider the motion of a particle in the field of force $F$ given by

$$F = \frac{1}{r^2}\left(1 - \frac{\dot{r}^2 - 2r\ddot{r}}{c^2}\right). \tag{3.33}$$

The corresponding potential $V$ is

$$V = \frac{1}{r}\left(1 + \frac{\dot{r}^2}{c^2}\right). \tag{3.34}$$

We have derived (3.34) in chapter I. The Lagrangian is then

$$L = T - V = \frac{1}{2}m\dot{r}^2 - \frac{1}{r}\left(1 + \frac{\dot{r}^2}{c^2}\right).$$

The conjugate momentum. $p_r$ is

$$p_r = \frac{\partial L}{\partial \dot{r}} = m\dot{r} - \frac{2\dot{r}}{rc^2}.$$

The Hamiltonian $H$ is (in one dimension)

$$H = p_r \cdot \dot{r} - L = m\dot{r}^2 - \frac{2\dot{r}^2/c^2}{r} - \frac{1}{2}m\dot{r}^2 + \frac{1}{r}\left(1 + \frac{\dot{r}^2}{c^2}\right)$$

$$H = \frac{1}{2}m\dot{r}^2 + \frac{1}{r}\left(1 - \frac{\dot{r}^2}{c^2}\right). \tag{3.35}$$

Last two examples were devoted to the construction of Hamiltonians for velocity dependent potentials. For a conservative system, consider the Lagrangian of a particle moving in a potential $V(r)$. The Lagrangian $L$ is

$$L = T - V(r)$$

and $T$ (given in polar co-ordinates) is

$$\frac{1}{2}m(\dot{r}^2 + r^2\dot{\theta}^2 + r^2\sin^2\theta \cdot \dot{\phi}^2).$$

Notice that with the form of $T$ and $V$ being a function of distance $r$, the Lagrangian does not depend on $\phi$-variable. That is $\phi$ is a cyclic variable. Therefore, by symmetry argument the corresponding conjugate momentum $p_\phi$ is a constant of motion ($\phi$ is cyclic). From $L = T - V$, we have

$$p_r = \frac{\partial L}{\partial \dot{r}} = m\dot{r}; \quad p_\theta = \frac{\partial L}{\partial \dot{\theta}} = mr^2\dot{\theta};$$

$$p_\phi = \frac{\partial L}{\partial \dot{\phi}} = mr^2\sin^2\theta \cdot \dot{\phi} = l = \text{constant}. \tag{3.36}$$

The Hamiltonian $H$ is

$$\begin{aligned} H &= p_\theta \cdot \dot{\theta} + p_r \cdot \dot{r} + p_\phi \cdot \dot{\phi} - L \\ &= m\dot{r}^2 + mr^2\dot{\theta}^2 + mr^2\sin^2\theta\dot{\phi}^2 - \frac{1}{2}m(\dot{r}^2 + r^2\dot{\theta}^2 + r^2\sin^2\theta\dot{\phi}^2) + V \end{aligned}$$

$$H = \frac{1}{2}m(\dot{r}^2 + r^2\dot{\theta}^2 + r^2\sin^2\theta\dot{\phi}^2) + V = T + V \tag{3.37}$$

$$= \frac{1}{2m}\left(p_r^2 + \frac{1}{r^2}p_\theta^2 + \frac{1}{r^2\sin^2\theta}p_\phi^2\right) + V. \tag{3.38}$$

We will discuss further the use of (3.37) in the next chapter in connection with central orbit problem.
To conclude this section, we consider the following Hamilton's equation of motion. Consider, the Hamiltonian $H$.

$$H = \frac{1}{2}\left(Zp^2 + 2Ypq + Xq^2\right). \tag{3.39}$$

where $X, Y$ and $Z$ are functions of time. This Hamiltonian represents a *generalised Harmonic Oscillator*. The equations of motion are

$$\dot{p} = -\frac{\partial H}{\partial q} = -(Yp + Xq) \quad \text{and} \quad \dot{q} = \frac{\partial H}{\partial p} = (Zp + Yq). \tag{3.40}$$

This pair of Hamilton's equation can be put as

$$\dot{C}(t) = M(t)\cdot C(t).$$

where, $C(t)$ is a column-vector $\binom{p}{q}$ and $M(t)$ is the $2\times 2$ matrix operator.

$$M = \begin{pmatrix} -Y & -X \\ Z & Y \end{pmatrix}. \tag{3.41}$$

This matrix-equation poses, however, a non-trivial problem.
We will discuss the problem once again later in this text.

## 3.6 Poisson Bracket

Let $G(q_i, p_i; t)$ be any dynamical variable of a system represented by the canonical conjugal variables $(q_r, p_r)$. The temporal development of this dynamical quantity is given by

$$\frac{dG}{dt} = \frac{\partial G}{\partial t} + \sum_r \left(\frac{\partial G}{\partial q_r}\dot{q}_r + \frac{\partial G}{\partial p_r}\dot{p}_r\right) \tag{3.42}$$

$$= \frac{\partial G}{\partial t} + \sum_r \left(\frac{\partial G}{\partial q_r}\frac{\partial H}{\partial p_r} - \frac{\partial G}{\partial p_r}\frac{\partial H}{\partial q_r}\right). \tag{3.43}$$

In obtaining the last expression we have used Hamiltons equation given by (3.16). The expression in the first bracket of equation (3.43) is defined as Poisson Bracket, namely:

$$[G, H]_{\text{P.B.}} = \sum_r \left(\frac{\partial G}{\partial q_r}\frac{\partial H}{\partial p_r} - \frac{\partial G}{\partial p_r}\frac{\partial H}{\partial q_r}\right). \tag{3.44}$$

Equation (3.44) defines the Poisson Bracket of the dynamical variable $G$ with $H$. In the same way, Poisson Bracket of two dynamical variables $A$ and

$B$ is given by

$$[A, B]_{\text{P.B.}} = \sum_r \left( \frac{\partial A}{\partial q_r} \frac{\partial B}{\partial p_r} - \frac{\partial A}{\partial p_r} \cdot \frac{\partial B}{\partial q_r} \right) \tag{3.45}$$

$$= -\sum_r \left( \frac{\partial B}{\partial q_r} \frac{\partial A}{\partial p_r} - \frac{\partial B}{\partial p_r} \cdot \frac{\partial A}{\partial q_r} \right) = -[B, A]. \tag{3.46}$$

Then

$$[A, B]_{\text{P.B.}} = -[B, A]_{\text{P.B.}} \tag{3.47}$$

If $c$ is a constant, then

$$[cA, B] = c[A, B]. \tag{3.48}$$

Let us calculate $[A + B, C]_{\text{P.B.}}$:

$$[A + B, C]_{\text{P.B.}} = \sum_r \left[ \frac{\partial(A + B)}{\partial q_r} \cdot \frac{\partial C}{\partial p_r} - \frac{\partial(A + B)}{\partial p_r} \frac{\partial C}{\partial q_r} \right].$$

That is,

$$[A + B, C]_{\text{P.B.}} = \sum_r \left( \frac{\partial A}{\partial q_r} \frac{\partial C}{\partial p_r} - \frac{\partial A}{\partial p_r} \frac{\partial C}{\partial q_r} \right) + \sum_r \left( \frac{\partial B}{\partial q_r} \frac{\partial C}{\partial p_r} - \frac{\partial C}{\partial q_r} \cdot \frac{\partial B}{\partial p_r} \right)$$
$$= [A, C]_{\text{P.B.}} + [B, C]_{\text{P.B.}}$$

Then

$$[A + B + X, C] = [A, C]_{\text{P.B.}} + [B, C]_{\text{P.B.}} + [X, C]_{\text{P.B.}} \tag{3.49}$$

Since

$$[A, B] = -[B, A],$$

we put $B = A$ in it then.
So

$$[A, A] = -[A, A], \quad \text{i.e.,} \quad [A, A] = 0. \tag{3.50}$$

If $C$ is a constant, then the P.B. (Poisson Bracket) of $C$ with any dynamic variable, $X$ is

$$[C, X] = \sum_r \left( \frac{\partial C}{\partial q_r} \frac{\partial X}{\partial p_r} - \frac{\partial C}{\partial q_r} \frac{\partial X}{\partial p_r} \right) = 0. \tag{3.51}$$

This is because

$$\frac{\partial C}{\partial q_r} = 0 = \frac{\partial C}{\partial p_r},$$

where $C$ is a constant. Again for $A, B, C$ dynamical variables,

$$[AB, C]_{\text{P.B.}} = \sum_r \left[ \frac{\partial(AB)}{\partial q_r} \frac{\partial C}{\partial p_r} - \frac{\partial(AB)}{\partial p_r} \frac{\partial C}{\partial q_r} \right]$$
$$= \sum_r \left( \frac{\partial A}{\partial q_r} \frac{\partial C}{\partial p_r} - \frac{\partial A}{\partial p_r} \frac{\partial C}{\partial q_r} \right) B + \sum_r A \left( \frac{\partial B}{\partial q_r} \frac{\partial C}{\partial p_r} - \frac{\partial B}{\partial p_r} \frac{\partial C}{\partial q_r} \right)$$
$$= [A, C]_{\text{P.B.}} B + A[B, C]_{\text{P.B.}} \tag{3.52}$$

Then

$$
\begin{aligned}
[ABC, D]_{\text{P.B.}} &= [XC, D]_{\text{P.B.}}, \quad \text{where } X = AB \\
&= X[C, D]_{\text{P.B.}} + [X, D]_{\text{P.B.}}C \\
&= AB[C, D]_{\text{P.B.}} + \{A[B, D]_{\text{P.B.}} + [A, D]_{\text{P.B.}}B\}C \\
&= AB[C, D]_{\text{P.B.}} + A[B, D]_{\text{P.B.}}C + [A, D]_{\text{P.B.}}BC. \quad (3.53)
\end{aligned}
$$

Note in (3.52) and (3.53), we have kept the order of the variables. In our present case it does not matter as they are $C$-numbers. But in operator theory they matter for instance when P.B. is replaced by commutator in quantum theory.

Let us find $[q_i, q_j]$.

$$
[q_i, q_j]_{\text{P.B.}} = \sum_r \left( \frac{\partial q_i}{\partial q_r} \cdot \frac{\partial q_j}{\partial p_r} - \frac{\partial q_i}{\partial p_r} \cdot \frac{\partial q_j}{\partial q_r} \right) = 0 \qquad (3.54)
$$

as $(\partial q/\partial p) = 0$.

Similarly,

$$
[p_i, p_j]_{\text{P.B.}} = 0. \qquad (3.55)
$$

For $[q_i, p_j]_{\text{P.B.}}$ we find, however,

$$
[q_i, p_j]_{\text{P.B.}} = \sum_r \left( \frac{\partial q_i}{\partial q_r} \frac{\partial p_j}{\partial p_r} - \frac{\partial q_i}{\partial p_r} \frac{\partial p_j}{\partial q_r} \right) = \sum_r \delta_{ir} \cdot \delta_{jr} = \delta_{ij}. \qquad (3.56)
$$

Thus

$$
[q_i, q_j]_{\text{P.B.}} = 0 = [p_i, p_j]_{\text{P.B.}}; \; [q_i, p_j]_{\text{P.B.}} = \delta_{ij}.
$$

These three Poisson Brackets are called *Fundamental Poisson Brackets* as with the help of these, one can calculate all P.B.'s of daily use.

From

$$
[q_i, H]_{\text{P.B.}}
$$

we get,

$$
\begin{aligned}
[q_i, H]_{\text{P.B.}} &= \sum_r \left( \frac{\partial q_i}{\partial q_r} \frac{\partial H}{\partial p_r} - \frac{\partial q_i}{\partial p_r} \frac{\partial H}{\partial q_r} \right) = \sum_r \delta_{ir} \frac{\partial H}{\partial p_r} \\
\text{or,} \quad [q_i, H]_{\text{P.B.}} &= \frac{\partial H}{\partial p_i}. \qquad (3.57)
\end{aligned}
$$

Likewise,

$$
[p_i, H]_{\text{P.B.}} = \sum_r \left( \frac{\partial p_i}{\partial q_r} \frac{\partial H}{\partial p_r} - \frac{\partial p_i}{\partial p_r} \frac{\partial H}{\partial q_r} \right) = -\sum_r \delta_{ir} \frac{\partial H}{\partial q_r} = -\frac{\partial H}{\partial q_i}. \qquad (3.58)
$$

Thus.

$$\dot{p}_i = -\frac{\partial H}{\partial q_i} = [p_i. H]_{\text{P.B.}} \tag{3.59}$$

$$\dot{q}_i = \frac{\partial H}{\partial p_i} = [q_i. H]_{\text{P.B.}} \tag{3.60}$$

(3.59. 3.60) are alternate way of putting Hamilton's equation in terms of Poisson Bracket. This is of fundamental importance in physics. Equation (3.59) tells us. if $p_i$ is a constant of motion. then the Poisson Bracket of $p_i$ with $H$ is zero. Thus

$$[p_i. H]_{\text{P.B.}} = 0.$$

implies $p_i$ is a constant of motion.
If $G = G(q.p)$. i.e.. time does not occurs explicitly in the expression then

$$\frac{dG}{dt} = [G. H]_{\text{P.B.}}.$$

i.e.. $G$ is a constant of motion if P.B of $G$ and $H$ is zero. when $G$ does not depend on time explicitly. In general.

$$\frac{dG}{dt} = 0$$

will imply

$$\frac{\partial G}{\partial t} = -[G. H] = [H. G].$$

For the Hamiltonian $H$. we can write

$$\frac{dH}{dt} = \frac{\partial H}{\partial t} + [H. H]_{\text{P.B.}}:$$

but $[H. H]_{\text{P.B.}} = 0$. so

$$\frac{dH}{dt} = \frac{\partial H}{\partial t}.$$

This relation has already been obtained.
Consider. the angular momentum $L_i$. We have

$$\mathbf{L} = \mathbf{q} \times \mathbf{p}. \quad \text{i.e..} \quad L_i = \varepsilon_{ijk} q_j \cdot p_k. \tag{3.61}$$

Therefore.

$$L_1 = q_2 p_3 - q_3 p_2.$$

Note. $\varepsilon_{ijk}$ is the alternating tensor. $\varepsilon_{ijk} = -\varepsilon_{jik}$. where $i. j$ and $k$ all run for 1. 2. 3. $\varepsilon_{123} = \varepsilon_{231} = \varepsilon_{312} = 1$. $\varepsilon_{112} = 0$. etc.

Thus,

$$\begin{aligned}[L_1, L_2]_{\text{P.B.}} &= [q_2p_3 - q_3p_2, q_3p_1 - q_1p_3]_{\text{P.B.}} \\ &= [q_2p_3, q_3p_1] - [q_2p_3, q_1p_3] - [q_3p_2, q_3p_1] + [q_3p_2, p_3q_1] \\ &= q_2[p_3, q_3]p_1 - 0 - 0 + p_2[q_3, p_3]q_1 \\ &= q_1p_2 - p_1q_2 = L_3. \end{aligned} \tag{3.62}$$

Thus,

$$[L_1, L_2]_{\text{P.B.}} = L_3. \tag{3.63}$$

In deriving (3.62) we used the formulae for $[AB, C]$ etc. and also used fundamental P.B.'s, i.e.,

$$[q_i, q_j]_{\text{P.B.}} = 0 = [p_i, p_j]_{\text{P.B.}} \quad \text{and} \quad [q_i, p_j] = \delta_{ij},$$
$$\text{i.e., } [q_1, p_1] = 1 = [q_2, p_2] = [q_3, p_3]$$

but $[q_1, p_2] = 0 = [q_2, p_1]$ etc., i.e., $[q_i, p_j] = 0$, wherever $i \neq j$.
Likewise

$$[L_2, L_3]_{\text{P.B.}} = L_1; \ [L_3, L_1]_{\text{P.B.}} = L_2.$$

We can combine all these to form

$$[L_i, L_j]_{\text{P.B.}} = \varepsilon_{ijk}L_k. \tag{3.64}$$

We check that for $i = 1$, $j = 2$, $k$ must be 3, so that $\varepsilon_{123} = 1$, and we find

$$[L_1, L_2]_{\text{P.B.}} = L_3,$$

as we have shown. The Poisson Bracket of $L_1$ with $H$ is

$$\begin{aligned}[L_1, H]_{\text{P.B.}} &= [q_2p_3 - p_2q_3, H]_{\text{P.B.}} = [q_2p_3, H]_{\text{P.B.}} - [p_2q_3, H]_{\text{P.B.}} \\ &= q_2[p_3, H]_{\text{P.B.}} + p_3[q_2, H]_{\text{P.B.}} - p_2[q_3, H]_{\text{P.B.}} - q_3[p_2, H]_{\text{P.B.}} \\ &= q_2\dot{p}_3 + p_3\dot{q}_2 - p_2\dot{q}_3 - q_3\dot{p}_2 = \frac{d}{dt}(q_2p_3 - q_3p_2) = \frac{d}{dt}L_1. \end{aligned} \tag{3.65}$$

That is

$$[L_i, H] = \frac{dL_i}{dt}. \tag{3.66}$$

In deriving (3.63) we used

$$[p, H]_{\text{P.B.}} = \dot{p}; \ [q, H]_{\text{P.B.}} = \dot{q}.$$

Thus, from (3.63) we find if $\mathbf{L}$ is conserved then Poisson Bracket of $\mathbf{L}$ with Hamiltonian $H$ is zero.
The angular momentum $L_i$ satisfy an interesting relation, namely

$$[[L_i, L_j]_{\text{P.B.}}, L_k]_{\text{P.B.}} + [[L_j, L_k]_{\text{P.B.}}, L_i]_{\text{P.B.}} + [[L_k, L_i]_{\text{P.B.}}, L_j]_{\text{P.B.}} = 0. \tag{3.67}$$

This is known as *Jacobi identity*. Using

$$[L_i, L_j]_{\text{P.B.}} = \varepsilon_{ijk} L_k$$

we write for the left-hand side of (3.67) the following

$$\begin{aligned}&\{\varepsilon_{ijl}\varepsilon_{lkm} + \varepsilon_{jkl}\varepsilon_{lim} + \varepsilon_{kil}\varepsilon_{ljm}\} L_m \\ &= \{\delta_{ik}\delta_{jm} - \delta_{im}\delta_{jk} + \delta_{ji}\delta_{km} - \delta_{jm}\delta_{ki} + \delta_{kj}\delta_{im} - \delta_{km}\delta_{ij}\} L_m = 0.\end{aligned}$$

Hence, Jacobi Identity for $L_i$'s is satisfied.

## 3.7 Jacobi Identity

For any three dynamical quantities $X, Y$ and $Z$, the following double P.B. relation must be true namely:

$$[X, [Y, Z]_{\text{P.B.}}]_{\text{P.B.}} + [Y, [Z, X]_{\text{P.B.}}]_{\text{P.B.}} + [Z, [X, Y]_{\text{P.B}}]_{\text{P.B.}} = 0. \qquad (3.68)$$

The relation (3.68) is called *Jacobi Identity*. (Below we will drop the suffix P.B. in the square bracket). The proof is as follows: consider the first two terms of (3.68), i.e.,

$$[X, [Y, Z]] + [Y, [Z, X]].$$

This equals

$$\begin{aligned}&[X, [Y, Z]] - [Y, [X, Z]] \\ &= \sum_i \left\{ \left[X, \frac{\partial Y}{\partial q_i}\frac{\partial Z}{\partial p_i}\right] - \left[X, \frac{\partial Z}{\partial q_i}\frac{\partial Y}{\partial p_i}\right] - \left[Y, \frac{\partial X}{\partial q_i}\frac{\partial Z}{\partial p_i}\right] + \left[Y, \frac{\partial Z}{\partial q_i}\frac{\partial X}{\partial p_i}\right] \right\} \\ &= \sum_i \left\{ \frac{\partial Y}{\partial q_i}\left[X, \frac{\partial Z}{\partial p_i}\right] + \left[X, \frac{\partial Y}{\partial q_i}\right]\frac{\partial Z}{\partial p_i} - \frac{\partial Z}{\partial q_i}\left[X, \frac{\partial Y}{\partial p_i}\right] - \left[X, \frac{\partial Z}{\partial q_i}\right]\frac{\partial Y}{\partial p_i} \right. \\ &\quad \left. - \frac{\partial X}{\partial q_i}\left[Y, \frac{\partial Z}{\partial p_i}\right] - \left[Y, \frac{\partial X}{\partial q_i}\right]\frac{\partial Z}{\partial p_i} + \frac{\partial Z}{\partial q_i}\left[Y, \frac{\partial X}{\partial p_i}\right] + \left[Y, \frac{\partial Z}{\partial q_i}\right]\frac{\partial X}{\partial p_i} \right\}. \qquad (3.69)\end{aligned}$$

Of the expression in equation (3.69) we collect the following four terms, the remaining terms will be considered later.

$$\begin{aligned}&\sum_i \left\{ -\frac{\partial Z}{\partial q_i}\left[X, \frac{\partial Y}{\partial p_i}\right] - \frac{\partial Z}{\partial q_i}\left[\frac{\partial X}{\partial p_i}, Y\right] + \left[X, \frac{\partial Y}{\partial q_i}\right]\frac{dZ}{\partial p_i} + \left[\frac{\partial X}{\partial q_i}, Y\right]\frac{\partial Z}{\partial p_i} \right\} \\ &= \sum_i \left\{ \frac{\partial}{\partial q_i}[X, Y]\frac{\partial Z}{\partial p_i} - \frac{\partial Z}{\partial q_i}\frac{\partial}{\partial p_i}[X, Y] \right\} = [[X, Y], Z]. \qquad (3.70)\end{aligned}$$

In deriving (3.70), we need

$$\frac{\partial}{\partial q}[X, Y] = \left[\frac{\partial X}{\partial q}, Y\right] + \left[X, \frac{\partial Y}{\partial q}\right].$$

This can be easily verified. The remaining terms of (3.69) are given by $R$, where

$$R = \sum_i \left\{ \frac{\partial Y}{\partial q_i}\left[X, \frac{\partial Z}{\partial p_i}\right] - \left[X, \frac{\partial Z}{\partial q_i}\right]\frac{\partial Y}{\partial p_i} - \frac{\partial X}{\partial q_i}\left[Y, \frac{\partial Z}{\partial p_i}\right] + \left[Y, \frac{\partial Z}{\partial q_i}\right]\frac{\partial X}{\partial p_i} \right\}.$$

We now expand each of the P.B's and write $R$ as

$$R = \sum_i \sum_j \left[ \frac{\partial^2 Z}{\partial p_i \partial p_j}\left(\frac{\partial Y}{\partial q_i}\frac{\partial X}{\partial q_j} - \frac{\partial X}{\partial q_i}\frac{\partial Y}{\partial q_j}\right) + \frac{\partial^2 Z}{\partial q_i \partial q_j}\left(\frac{\partial X}{\partial p_j}\frac{\partial Y}{\partial p_i} - \frac{\partial X}{\partial p_i}\frac{\partial Y}{\partial p_j}\right) \right.$$
$$\left. + \frac{\partial^2 Z}{\partial q_j \partial p_i}\left(-\frac{\partial Y}{\partial q_i}\frac{\partial X}{\partial p_j} + \frac{\partial X}{\partial q_i}\frac{\partial Y}{\partial p_j}\right) + \frac{\partial^2 Z}{\partial q_i \partial p_j}\left(-\frac{\partial X}{\partial q_j}\frac{\partial Y}{\partial p_i} + \frac{\partial Y}{\partial q_j}\frac{\partial X}{\partial p_i}\right) \right].$$

The first two terms vanish became they appear as product of a symmetric and an anti-symmetric tensor. The third term became opposite to the last term when we interchange $i \leftrightarrow j$ in the 3rd term as $i$ and $j$ occur as dummy indices. Hence, the remaining terms denoted by $R$ become equal to zero. Hence, the first two terms of equation (3.68) becomes,

$$[[X, Y], Z]$$

as given by (3.70). Adding this to the last term of (3.68) we get

$$[[X, Y], Z] + [Z, [X, Y]] \tag{3.71}$$

which is zero. Thus, we obtain the Jacobi Identity.
As a simple example consider the variables $q$, $p$, and $H$. Then, Jacobi Identity gives

$$[[q, p], H] + [[p, H], q] + [[H, q], p] = 0,$$
$$\text{or,}\quad [[q, p], H] + [\dot{p}, q] - [\dot{q}, p] = 0,$$
$$\text{or,}\quad [1, H] = [\dot{q}, p] + [q, \dot{p}] = \frac{d}{dt}[q, p] = 0 \tag{3.72}$$

as $[1, H] = 0$; we used $[q, p]_{\text{P.B.}} = 1$. The result in (3.72) is of course, obvious. We may note that the Poisson Bracket of

$$[q_i, p_j] = \delta_{ij} \tag{3.73}$$

tells us that the left-hand side of (3.73) is time-dependent as both $p$ and $q$ are time-dependent, whereas the right-hand-side is not time-dependent. Then, we should expect

$$[q_i(t + \delta t), p_j(t + \delta t)] = \delta_{ij}$$

as $[q_i(t), p_j(t)] = \delta_{ij}$.

Let us calculate

$$\begin{aligned}[q_i(t+\delta t), p_j(t+\delta t)] &= [q_i(t)+\dot{q}_i(t)\delta t, p_j(t)+\dot{p}_j(t)\delta t] \\ &= [q_i, p_j] + \delta t\,\{[\dot{q}_i, p_j] + [q_i, \dot{p}_j]\} + (\delta t)^2 \\ &= \delta_{ij} + \delta t\cdot \frac{d}{dt}[q_i, p_j] = \delta_{ij}.\end{aligned}$$

where we used

$$\frac{d}{dt}[q_i, p_j] = 0.$$

Physically this result that $[q_i, p_j]_{\text{P.B.}}$ is equal to $\delta_{ij}$ for all time is correct. If the value changes with time, then physical results will change. As we have mentioned, already that this P.B. is called the *fundamental Poisson Bracket*, with the help of this we can calculate any physical result.

## 3.8 Poisson's Theorem

If the dynamical variables $f$ and $g$ are constant of motion, then their Poisson Bracket is also a constant of motion.

We have

$$\frac{df}{dt} = \frac{\partial f}{\partial t} + [f, H] \qquad (3.74)$$

and

$$\frac{dg}{dt} = \frac{\partial g}{\partial t} + [g, H]. \qquad (3.75)$$

But,

$$\frac{df}{dt} = 0 = \frac{dg}{dt},$$

as they are constant of motion. We have for $(d/dt)[f, g]$ the following

$$\frac{d}{dt}[f, g] = \frac{\partial}{\partial t}[f, g] + [[f, g], H]\,. \qquad (3.76)$$

For any 3 variables $f, g, H$, we have the Jacobi Identity,

$$[[f, g], H] + [[g, H], f] + [[H, f], g] = 0. \qquad (3.77)$$

Hence

$$\frac{d}{dt}[f, g] = \frac{\partial}{\partial t}[f, g] - [[g, H], f] - [[H, f], g]\,. \qquad (3.78)$$

In deriving (3.78) we used (3.77).

We write (3.78) as

$$\begin{aligned}\frac{d}{dt}[f, g] &= \left[\frac{\partial f}{\partial t}, g\right] + \left[f, \frac{\partial g}{\partial t}\right] + [f, [g, H]] + [[f, H], g] \\ &= \left[\frac{\partial f}{\partial t} + [f, H], g\right] + \left[f, \frac{\partial g}{\partial t} + [g, H]\right] \\ &= \left[\frac{df}{dt}, g\right] + \left[f, \frac{dg}{dt}\right] = 0\end{aligned}$$

as

$$\frac{df}{dt} = 0 \quad \text{and} \quad \frac{dg}{dt} = 0.$$

Hence,

$$\frac{d}{dt}[f, g] = 0.$$

i.e., P.B. of $f$ and $g$ is a constant of motion.
So we proved the theorem that the Poisson Bracket of two constant of motion is also a constant of motion.

## 3.9 Lagrange Bracket

Analogous to Poisson Bracket there exists in the literature another bracket, known as *Lagrange Bracket* and it is defined as,

$$[u, v]_{\text{L.B.}} = \sum_i \left( \frac{\partial q_i}{\partial u}\frac{\partial p_i}{\partial v} - \frac{\partial q_i}{\partial v}\frac{\partial p_i}{\partial u} \right).$$

From which we also find $[u, v]_{\text{L.B.}} = -[v, u]_{\text{L.B.}}$ For the fundamental L.B.,

$$[q_i, q_j]_{\text{L.B.}} = \sum_u \left( \frac{\partial q_k}{\partial q_i}\frac{\partial p_k}{\partial q_j} - \frac{\partial q_k}{\partial q_j}\frac{\partial p_k}{\partial q_i} \right) = 0.$$

as

$$\frac{\partial p_k}{\partial q_j} = 0 = \frac{\partial p_k}{\partial q_i}.$$

Likewise

$$[p_i, p_j]_{\text{L.B.}} = 0$$
$$\text{and } [q_i, p_j]_{\text{L.B.}} = \delta_{ij}.$$

Thus, the fundamental brackes have the same value in both the bracket systems.
There exists a direct relationship between the Lagrange and the Poisson Bracket namely

$$\sum_{l=1}^{2n} [u_i, u_l]_{\text{L.B.}} [u_l, u_j]_{\text{P.B.}} = \delta_{ij}.$$

The proof is straight-forward, but involves a bit a algebra. We do not continue discussion with this further because the use of Lagrange bracket in physics is rare.

## 3.10 Routhian

Let us consider a Lagrangian of a system given by

$$L = L(q_1, \cdots, q_n; \dot{q}_1, \cdots, \dot{q}_n; t).$$

Suppose $q_1$ is a cyclic variable. Then

$$\frac{\partial L}{\partial q_1} = 0, \quad \text{or,} \quad p_1 = \text{constant},$$

where $p_1$ is the conjugate momentum of $q_1$. Since, Hamiltonian $H$ is constructed as

$$H(q_1, \cdots; p_1, \cdots; t)$$

from

$$\sum p\dot{q} - L(q_1, \cdots; \dot{q}_1, \cdots; t),$$

therefore $q_1$ also does not occur explicitly and $p = L_1$ (a constant). Then

$$H = H(q_2, \cdots; L_1, p_2, \cdots; t).$$

Neither $q_1$ nor $p_1$ occur in $H$. But

$$\dot{q}_1 = \frac{\partial H}{\partial p_1} = \frac{\partial H}{\partial L_1}.$$

This equation can be used to find $q_1 = q_1(t)$, giving the trajectory. Consider now that of the generalised co-ordinates $q_1, \cdots, q_n$ let $q_1, \cdots, q_s$ are cyclic. Then

$$\frac{\partial L}{\partial q_i} = 0 \text{ for } i = 1, \cdots, s \tag{3.79}$$

$$\text{and } p_i = \frac{\partial L}{\partial \dot{q}_i} \tag{3.80}$$

for $i = 1, \cdots, s$, are *constants*.
As

$$L = L(q_1, \cdots, q_s, q_{s+1}, \cdots, q_n; \dot{q}_1, \cdots, \dot{q}_n; t),$$

so by Legendre transformation

$$\sum_{i=1}^{s} p_i\dot{q}_i - L = R(q_1, \cdots, q_s; p_1, \cdots, p_s; q_{s+1}, \cdots, q_n, \dot{q}_{s+1}, \cdots, \dot{q}_n; t). \tag{3.81}$$

We now calculate the differential of both sides of (3.81), then

$$d\left(\sum_{i=1}^{s} p_i \dot{q}_l - L\right)$$
$$= \sum_{i=1}^{s}(dp_i\dot{q}_i + p_i d\dot{q}_i) - \sum_{i}^{s}\frac{\partial L}{\partial q_i}dq_i - \sum_{s+1}^{n}\frac{\partial L}{\partial q_i}dq_i - \sum_{1}^{n}\frac{\partial L}{\partial \dot{q}_i}d\dot{q}_i - \frac{\partial L}{\partial t}dt$$
$$= \sum_{1}^{s}(dp_i\dot{q}_i + p_i d\dot{q}_i) - \sum_{1}^{s}\dot{p}_i dq_i - \sum_{s+1}^{n}\frac{\partial L}{\partial q_i}dq_i - \sum_{1}^{s}p_i d\dot{q}_i - \sum_{s+1}^{n}\frac{\partial L}{\partial \dot{q}_i}d\dot{q}_i - \frac{\partial L}{\partial t}dt$$
$$- \sum_{1}^{s}(dp_i\dot{q}_i - dq_i\dot{p}_i) \quad \sum_{s+1}^{n}\frac{\partial L}{\partial q_i}dq_i - \sum_{s+1}^{n}\frac{\partial L}{\partial \dot{q}_i}d\dot{q}_i - \frac{\partial L}{\partial t}dt.$$

The left-hand side of (3.81) gives

$$dR(q_1,\cdots,q_s;p_1,\cdots,p_s;q_{s+1},\cdots,q_n;\dot{q}_{s+1},\cdots,\dot{q}_n;t) \tag{3.82}$$
$$= \sum_{1}^{s}\left(\frac{\partial R}{\partial q_i}dq_i + \frac{\partial R}{\partial p_i}dp_i\right) + \sum_{s+1}^{n}\frac{\partial R}{\partial q_i}dq_i + \sum_{s+1}^{n}\frac{dR}{d\dot{q}_i}d\dot{q}_i + \frac{\partial R}{\partial t}dt. \tag{3.83}$$

Hence equating (3.82) and (3.83), we obtain

$$\frac{\partial R}{\partial q_i} = -\dot{p}_i, \quad \frac{dR}{dp_i} = \dot{q}_i \tag{3.84}$$

when $i = 1, 2, \cdots, S$.

$$\frac{\partial R}{\partial q_i} = -\frac{\partial L}{\partial q_i}; \quad \frac{\partial R}{\partial \dot{q}_i} = -\frac{\partial L}{\partial \dot{q}_i}; \quad \frac{\partial R}{\partial t} = -\frac{\partial L}{\partial t} \tag{3.85}$$

when $i - s + 1, \cdots, n$.

This function $R$ is called the *Routhian*. $R$ takes place of $H$ for the cyclic variables.

$$\frac{\partial R}{\partial q_i} = 0 \text{ implying } p_i = \text{ constant.}$$

Therefore, taking the constants, $L_i$,

$$\dot{q}_i = \frac{\partial R}{\partial L_i} \quad \text{for } i = 1, 2, \cdots, s.$$

For the rest of $q_i$'s and $\dot{q}_i$'s when $i = s + 1, \cdots, n$.

$R$ satisfies the same equation as the negative of Lagrangian. $L$ satisfies Euler-Lagrange equation for $i = s + 1, \cdots, n$. Then $R$ also satisfies

$$\frac{\partial R}{\partial q_i} = \frac{d}{dt}\left(\frac{\partial R}{\partial \dot{q}_i}\right); \quad i = s + 1, \cdots, n.$$

Thus when $L$ contains both cyclic and non-cyclic variables we have *Routhian* behaves as Hamiltonian for cyclic variables and $R$ behaves as Lagrangian for the non-cyclic variables.

## 3.11 Some Simple Applications

**1.** As a simple example consider the motion of an electron in a potential.

$$V = \frac{\mu}{q}$$

in one-dimension ($\mu$ is a constant). The Hamiltonian, $H = T + V$.

$$H = \frac{p^2}{2m} + V(q), \quad \text{where } V(q) = \mu/q. \tag{3.86}$$

The Hamilton's equations are, in the Poisson Bracket form are

$$\dot{p} = [p, H]_{\text{P.B.}}; \quad \dot{q} = [q, H].$$

Now,

$$\begin{aligned}[p, H] &= \left[p, \frac{p^2}{2m}\right]_{\text{P.B.}} + [p, V(q)]_{\text{P.B.}} = 0 + [p, V(q)]_{\text{P.B.}} \\ &= \frac{\partial p}{\partial q} \cdot \frac{\partial V}{\partial p} - \frac{\partial p}{\partial p} \cdot \frac{\partial V}{\partial q} = -\frac{\partial V}{\partial q}.\end{aligned}$$

Then,

$$\left[p, +\frac{\mu}{q}\right] = +\frac{\mu}{q^2}.$$

Hence,

$$\dot{p} = \frac{\mu}{q^2}.$$

Again,

$$\begin{aligned}\dot{q} = [q, H] &= \left[q, \frac{p^2}{2m}\right] + [q, V(q)] = \frac{1}{2m}[q, p^2] \\ &= \frac{1}{2m}\{p[q, p] + [q, p]p\} = \frac{2p}{2m} = \frac{p}{m}.\end{aligned}$$

Hence

$$\dot{q} = \frac{p}{m} \text{ or, } \ddot{q} = \frac{\dot{p}}{m} = \frac{1}{m}\frac{\mu}{q^2} = \frac{k}{q^2}; \quad k = \frac{\mu}{m}.$$

The equation to be solved is

$$\ddot{q} = \frac{k}{q^2} \text{ or, } 2\dot{q}\ddot{q} = 2k\dot{q}\frac{1}{q^2}, \quad \text{or, } \quad \frac{d}{dt}(\dot{q})^2 = -2k\frac{d}{dt}\left(\frac{1}{q}\right).$$

$$\text{or, } (\dot{q})^2 + \frac{2k}{q} = C \text{ (a constant)}.$$

$$\dot{q}^2 = C - \frac{2k}{q}; \text{ or, } t = \int \frac{\sqrt{q}dq}{\sqrt{(Cq - 2k)}} + \text{Constant}. \tag{3.87}$$

Equation (3.87) gives the trajectory.

**2.** Consider the motion of a free particle with Hamiltonian

$$H = \frac{p^2}{2m_1}.$$

We can show that the quantity $F$ given by

$$F = x - \frac{pt}{m}.$$

is a constant of motion.

Because

$$\frac{dF}{dt} = \frac{\partial F}{\partial t} + [F, H]_{\text{P.B.}}$$

The use of this yields the result. We have

$$\frac{\partial F}{\partial t} = -\frac{p}{m} = -\frac{m\dot{x}}{m} = -\dot{x}.$$

Again, $[F, H]_{\text{P.B.}}$ is

$$[F, H]_{\text{P.B.}} = \left[x - \frac{pt}{m}, H\right]_{\text{P.B.}} = [x, H]_{\text{P.B.}} - \frac{t}{m}[p, H]_{\text{P.B.}}$$

But

$$[p, H]_{\text{P.B.}} = \left[p, \frac{p^2}{2m}\right]_{\text{P.B.}} = 0.$$

Therefore,

$$[F, H] = [x, H] = \dot{x}.$$

So,

$$\frac{\partial F}{\partial t} + [F, H] = -\dot{x} + \dot{x} = 0.$$

Hence,

$$\frac{dF}{dt} = 0.$$

In the discussion above we have found

$$[p, V(q)]_{\text{P.B.}} = -\frac{\partial V}{\partial q} \quad \text{and} \quad [q, V(p)]_{\text{P.B.}} = \frac{\partial V}{\partial p}.$$

These suggest, that the effects of calculating

P.B. of $p$ with $V(q)$ is to differential $V(q)$ by $-(\partial/\partial q)$

and the P.B. of $q$ with $V(p)$ is to differentiate $V(p)$ by $(\partial/\partial p)$.

These relations are very akin to those one finds in quantum theory, one, however, calculates Commutator Bracket and not Poisson Bracket.

**3.** Let us now apply the P.B. relations involving angular momentums, $L_i$. Consider $[p_i, L_j]_{\text{P.B.}}$.

$$[p_i, L_j] = [p_i, \varepsilon_{jkl}q_k p_l] = \varepsilon_{jkl}[p_i, q_k, p_l].$$

But

$$[p_i, q_k p_l] = [p_i, q_k]p_l = -p_l\delta_{ik}.$$

Therefore,

$$[p_l, L_j] = -\varepsilon_{jkl}p_l\delta_{ik} = -\varepsilon_{jil}p_l. \quad \text{or,} \quad [p_i, L_j] = \varepsilon_{ijl}p_l.$$

By similar argument

$$[q_i, L_j] = \varepsilon_{ijl}q_l.$$

Therefore,

$$[\mathbf{p}\cdot\mathbf{q}, L_j] = \sum_i [p_i q_i, L_j].$$

But

$$[p_i q_i, L_j] = p_i[q_i, L_j] + [p_i, L_j]q_i = \varepsilon_{ijl}p_i q_l + \varepsilon_{ijl}q_i p_l$$

(Summation over repeated indices underflood).

Hence,

$$[\mathbf{p}, \mathbf{q}, L_j] = \varepsilon_{ijl}(p_i q_l + p_l q_i) = 0.$$

because it is the product of symmetric $(p_i q_l + p_l q_i)$ term with anti-symmetric $\varepsilon_{ijl}$ tensor.

Then, if $\phi$ is any scalar, then $[\phi, L_j] = 0$. One can expand $\phi$ as terms involving $\mathbf{p}\cdot\mathbf{q}$, $\mathbf{p}^2\cdot\mathbf{q}^2$, etc. Since

$$L^2 = L_1^2 + L_2^2 + L_3^2.$$

where $L_i$ is angular momentum $L^2$ is scalar. Then $[L^2, L_j] = 0$, i.e., P.B. of components of angular momentum with the magnitude $L^2$, vanishes.

**4.** If the Hamiltonian, $H$ is a constant of motion, then

$$\frac{dH}{dt} = \frac{\partial H}{\partial t} = 0;$$

since $p$ does not contain time explicitly

$$\frac{\partial}{\partial t}[p, H] = 0.$$

Therefore,

$$\frac{d}{dt}[p, H] = [[p, H], H].$$

So,

$$\ddot{p} = [[p, H], H]$$

and similarly

$$\ddot{q} = [[q, H], H].$$

We can continue the higher order-derivatives to connect with higher order Poisson Brackets. We can therefore show that

$$p(t) = p(0) + t[p, H] + \frac{t^2}{2!}[[p, H], H] + \frac{t^3}{3!}[[[p, H], H], H] + \cdots,$$

where of course the P.B.'s are evaluated at $t = 0$. A similar expansion holds for $q(t)$.

**5.** Consider the value of $[p, [p, H]]$, where

$$H = \frac{p^2}{2m} + V(q).$$

We have

$$[p, H] = -\frac{\partial V}{\partial q}, \quad [p, [p, H]] = -\left[p, \frac{\partial V}{\partial q}\right] = \frac{\partial^2 V}{\partial q^2},$$

$$\text{or, } [H, p] = \frac{\partial V}{\partial q}, \quad [[H, p], p] = \frac{\partial^2 V}{\partial q^2}, \cdots.$$

Therefore, from

$$V(q) = V(0) + qV'(q)_{q=0} + \frac{q^2}{2!}\frac{\partial^2 V}{\partial q^2}(q)_{q=0} + \cdots,$$

we can write

$$V(q) = V(0) + q[H, p] + \frac{q^2}{2!}[[H, p], p] + \cdots$$

where all P.B.'s are evaluated at $t = 0$. The above expression gives expansion of any function in terms of P.B.'s.

**6.** To see the connection between Quantum and Classical mechanic we can show that two *operators*

$$(AB - BA) = \alpha[A, B],$$

where the right hand side expression is classical Poisson Bracket and $\alpha$ is a constant. We only verify this for $A = x$ and $B = p =$ constant time $(\partial/\partial x)$. Because if we calculate $[A, B]$ we find $[x, p] = 1$, whereas for operator relation $(xp - px)$ becomes

$$(xp - px)f(x) = x\frac{\partial}{\partial x}f - \frac{\partial}{\partial x}\cdot xf = x\frac{\partial f}{\partial x} - x\frac{\partial f}{\partial x} - f = -f.$$

i.e., $(xp - px)f = -f$, or, in operator-notation

$$(xp - px) = -1 \ (1 = \text{unit operator}).$$

So, $(xp - px) = \text{constant}[x, p]$ is satisfied.

**7.** As an use of cyclic variable we consider the solution of the equation of motion given by the Lagrangian for two variables $q_1$ and $q_2$. The

$$L = \frac{1}{2}\dot{q}_1^2 \cdot \frac{1}{a + bq_2^2} + \frac{1}{2}\dot{q}_2^2 - (c + dq_2^2).$$

where $a, b, c, d$ are constants. Note, $L$ does not contain $q_1$, i.e., $(\partial L/\partial q_1) = 0$. $q_1$ is cyclic so $(\partial L/\partial \dot{q}_1) = \text{constant}$.
That is

$$\dot{q}_1 \frac{1}{a + bq_2^2} = k \ (\text{constant}).$$

For $q_2$,

$$\frac{\partial L}{\partial q_2} = \frac{d}{dt}\frac{\partial L}{\partial \dot{q}_2} = \ddot{q}_2, \quad \text{i.e.,} \quad -\frac{b\dot{q}_1^2 q_2}{(a + bq_2^2)^2} = \ddot{q}_2, \quad \text{or,} \quad \ddot{q}_2 + bk^2 q_2 = 0.$$

where we have used

$$\frac{\dot{q}_1}{a + bq_2^2} = k.$$

i.e., $q_2$ satisfies Harmonic oscillator equation. We can use this result to find $q_1$ from

$$\dot{q}_1 = k(a + bq_2^2), \quad \text{i.e.,} \quad q_1 = \int (a + bq_2^2)\, dt + \text{ constant}.$$

where $q_2$ is $e^{\pm i\omega t}$ where $\omega^2 = bk^2$.

## 3.12 Action as a function of co-ordinates

We defined action $S$ by

$$S = \int L dt; \ \ L = \text{Lagrangian}. \tag{3.88}$$

The integral is taken along a path between points $q^{(1)}$ and $q^{(2)}$ at $t = t_1$ and $t = t_2$-respectively; thus

$$q^{(1)} = q(t_1) \quad \text{and} \quad q^{(2)} = q(t_2).$$

Of the many paths that can be found between these two points only one path is the actual one for which the action $S$ is minimum.

Let us now consider another aspect of $S$, when we regard $S$ as a quantity characterising the motion along the actual path, and compare the values of $S$ for paths having a common beginning at $q^{(1)}$, but passing through different points at $t = t_2$. That is end point variation at $t = t_2$ is different from zero but $\delta q = 0$ at $t = t_1$.
Therefore,

$$\delta S = \left[\frac{\partial L}{\partial \dot{q}}\delta q\right]_{t_1}^{t_2} + \int_{t_1}^{t_1}\left[\frac{\partial L}{\partial q} - \frac{d}{dt}\left(\frac{\partial L}{\partial \dot{q}}\right)\right]\delta q\,dt. \tag{3.89}$$

Since, for actual path

$$\frac{\partial L}{\partial q} - \frac{d}{dt}\left(\frac{\partial L}{\partial \dot{q}}\right) = 0$$

and again $\delta q = 0$ at $t = t_1$, therefore from (3.89), we get

$$\delta S = p\delta q(t_2).$$

where $p = (\partial L/\partial \dot{q})$.
Hence writing $\delta q$ for $\delta q(t_2)$ and for many co-ordinates

$$\delta S = \sum_i p_i \delta q_i$$

which immediately gives

$$\frac{\partial S}{\partial q_i} = p_i. \tag{3.90}$$

Again from $S = \int L dt$ we get

$$\frac{dS}{dt} = L. \tag{3.91}$$

Hence

$$\frac{dS}{dt} = \frac{\partial S}{\partial t} + \sum_i \frac{\partial S}{\partial q_i}\frac{dq_i}{dt} = \frac{\partial S}{\partial t} + \sum_i \frac{\partial S}{\partial q_i}\dot{q}_i = \frac{\partial S}{\partial t} + \sum_i p_i \dot{q}_i. \tag{3.92}$$

But $(dS/dt) = L$; so from (3.92)

$$L = \frac{\partial S}{\partial t} + \sum_i p_i \dot{q}_i \text{ but } L = \sum_i p_i \dot{q}_i - H.$$

Therefore,

$$H = -\frac{\partial S}{\partial t}.$$

Also for total differential

$$dS = \sum_i p_i dq_i - H dt.$$

If the end variations at $t = t_1$ and $t = t_2$. are different from zero. then

$$dS = \sum_i p_i^{(2)} dq_i^{(2)} - H^{(2)}\, dt^{(2)} - \sum_i p_i^{(1)} dq_i^{(1)} + H^{(1)} dt^{(1)}. \tag{3.93}$$

Equation (3.93) shows whatever the external forces be present in the system. the right-hand side of (3.93) is a perfect differential. Thus. only those motions are allowed for which right-hand side of (3.93) is a perfect differential.

## 3.13 Hamilton's Characteristic Function: Maupertuis Principle

Suppose we are interested only in the motion but not the trajectory. we can write a restricted form of veriational principle. Let us take the situation in which the initial and final positions are fixed and a given initial time. However. we allow the variation in final time.
Then from (3.93). we have by putting $\delta q^{(1)} = \delta q^{(2)} = 0$ and $dt^{(1)} = 0$

$$\delta S = -H\delta t. \tag{3.94}$$

where $\delta t$ is written for $dt^{(2)}$. If further we are interested in real motion in which energy is conserved then $H$ and $L$ do not involve time explicitly and $H = E$. Then (3.94) is written as

$$\delta S = -E\delta t. \quad \text{i.e..} \quad \delta S + E\delta t = 0.$$

Thus writing the action for real motion. we have for $S$.

$$S = \int_{t_0}^{t} \left(\sum p_i \dot{q}_i - H\right) dt = \int \sum p_i \dot{q}_i dt - E(t - t_0) \tag{3.95}$$

as $H$ is Energy. $E$.
Therefore,

$$S = S_0 - E(t - t_0). \tag{3.96}$$

where

$$S_0 = \sum_i \int p_i dq_i. \tag{3.97}$$

$S_0$ is called *Hamilton's characteristic action* or *abreviated action*. From (3.96).

$$\delta S = \delta S_0 - E\delta t.$$
$$\text{i.e.,} \quad \delta S + E\delta t = \delta S_0.$$

but $\delta S + E\delta t = 0$, so we have

$$\delta S_0 = 0. \tag{3.98}$$

Then (3.98) satisfies least action principle for all paths for which energy is conserved.

Since, $L$ does not depend on time explicitly,

$$p_i = \frac{\partial L}{\partial \dot{q}_i} = \frac{\partial T}{\partial \dot{q}_i}.$$

For the case when $t$ does not occur explicitly in $T$,

$$\sum p_i \dot{q}_i = \sum_{i,k} a_{ik} \dot{q}_i \dot{q}_k.$$

where

$$T = \frac{1}{2} \sum_{i,k} a_{ik} \dot{q}_i \dot{q}_k.$$

Again, $E = T + V$, gives

$$2[E - V(q)] = 2T = \sum_{i,k} a_{ik} \dot{q}_i \dot{q}_k.$$

$$\text{or, } dt = \sqrt{\left(\sum_{ik} a_{ik}\, dq_i dq_k\right) \Big/ 2[E - V(q)]}.$$

Since

$$S_0 = \int \sum_i p_i \dot{q}_i dt = \int \sum_{ik} a_{ik} \dot{q}_i \dot{q}_k dt = \int \sum a_{ik}\, dq_k\, dq_k\, (dt)^{-1}$$

$$= \int \sqrt{\left(\sum a_{ik}\, dq_i dq_k\right) 2[E - V(q)]}$$

hence the least action principle is

$$\delta S_0 = \delta \int \sqrt{\left[2(E - V) \sum_{ik} a_{ik}\, dq_i dq_k\right]} = 0.$$

For a single particle moving in a potential $V$, we get

$$S_0 = \int p dq = \int mv \cdot dx.$$

We have used configuration space. Since

$$T = \frac{1}{2} mv^2 \quad \text{and} \quad E - V = T.$$

$$S_0 = \int \sqrt{[2m(E - V)]}\, dx$$

and motion is given by

$$\delta \int \sqrt{[2m(E - V)]}\, dx = 0.$$

## Exercises

1. For the Lagrangian

$$L = \frac{1}{2Z}\dot{q}^2 - \frac{1}{2}\left(X - \frac{Y^2}{Z}\right)q^2 - \frac{Y}{Z}q \cdot \dot{q}.$$

where $X, Y, Z$ are time-dependent, show that the conjugate momentum $p$ is $p = (\dot{q} - Y_q)/Z$ and the Hamiltonian

$$H = \frac{1}{2}(Zp^2 + 2Ypq + Xq^2).$$

[*Hints:*

$$\begin{aligned}
\text{Lagrangian } L &= \frac{1}{2Z}\dot{q}^2 - \frac{1}{2}\left(X - \frac{Y^2}{Z}\right)q^2 - \frac{Y}{Z}q \cdot \dot{q} \\
p &= \frac{\partial L}{\partial \dot{q}} = \frac{1}{Z}\dot{q} - \frac{1}{Z}Y \cdot q \\
p &= \frac{1}{Z}(\dot{q} - Y \cdot q) \\
H = p\dot{q} - L &= \frac{1}{Z}(\dot{q} - Y \cdot q)\dot{q} - \frac{1}{2Z}\dot{q}^2 + \frac{1}{2}\left(X - \frac{Y^2}{Z}\right)q^2 + \frac{Y}{Z}q\dot{q} \\
&= \frac{1}{2Z}\dot{q}^2 + \frac{1}{2}\left(X - \frac{Y^2}{Z}\right)q^2 \\
&= \frac{1}{2Z}(pZ + Yq)^2 + \frac{1}{2}\left(X - \frac{Y^2}{Z}\right)q^2 \\
&= \frac{1}{2Z}\left(p^2z^2 + Y^2q^2 + 2YZpq\right) + \left(\frac{1}{2}Xq^2 - \frac{1}{2Z}Y^2q^2\right) \\
H &= \frac{1}{2}(Zp^2 + 2Ypq + Xq^2).]
\end{aligned}$$

2. If the canonical variables are not all independent but are connected by auxilliary conditions

$$\psi_k(q_i, p_i; t) = 0,$$

show that the canonical equations of motion are given by

$$\frac{\partial H}{\partial p_i} + \sum_k \lambda_k \frac{\partial \psi_k}{\partial p_i} = \dot{q}_i \quad \text{and} \quad \frac{\partial H}{\partial q_i} + \sum_i \lambda_k \frac{\partial \psi_k}{\partial q_i} = -\dot{p}_i.$$

where $\lambda_k$'s are Lagrange's undetermined multipliers.

3. Calculate the following Poisson Brackets

(*a*) $[p, [p, H]]$.

(b) $[[\dot{q}.\dot{p}].H]+[[\dot{p}.H].\dot{q}]+[[H.\dot{q}].\dot{p}]$. where $H=\dfrac{p^2}{2m}+V(q)$.

[*Hints:* $H=\dfrac{p^2}{2m}+V(q)$ $[p.[p.H]]$. Calculate.

$$[p.H]=\left[p.\frac{p^2}{2m}+V(q)\right]=[p.V(q)]=-\frac{\partial V}{\partial q}$$

Hence.

$$[p.[p.H]]=\left[p.-\frac{\partial V}{\partial q}\right]=-\left[p.\frac{\partial V}{\partial q}\right]=\frac{\partial^2 V}{\partial q^2}.$$

$[[\dot{q}.\dot{p}].H]+[[\dot{p}.H].\dot{q}]+[[H.\dot{q}].\dot{p}]=0$. because

$$[[\dot{q}.\dot{p}].H]=\frac{p}{m^2}\frac{\partial^3 V}{\partial q^3};[[\dot{p}.H].\dot{q}]=-\frac{p}{m^2}\frac{\partial^3 V}{\partial q^3};[[H.\dot{q}].\dot{p}]=0.$$

4. A particle of mass $m$ is constrained to more on an inclined rod in a vertical-plane under the influence of a central potential $V(r)$. The rod is making an angle $\theta$. with the $z$-axis and is rotating around the $z$-axis with a constant angular velocity $\omega$. Show that the Hamiltonian $H$ of the system can be written as

$$H=\frac{p_r^2}{2m}-a^2\omega^2 r^2+V(r).$$

where

$$a^2=(\sin^2\theta)\frac{m}{2}$$

obtain Hamilton's equation of motion and solve this equations for

$$V(r)=br^2;\ \ b>0.$$

[*Hints:* K.E. in spherical co-ordinate

$$T=\frac{1}{2}m(\dot{r}^2+r^2\dot{\theta}^2+r^2\sin^2\theta\dot{\phi}^2).$$

As $\theta$ is fixed. we have

$$T=\frac{1}{2}m(\dot{r}^2+r^2\sin^2\theta\dot{\phi}^2).$$

Again. $\phi=\omega t$. hence

$$T=\frac{1}{2}m(\dot{r}^2+r^2\sin^2\theta.\omega^2)$$

$$L=T-V(r)=\frac{1}{2}m(\dot{r}^2+r^2\sin^2\theta\omega^2)-V(r)$$

$$p_r=\frac{\partial L}{\partial \dot{r}}=m\dot{r}.$$

$$H=p_r\dot{r}-L=\left(\frac{p_r^2}{2m}-\frac{r^2}{2}\sin^2\theta.m\omega^2\right)+V(r)=\frac{p_r^2}{2m}-r^2a^2\omega^2+V(r).$$

$$\therefore a^2=\left(\frac{\sin^2\theta}{2}\right)m.$$

One now can obtain canonical equation of motion by putting $V = br^2$.]

5. (*a*) Is the constraint given by $x\dot{y} - y\dot{x} = k$ (a constant) a holonomic one?

   (*b*) A particle is constrained to move on a circle while the centre of the circle is pulled along the $x$-axis with constant velocity. Is the constraint on the system a holonomic one?

6. (*a*) Are the two Lagrangians $L_1$ and $L_2$ given by
$$L_1 = (q + \dot{q})^2 \text{ and } L_2 = (q^2 + \dot{q}^2) \text{ equivalent?}$$

   (*b*) Show that the Lagrangians
$$L_I = \frac{1}{2Z}\dot{q}^2 - \frac{1}{2}\left(X - \frac{Y^2}{Z}\right)q^2 - \frac{Y}{Z}q\dot{q}$$
$$L_{II} = \frac{1}{2Z}\dot{q}^2 - \frac{1}{2}\left[X - \frac{Y^2}{Z} - \frac{d}{dt}\left(\frac{Y}{Z}\right)\right]q^2$$
   are equivalent ($X, Y$ and $Z$) are time-dependent constants, what is the Hamiltonian for $L_{II}$?

[*Hints* (*b*): $L_I$ and $L_{II}$ differ by a total time-derivative
$$L_I = L_{II} - \frac{1}{2}\frac{d}{dt}\left(\frac{Y}{Z}q^2\right).]$$

7. For the Lagrangian
$$L = \sum_i f_i(q_i)\dot{q}_i^2 - \sum_i V_i(q_i)$$
obtain Lagrangian and Hamiltonian equation of motion. Solve the Lagrangian equations.

8. A pendulum bob is suspended by a coil spring and allowed to swing in a vertical-plane. Obtain Hamilton's canonical equations.

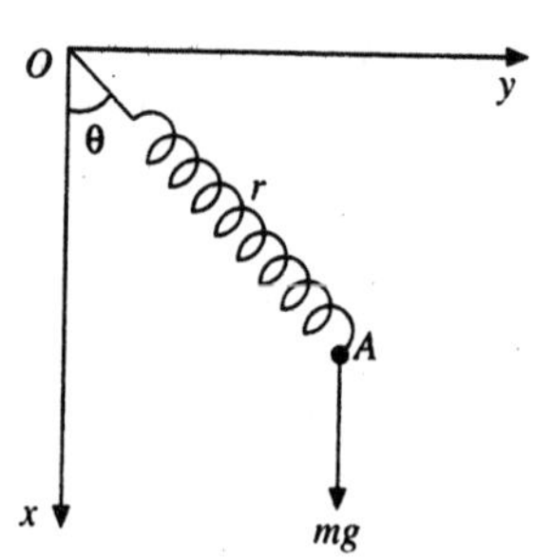

Fig 3.2

[*Hints:* $OA$ = coil spring of length $r$, ($r$ = elongated length) $r_0$ = initial length of the spring; $k$ is the spring constant. Hence the force, $F$ is
$$F = -k(r - r_0).$$
Therefore, the potential, $V$ is obtained from
$$F = -\frac{\partial V}{\partial r}.$$
Hence
$$V = \frac{1}{2}k(r - r_0)^2.$$

Let $(r, \theta)$ be the co-ordinate of the bob, then

$$\text{K.E.} = T = \frac{1}{2}m(\dot{r}^2 + r^2\dot{\theta}^2).$$

Potential energy

$$= -mgr\cos\theta + \frac{1}{2}k(r - r_0)^2.$$

$$L = T - V;\ p_r = \frac{\partial L}{\partial \dot{r}} = m\dot{r};$$

$$p_\theta = \frac{\partial L}{\partial \dot{\theta}} = mr^2\dot{\theta};\ H = p_r\dot{r} + p_\theta\dot{\theta} - L.$$

Hence

$$H = \frac{p_r^2}{2m} + \frac{p_\theta^2}{2mr^2} - mgr\cos\theta + \frac{1}{2}k(r - r_0)^2.$$

Canonical equation

$$-\dot{p}_r = \frac{\partial H}{\partial r};\ \dot{r} = \frac{\partial H}{\partial p_r};\ -\dot{p}_\theta = \frac{\partial H}{\partial \theta};\ \dot{\theta} = \frac{\partial H}{\partial p_\theta};$$

these are easily found from $H$.]

9. A particle is constrained to move on the arc of a parabola $x^2 = 4ay$ ($y$-axis vertical), under the action of gravity. Show that the Hamiltonian of the system is

$$H = \frac{2a^2p_x^2}{m(4a^2 + x^2)} + \frac{mg}{4a}x^2,$$

where $p_x$ = conjugate momentum of $x$, $m$ = mass, $g$ = acceleration due to gravity.

10. Two masses $m_1$ and $m_2$ are connected by a coil spring of spring constant $k$ and of unstretched length $l_0$. They are constrained to move along a vertical line only under gravity. Show that the Hamiltonian is

$$H = \frac{1}{2m_1}(p_1 + p_3)^2 + \frac{p_3^2}{2m_2} + m_1gy_1 + m_2g(y_1 - y_3) + \frac{1}{2}k(y_3 - l_0)^2;$$

$m_1$ is at a height $y_1$, $m_2$ at a height $y_2$; $y_3 = y_1 - y_2$ = distance between $m_1$ and $m_2$; $p_3, p_1$ conjugate to $y_3$ and $y_1$.

[*Hints:* $m_1$ is at a height $y_1$; $m_2$ is at $y_2$ and $y_3$ is the distance between $m_1$ and $m_2$.

The constraint: $y_1 - y_2 = y_3$.

$$P.E. \text{ of } m_1 = mg\,y_1.$$

Potential energy of $m_2 = m_2g(y_1 - y_3)$.

Force between $m_1$ and $m_2 = F = -k(y_3 - l_0)$, where $l_0$ is the length of the spring, where unstretched. Hence

$$V = \frac{1}{2}k(y_3 - l_0)^2.$$

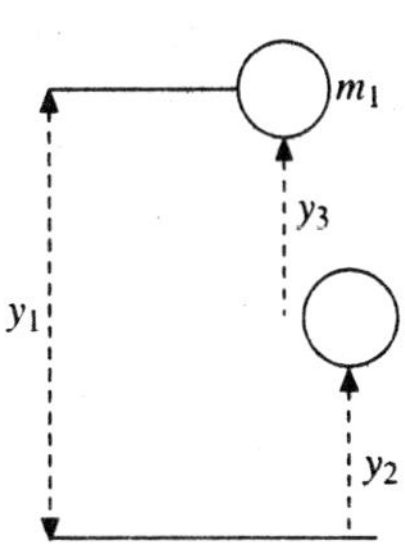


Fig 3.3

Hence

$$V_{Tot} = m_1 g y_1 + m_2 g(y_1 - y_3) + \frac{1}{2}k(y_3 - l_0)^2.$$
$$T = \text{K.E.} = \frac{1}{2}m_1\dot{y}_1^2 + \frac{1}{2}m_2(\dot{y}_1 - \dot{y}_3)^2.$$

Hence

$$\begin{aligned} L &= T - V \\ &= \frac{1}{2}m_1\dot{y}_1^2 + \frac{1}{2}m_2(\dot{y}_1^2 + \dot{y}_3^2 - 2\dot{y}_1\dot{y}_3) - m_1 g y_1 - m_2 g(y_1 - y_3) - \frac{1}{2}k(y_3 - l_0)^2. \end{aligned}$$

Hence

$$\begin{aligned} p_1 &= \frac{\partial L}{\partial \dot{y}_1} = m_1\dot{y}_1 + m_2\dot{y}_1 - m_2\dot{y}_3 \\ p_3 &= \frac{\partial L}{\partial \dot{y}_3} = m_2\dot{y}_3 - m_2\dot{y}_1. \quad p_3 = m_2(\dot{y}_3 - \dot{y}_1). \\ p_1 &= (m_1 + m_2)\dot{y}_1 - m_2\dot{y}_3; \quad p_1 + p_3 = m_1\dot{y}_1. \end{aligned}$$

Now we easily find

$$H = p_1\dot{y}_1 + p_3\dot{y}_3 - L$$

reduces to

$$\begin{aligned} H &= \frac{1}{2}(m_1 + m_2)\dot{y}_1^2 - m_2\dot{y}_3\dot{y}_1 + \frac{1}{2}m_2\dot{y}_1^2 + V \\ &= \frac{1}{2}m_1\dot{y}_1^2 + \frac{1}{2}m_2(\dot{y}_1 - \dot{y}_3)^2 + V \\ &= \frac{1}{2}m_1(p_1 + p_3)^2/m_1^2 + \frac{1}{2}m_2 p_3^2/m_2^2 + V \\ &= \frac{1}{2}(p_1 + p_3)^2/m_1 + p_3^2/2m_2 + m_1 g y_1 + m_2 g(y_1 - y_3) + \frac{1}{2}k(y_3 - l_0)^2. \end{aligned}$$

11. Use $H = p\dot{q} - L$ to calculate the Poisson Bracket of $[p, H]_{\text{P.B.}}$ and show that it is $\dot{p}$.

[*Hints:* The Poisson Bracket $[p, H]$ is

$$[p, H] = [p, p\dot{q}] - [p, L(q, \dot{q})] = p[p, \dot{q}] - [p, L(q, Q)]; \quad \dot{q} = Q \text{ is used.}$$

To find $[p, L]$, expand $L$ as a

$$\sum_{m,n} a_{mn} q^m Q^n.$$

$$[p, L(q, Q)] = [p, q]\frac{\partial L}{\partial q} + [p, Q]\frac{\partial L}{\partial Q} = -\frac{\partial L}{\partial q} + [p, \dot{q}]\frac{\partial L}{\partial \dot{q}} = -\dot{p} + [p, \dot{q}]p$$

(we use $p = \frac{\partial L}{\partial \dot{q}}$ and $[q, p] = 1$.)

$$[p, H] = p[p, \dot{q}] - [p, L] = p[p, \dot{q}] + \dot{p} - [p, \dot{q}]p = \dot{p}.]$$

12. (*a*) For the Lagrangian

$$L = \dot{q}_1^2 + \dot{q}_2^2 + \dot{q}_1\dot{q}_2 - q_1^2 - q_2^2.$$

find the conjugate momenta $p_1$ and $p_2$ and obtain the Hamiltonian. Write down the Hamilton's canonical equations and solve them.

(*b*) For the Hamiltonian

$$H = q_1p_1 - q_2p_2 - aq_1^2 + bq_2^2$$

solve the Hamilton's canonical equation of motion.

13. Is the angular momentum conserved for a system whose Hamiltonian is given by

$$H = a\mathbf{p}^2 + b\mathbf{p}\cdot\mathbf{q} + c\mathbf{q}^2.$$

where $a, b, c$ are constants.

[*Hints:* $a, b, c$ are constants, and they are scalar and are invariant under rotations.

Further $\mathbf{p}^2$, $\mathbf{p}.\ \mathbf{q}$ and $\mathbf{q}^2$ are all scalars, so they are also invariant under rotation. Hence the Hamiltonian $H$ is invariant under rotation. Therefore, angular momentum is conserved.]

14. Find The Routhian for the system with the Lagrangian

$$\frac{1}{2}m\dot{r}^2 + r^2\dot{\theta}^2 + r^2\sin^2\theta\dot{\phi}^2) - V(r)$$

by eliminating the cyclic variable.

[*Hints:* The cyclic variable is $\phi$. Hence

$$p_\phi = \frac{\partial L}{\partial\dot{\phi}}; \quad p_\phi = mr^2\sin^2\theta\cdot\dot{\phi}. \quad \text{or.} \quad \dot{\phi} = \frac{p_\phi}{mr^2\sin^2\theta}.$$

Therefore, the Routhian $R$ given by $R = p_\phi\,\dot{\phi} \quad L$ becomes

$$\frac{1}{2}\frac{p_\phi^2}{mr^2\sin^2\theta} - \frac{1}{2}m(\dot{r}^2 + r^2\dot{\theta}^2) + V(r)]$$

15. Find an expression for energy in terms of Routhian.

[*Hints:* By definition

$$R(q, p; \xi, \dot{\xi}) = p\dot{q} - L.$$

where $L = L(q, \dot{q}, \xi, \dot{\xi})$ and $p = \frac{\partial L}{\partial \dot{q}}$. Energy, $E$, is defined as

$$E = \dot{q}\frac{\partial L}{\partial\dot{q}} + \dot{\xi}\frac{\partial L}{\partial\dot{\xi}} - L. \quad \text{or.} \quad E = p\dot{q} - L + \dot{\xi}\frac{\partial L}{\partial\dot{\xi}}$$

and from definition of $R$ and

$$\frac{\partial R}{\partial \dot{\xi}} = -\frac{\partial L}{\partial \dot{\xi}}; \quad E = R - \dot{\xi}\frac{\partial R}{\partial \dot{\xi}}]$$

16. For the Lagrangian

$$L = \frac{1}{2}\dot{q}^2 - q\dot{q} + q^2.$$

find $p$ in terms of $q$ and $\dot{q}$ and obtain the value of the $[p, \dot{q}^2]_{P.B.}$.

[*Hints:*

$$p = \frac{\partial L}{\partial \dot{q}} = \dot{q} - q; \text{ so } \dot{q} = p + q.$$

$$[p, \dot{q}^2] = 2\dot{q}[p, \dot{q}] = 2\dot{q}[p, p + q] = 2\dot{q}[p, q] = -2\dot{q}.]$$

17. Given

$$T = \frac{1}{2}m\dot{r}^2; \quad V = \frac{1}{r}\left(1 + \frac{\dot{r}^2}{c^2}\right) \text{ and } L = T - V.$$

Find the Hamiltonian and determine:

(*a*) whether $H = T + V$;

(*b*) whether $dH/dt = 0$.

Chapter IV

# Central Orbit and Theory of Scattering

## 4.1 Two-body Problem under Central Force

Consider the interaction of two particles under a mutually attractive force characterised by a potential $V$, which is a function of the distance between the two particles. Let $\mathbf{r}_1$ and $\mathbf{r}_2$ are the position vectors of the two particles of masses $m_1$ and $m_2$ respectively. The kinetic energy $T$ is

$$\frac{1}{2}m_1\dot{\mathbf{r}}_1^2 + \frac{1}{2}m_2\dot{\mathbf{r}}_2^2$$

and potential

$$V = V(|\mathbf{r}_1 - \mathbf{r}_2|).$$

The Lagrangian $L$ is $T - V(r)$ where $r = |\mathbf{r}_1 - \mathbf{r}_2|$. We use centre of mass and relative variables $\mathbf{R}$ and $\mathbf{r}$, where

$$\mathbf{R} = \frac{(m_1\mathbf{r}_1 + m_2\mathbf{r}_2)}{(m_1 + m_2)} \quad \text{and} \quad \mathbf{r} = \mathbf{r}_1 - \mathbf{r}_2.$$

Therefore,

$$\mathbf{r}_1 = \mathbf{R} + \frac{(m_2\mathbf{r})}{(m_1 + m_2)} \quad \text{and} \quad \mathbf{r}_2 = \mathbf{R} - \frac{(m_1\mathbf{r})}{(m_1 + m_2)}.$$

In terms of these two variables $\mathbf{R}$ and $\mathbf{r}$, $L$ becomes

$$L = \frac{1}{2}M\dot{\mathbf{R}}^2 + \frac{1}{2}\mu\dot{\mathbf{r}}^2 - V(r). \tag{4.1}$$

where

$$\mu = \frac{m_1 m_2}{(m_1 + m_2)}$$

is the reduced mass of the system.

7

In equation (4.1), the first term represents the free motion of the centre of mass ($M = m_1 + m_2$) as $V$ is a function of the relative co-ordinate only. Thus, dropping the centre of mass motion, the Lagrangian $L$ becomes

$$L = \frac{1}{2}\mu\dot{\mathbf{r}}^2 - V(r). \tag{4.2}$$

Transforming $T$ to spherical polar co-ordinates $(r, \theta, \phi)$ we get,

$$T = \frac{1}{2}\mu\left(\dot{r}^2 + r^2\dot{\theta}^2 + r^2\sin^2\theta\dot{\phi}^2\right).$$

Therefore,

$$L = \frac{1}{2}\mu\left(\dot{r}^2 + r^2\dot{\theta}^2 + r^2\sin^2\theta\dot{\phi}^2\right) - V(r). \tag{4.3}$$

From (4.3) we see that $\phi$, does not occur in $L$ so $p_\phi$, the conjugate momentum to $\phi$ must be a constant of motion, i.e.,

$$p_\phi = \frac{\partial L}{\partial\dot{\phi}} = \mu r^2\sin^2\theta\cdot\dot{\phi} = \text{constant}.$$

But,

$$\mu r^2\sin^2\theta\dot{\phi} = L_z,$$

where $\mathbf{L}$ is the angular momentum vector. Since $L_z$ is a constant, we thus choose that the motion is confined to the $xy$-plane. Hence the Lagrangian can be chosen as,

$$L = \frac{1}{2}\mu\left(\dot{r}^2 + r^2\dot{\theta}^2\right) - V(r). \tag{4.4}$$

Since the Lagrangian is invariant under rotation angular momentum vector $\mathbf{L}$ is constant. As $\mathbf{L} = \mathbf{r}\times\mathbf{p}$, therefore $\mathbf{r}\times\mathbf{p}$ is constant. The plane containing $\mathbf{r}$ and $\mathbf{p}$ is always perpendicular to $\mathbf{L}$. Hence the components of $\mathbf{L}$ along any axis through the centre of force is constant. As there is no component of force perpendicular to the plane containing $\mathbf{r}$ and $\mathbf{p}$ the motion will therefore be confined to a plane of $\mathbf{r}$ and $\mathbf{p}$.

## 4.2 Equation of the Orbit under Central Force

From equation (4.4) we have for the Hamiltonian $H$, the following

$$H = T + V = \frac{1}{2}\mu\left(\dot{r}^2 + r^2\dot{\theta}^2\right) + V(r). \tag{4.5}$$

where $\dot{r}^2 + r^2\dot{\theta}^2$ is the expression for $T$ in terms of the polar variable $(r, \theta)$ in two-dimensional $xy$-plane. As $H$ is conserved owing to the fact that $t$ does not occur explicitly in $H$, then total energy $E$ is

$$E = \frac{1}{2}\mu\left(\dot{r}^2 + r^2\dot{\theta}^2\right) + V(r). \tag{4.6}$$

From the Lagrangian

$$L = \frac{1}{2}\mu\left(\dot{r}^2 + r^2\dot{\theta}^2\right) - V(r) \tag{4.7}$$

we find $\partial L/\partial\theta = 0$, which implies

$$p_\theta = \frac{\partial L}{\partial\dot{\theta}} = \mu r^2\dot{\theta} = \text{constant}.$$

Thus

$$\mu r^2\dot{\theta} = l \quad \text{(a constant)}, \quad \text{or,} \quad \dot{\theta} = \frac{l}{\mu r^2}. \tag{4.8}$$

Therefore $E$ becomes

$$E = \frac{1}{2}\mu\left(\dot{r}^2 + \frac{l^2}{\mu^2 r^2}\right) + V, \quad \text{or,} \quad \dot{r}^2 = \frac{2(E-V)}{\mu} - \frac{l^2}{\mu^2 r^2}. \tag{4.9}$$

We write

$$\dot{r} = \frac{dr}{d\theta}\cdot\frac{d\theta}{dt}$$

and use $\dot{\theta} = l/\mu r^2$. We then find

$$\dot{r} = \frac{l}{\mu r^2}\cdot\frac{dr}{d\theta}. \tag{4.10}$$

Hence from (4.9)

$$d\theta = \frac{l}{\mu r^2}dr\cdot\frac{1}{\{[2(E-V)/\mu] - [l^2/(\mu^2 r^2)]\}^{1/2}}. \tag{4.11}$$

We use

$$M^2 = \frac{l^2}{2\mu} \quad \text{and} \quad r = \frac{1}{u}.$$

then

$$\theta + \text{constant} = -M\int\frac{du}{[E - V(u) - M^2u^2]^{1/2}}. \tag{4.12}$$

Equation (4.11) is the equation for the orbit in integral form. From the Lagrangian (4.7) we get

$$\frac{\partial L}{\partial\theta} = 0 = \frac{d}{dt}\left(\frac{\partial L}{\partial\dot{\theta}}\right), \quad \text{i.e.,} \quad \mu r^2\dot{\theta} = l \text{ (constant)}$$

and the equation

$$\frac{\partial L}{\partial r} = \frac{d}{dt}\left(\frac{\partial L}{\partial\dot{r}}\right)$$

yields

$$\mu\ddot{r} = \mu r\dot{\theta}^2 - \frac{\partial V}{\partial r}. \tag{4.13}$$

Since

$$\mu r^2\dot{\theta} = l;$$

equation (4.13) reduces to

$$\mu\ddot{r} = \frac{l^2}{r^3\mu} - \frac{\partial V}{\partial r}. \tag{4.14}$$

Put $r = 1/u$ as before, then

$$\frac{dr}{dt} = -\frac{1}{u^2}\frac{du}{d\theta}\cdot\frac{d\theta}{dt} = -\frac{l}{\mu}\frac{du}{d\theta}.$$

Therefore,

$$\frac{d^2r}{dt^2} = -\frac{l}{\mu}\frac{d^2u}{d\theta^2}\cdot\frac{d\theta}{dt} = -\frac{l^2}{\mu^2}u^2\frac{d^2u}{d\theta^2}. \tag{4.15}$$

We use (4.15) in (4.14).
Then

$$-\frac{l^2}{\mu}u^2\frac{d^2u}{d\theta^2} = \frac{l^2}{\mu}u^3 + F. \tag{4.16}$$

where

$$F = -\frac{\partial V}{\partial r}.$$

Equation (4.16) gives

$$-\frac{l^2}{\mu}\left(u^2\frac{d^2u}{d\theta^2} + u^3\right) = F. \tag{4.17}$$

From this equation we can find the orbit if the law of force is given. Alternatively if the orbit is known, we can easily find the law of force. Equation (4.17) is the *differential equation* of the orbit. As an application of (4.17) suppose the orbit is $r = e^{k\theta}$, where $k$ is a constant.
Then from (4.17) we get

$$-\frac{l^2}{\mu}\left(u^2\frac{d^2u}{d\theta^2} + u^3\right) = -\frac{l^2(k^2+1)}{\mu}e^{-3k\theta}.$$

So the force is porportional to $-1/r^3$.

## 4.3 Planetary Motion: Kepler's Laws

We discuss the motion of a planet around the sun. The attractive potential $V$ is given by

$$V = -\frac{a}{r}.$$

where $a$ is a constant signifying the strength of the attractive force between the Sun and the planet. The orbit of the planet can be easily found from equation (4.12). Substituting $V = -au$, where $u = 1/r$, we get from (4.12),

$$k + \theta = -M \int \frac{du}{\sqrt{(E + au - M^2u^2)}}, \quad \text{where } M^2 = \frac{l^2}{2\mu}. \tag{4.18}$$

$l$ being a constant and $\mu$ stands for the reduced mass of sun and the planet. We write the denominator of the integrand in (4.18) as

$$\sqrt{\left[\left(E + \frac{a^2}{4M^2}\right) - \left(M^2u^2 + \frac{a^2}{4M^2} - au\right)\right]}$$

$$= \sqrt{\left[L^2 - \left(Mu - \frac{a}{2M}\right)^2\right]}. \tag{4.19}$$

where

$$L^2 = E + \frac{a^2}{4M^2}.$$

Using (4.19) in (4.18) we find

$$k + \theta = -M \int \frac{du}{\sqrt{[L^2 - (Mu - a/2M)^2]}}.$$

which can be easily integrated by putting

$$Mu - \frac{a}{2M} = z.$$

The result of integration gives us

$$\cos(\theta + k) = \frac{z}{L}. \tag{4.20}$$

We now absorb the constant $k$ in the definition of angle $\theta$ in (4.21) and write the equation of the orbit as

$$r = M/\left(L\cos\theta + \frac{a}{2M}\right) \tag{4.21}$$

which is an equation of a conic as (4.21) is of the form

$$r = \frac{\tilde{l}}{1 + e\cos\theta}. \tag{4.22}$$

where $\tilde{l}$ = latus rectum and $e$ = eccentricity.

We write (4.21) as

$$r = \frac{2M^2/\alpha}{1 + \sqrt{\left(1 + \frac{4EM^2}{\alpha^2}\right)} \cdot \cos\theta}. \tag{4.23}$$

Hence $\tilde{l} = 2M^2/\alpha$; and the eccentricity, $e$ is

$$e = \sqrt{\left(1 + \frac{4EM^2}{\alpha^2}\right)}. \tag{4.24}$$

Hence (a) when $E > 0, e > 1$, the conic is hyperbola.
(b) when $E < 0, e < 1$, the conic is an ellipse.
(c) when $E = 0, e = 1$, just bound system, conic is a parabola.

For $e = 0$, the orbit becomes a circle. The energy of the system is then given by

$$E = -\frac{\alpha^2}{4M^2} = -\frac{2\alpha^2\mu}{4l^2} = -\frac{\mu\alpha^2}{2l^2}.$$

Putting $l^2 = n^2\hbar^2$ we get the Bohr-energy-levels ($l$ = angular momentum in two-dimension). When $E < 0$, the system is bound as the earth is bound to sun. Hence we get

(*a*) One of the laws of Kepler, stating that planet is revolving round the sun in an ellipse keeping sun at one of its focii.

(*b*) Since $\mu r^2\dot{\theta} = l$ (a constant) we find

$$\frac{\left(\frac{1}{2}r \cdot r d\theta\right)}{dt} = \text{a constant}, \quad \text{i.e.,} \quad \frac{dA}{dt} = \text{constant}.$$

where $dA$ is an elementary area $r^2 d\theta/2$ covered in time $dt$. Thus the *areal-velocity* of the planet is a constant.

The total time period is obtained from the total area covered in one revolution. From the property of the ellipse, we know that if $\tilde{a}$ and $\tilde{b}$ are the semi-major and semi-minor axes, then

$$\tilde{l} = \tilde{a}(1 - e^2)$$
$$\tilde{b}^2 = \tilde{a}^2(1 - e^2) \quad e = \text{eccentricity}.$$
$$\text{and} \quad A = \pi\tilde{a}\tilde{b};$$

the time period, $\tau = A/$ areal velocity.

The areal velocity

$$\frac{dA}{dt} = \frac{l}{2\mu}.$$

We then have

$$\tilde{b} = (\tilde{a}\tilde{l})^{1/2},$$

where

$$\tilde{l} = \frac{2M^2}{a} = \text{constant}.$$

Thus $A$ is proportional to

$$\tilde{a}\tilde{b} = (\tilde{a})^{3/2}(\tilde{l})^{1/2}.$$

Hence

$$\tau = \pi\tilde{a}^{3/2}\cdot\left(\frac{2M^2}{a}\right)^{1/2}\cdot\frac{2\mu}{l}.$$

This implies that the time period $\tau$ is proportional to $(\tilde{a})^{3/2}$, i.e., square of the time period is proportional to the cube of the semi major axis of the ellipse.

The understanding of Kepler's laws on the basis of Newton's laws of motion was a great achievement. Kepler published his laws purely from planetary observations during 1610.

## 4.4 Some Simple Applications

In this section, we discuss further application of the orbit equation (4.12). There are not many potentials for which the result of integration is obtained in closed form. Quite generally it leads to elliptic integrals. For potentials like $e^{-\mu r}/r$ or $e^{-\mu r}$, the integrals become extremely involved. We discuss a few cases where integration can be performed.

**Example 1.** $V(r) = ar^2 = a/u^2$. *From the orbit equation*

$$\theta = -M\int\frac{du}{\sqrt{(E - V - M^2u^2)}}$$

*we get by putting* $V = a/u^2$,

$$\begin{aligned}\theta &= -M\int\frac{u\,du}{\sqrt{(Eu^2 - a - M^2u^4)}}\\ &= -\frac{M}{2}\int\frac{2u\,du}{\sqrt{\left[\left(\frac{E^2}{4M^2} - a\right) - \left(Mu^2 - \frac{E}{2M}\right)^2\right]}}.\end{aligned}$$

*Solution:* Putting

$$L^2 = \frac{E^2}{4M^2} - a \quad\text{and}\quad z = Mu^2 - \frac{E}{2m}$$

the orbit equation becomes

$$\theta = -\frac{1}{2}\int \frac{dz}{\sqrt{(L^2 - z^2)}} = \frac{1}{2}\cos^{-1}\left(\frac{z}{L}\right).$$

Thus,

$$z = L\cos 2\theta.$$

This can be put in the form,

$$r^2 = M \Bigg/ \left[\frac{E}{2M} + \sqrt{\left(\frac{E^2}{4M^2} - a\right)}\cos 2\theta\right].$$

**Example 2.** *The case when*

$$V = -\frac{a}{r^2} = -au^2.$$

***Solution:*** In this case, we find the orbit equation is

$$\theta = -M\int \frac{du}{\sqrt{(E + au^2 - M^2u^2)}} = -M\int \frac{du}{\sqrt{(E - (M^2 - a)u^2)}}$$
$$= -\frac{M}{M'}\int \frac{du}{\sqrt{(L^2 - u^2)}},$$

where $M'^2 = M^2 - a$ and $L^2 = E/M'^2$. The last integral gives

$$\theta = \left(\frac{M}{M'}\right)\cos^{-1}\frac{u}{L}, \quad \text{i.e.,} \quad L\cos\left(\frac{M'}{M}\theta\right) = u$$

as the equation for the orbit.

**Example 3.** *As an use of the differential form of the orbit equation we want to find the law of force for the path*

$$u^n a^n \cos n\theta = 1.$$

***Solution:*** From $u^n a^n \cos n\theta = 1$, we find

$$n\log u + n\log a + \log\cos n\theta = 0.$$

Therefore,

$$n\frac{1}{u}\frac{du}{d\theta} - n\tan n\theta = 0 \quad \text{and} \quad \frac{d^2u}{d\theta^2} = u(\tan^2 n\theta + n\sec^2 n\theta).$$

Therefore,

$$F = -u^2\left(\frac{d^2u}{d\theta^2} + u\right) = -u^3(n+1)\sec^2 n\theta$$
$$= -(n+1)a^{2n}u^{2n+3} = -(n+1)a^{2n}\frac{1}{r^{2n+3}}.$$

Hence $F$ is proportional to $1/r^{2n+3}$.

**Example 4.** *Consider an object moving in a circular orbit in such a way that its centre of force is on the circumference of the circle. To find the law of force.*

***Solution:*** The path described by the object can be shown by the diagram given below (fig. 4.1).
$O$ is the origin, the centre of force. $P$ any position of the object $(x, y)$ in polar co-ordinate $x = r\cos\theta$, $y = r\sin\theta$. Centre of the circle is at $C$. The point $O$ is on the circumference such that $OY$ is the tangent at $O$. $OC$ = radius = $a$, of the circle.

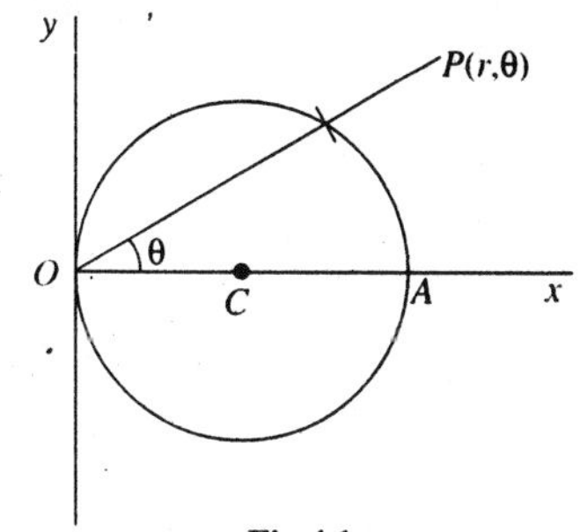

Fig 4.1

The equation of the circle is

$$(x-a)^2 + y^2 = a^2, \quad \text{i.e.,} \quad r^2 - 2ax = 0,$$

$$\text{i.e., } r^2 - 2ar\cos\theta = 0, \quad \text{i.e.,} \quad r = 2a\cos\theta = 0, \quad \text{i.e.,} \quad 2au\cos\theta = 1.$$

This equation is a particular case of the previous example with $n = 1$. Hence the law of force is proportional to $1/r^5$.

## 4.5 Infinite and Finite Motion; Condition for Closed Path

We have the following Hamiltonian for the 2-dimensional motion under a potential $V(r)$, namely

$$H = \frac{1}{2}\mu(\dot{r}^2 + r^2\dot{\theta}^2) + V(r).$$

As angular momentum $\mu r^2\dot{\theta}$ = constant, which follows from the fact that $p_\theta$ is constant as $\theta$ is a cyclic co-ordinate we have

$$H = E = \frac{1}{2}\mu\left(\dot{r}^2 + \frac{l^2}{\mu^2 r^2}\right) + V(r).$$

Hence

$$\frac{1}{2}\mu\dot{r}^2 = E - \left(V + \frac{l^2}{2\mu r^2}\right). \tag{4.25}$$

$$= E - U_{eff}(r). \tag{4.26}$$

This expression shows that the radial motion is in one-dimension under an effective potential $U_{eff}$, where

$$U_{eff} = V(r) + \frac{l^2}{2\mu r^2}.$$

The quantity $\frac{l^2}{2\mu r^2}$ is called *Centrifugal energy*. The values of $r$ for which

$$E = V(r) + \frac{l^2}{2\mu r^2}$$

is satisfied determine the *limits of motion* as regards distances from the centre. When

$$E = V(r) + \frac{l^2}{2\mu^2}.$$

then from (4.25), we get radial velocity, $\dot{r} = 0$. The value $\dot{r} = 0$ indicates a *turning point* of the path. If the ranges of $r$ is such that $r \geq r_{\min}$, then the motion is infinite. The particle comes from infinity and goes back to infinity. If the values of $r$ lie between $r_{\min}$ and $r_{\max}$, then the motion is finite. The finite motion does not imply that the path is a closed curve. During the period when radius vector varies from $r_{\min}$ to $r_{\max}$ and back, the angle deflected $\Delta\theta$ is

$$\Delta\theta = 2\int_{r_{\min}}^{r_{\max}} \frac{M\,du}{\sqrt{(E - V - M^2u^2)}}; \quad M^2 = \frac{l^2}{2\mu}.$$

The condition for the path to be closed is $\Delta\theta = 2\pi\cdot(m/n)$, where $m$ and $n$ are positive integers. This means after $n$-number of periods, the particle makes $m$-number of complete revolutions and comes back to its original position.

There are only two potentials: (1) $V = -L/r$ and (2) $V = ar^2$, for which we know the orbits are closed.

## 4.6 Perihelion Advance of Mercury

It has been experimentally found that so called *elliptical orbit of the planet* Mercury is not exact. In fact, one observes that the semi-major axis after one revolution does not come back to its original position. The axis advances at the rate of $40''$ of an arc in a century. This is known as *perihelion advance*. The word perihelion is loosely connected here with the turning point of the motion. To understand this *advance* one assumes the presence of a small $r^{-3}$ force in addition to the inverse-square-law of force. The effective potential $V$ is then

$$V = -\frac{a}{r} + \frac{\beta}{r^2} = -au + \beta u^2. \tag{4.27}$$

where $\beta$ is very small. This $r^{-2}$ potential is repulsive in nature.

The general orbit equation is

$$\theta = -M\int \frac{du}{\sqrt{(E - V - M^2u^2)}}.$$

Substituting (4.27) in this equation we get

$$\theta = -M \int \frac{du}{\sqrt{(E + au - M'^2u^2)}}, \tag{4.28}$$

where $M'^2 = (M^2 + \beta)$.
We rewrite (4.28) as

$$\frac{M'}{M}\theta = -M' \int \frac{du}{\sqrt{(E - au - M'^2u^2)}}. \tag{4.29}$$

Equation (4.29) is again of the standard form of orbit equation. Where now the angle $\theta$ becomes $(M'/M)\theta$, i.e., $\alpha\theta$, where $\alpha = M'/M$ and $M'^2$ is the new conserved quantity and related to $M$ by $M'^2 = M^2 + \beta$.
We write (4.29) as

$$\alpha\theta = -\int \frac{dz}{\sqrt{(L^2 - z^2)}} \tag{4.30}$$

where

$$L^2 = E + \frac{a^2}{4M'^2} \quad \text{and} \quad z = M'u - \frac{a}{2M'}.$$

Equation (4.30) gives us a conic and we get the equation of the conic from (4.30) in terms of polar co-ordinate

$$r = \frac{2M'^2/a}{1 + e\cos\alpha\theta}, \tag{4.31}$$

where

$$\frac{2M'^2}{a} = \text{latus-rectum} = \tilde{a}(1 - e^2);$$

$\tilde{a}$, being the semi-major axis. The eccentricity $e$ is

$$e = \sqrt{\left(1 + \frac{4M'^2E}{a^2}\right)}.$$

Hence for $E < 0$, $e < 1$. But as $\theta \to \alpha\theta$, here (see equation 4.31), the *ellipse* does not become a closed orbit, we see, as

$$\theta \to \theta + 2\pi.$$

$\cos\alpha\theta$ does not become $\cos\alpha\theta$ again, showing the major axis shifts. The shift is given by

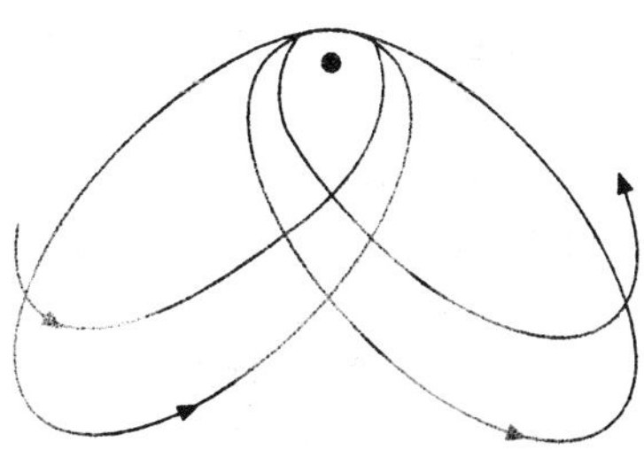

Fig 4.2

$$\theta \to \alpha\theta = \theta\left(\frac{M'}{M}\right) = \theta\sqrt{\left(\frac{M^2 + \beta}{M^2}\right)}.$$

Thus,

$$\theta \to \theta + \Delta\theta,$$

where

$$\Delta\theta = \frac{\beta\theta}{2M^2} = \frac{\mu\beta\theta}{l^2}.$$

Hence perihelion shift is proportional to the strength of repulsive $r^{-3}$ force. Thus the orbit is a *precessing ellipse* (fig. 4.2).

## 4.7 Definition of Scattering Cross-section

It is well known that the scattering experiments are of fundamental importance in physics. It gives information about the structure of small objects of atomic size and also it tells us the nature of forces acting between the incident and target particles. In the scattering experiment a beam of incident particles hits a target and gets scattered in various directions. One measures the angular distribution of the scattered particles to extract information about the nature of the force field of the target particle. In order to obtain this information we must be able to calculate the angular distribution when the force or the potential field is given. For this we first introduce the concept of *impact parameter.*

Consider the following situation, where a particle is coming from a long distance with a given positive energy, under a potential $-a/r$ $(a > 0)$. We know the path is a hyperbola as $E > 0$. $O$ is the origin, and the particle is following the hyperbola, $PO$ is along the asymptote of the hyperbola. This gives the incident direction of the path, $C$ is the focus, where the centre of the force-field is present. Then the perpendicular distance $CP' = b$ is called the *impact parameter.* Impact parameter is thus the distance of closest approach of the incident particle from the target. Physically one may imterpret the impact parameter as equal to the range of the potential which the incident particle experiences while coming close to the target particle. Obviously, the incident particle will get scattered when they are within the range of the potential.

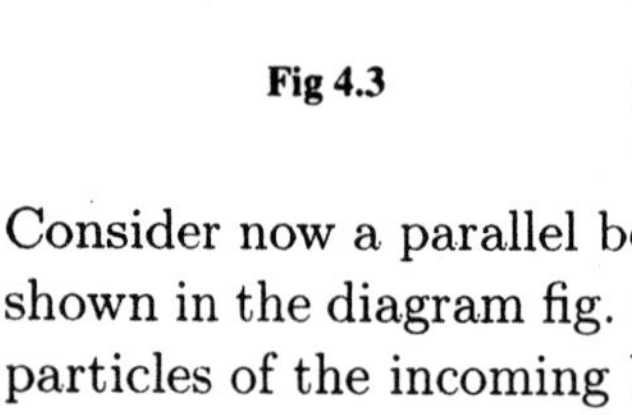

Fig 4.3

Consider now a parallel beam of particle coming along the positive $z$-axis as shown in the diagram fig. 4.4 and approaching the target $T$. Let, one of these particles of the incoming beam with impact parameters $b$, hits the target at $P$, within the range of the potential field of the target and gets scattered in a direction specified by polar angles $(\theta, \phi)$ within an angular range $d\theta$ and $d\phi$. The particles scattered through angles between $\theta$ and $\theta + d\theta$, $\phi$ and $\phi + d\phi$ are those which came in within impact parameters $b$ and $b + db$.

Let $ds$ be an elementary area at a distance $L$ from the point $P$ where incident

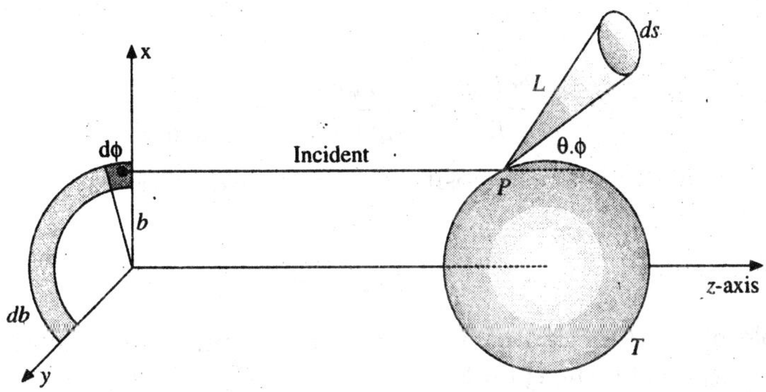


Fig 4.4

beam hits the target and is in the direction which makes angles $\theta.\ \phi$ with incident beam direction. Then,

$$ds = L^2 \sin\theta\ d\theta \cdot d\phi.$$

If $N_s$ denotes the number of particles scattered by the field of potential of the target, then $N_s$ is proportional to

$$L^2 \sin\theta\ d\theta\ d\phi = L^2 d\Omega, \tag{4.32}$$

where $d\Omega$ is the solid angle subtended at $P$ by $dS$.

$$N_s \text{ is also proportional to } 1/L^2, \tag{4.33}$$

where $L$ is the distance of $dS$ from $P$.
If $N$ stands for the number of particles incoming through the shaded region per unit time per unit area. Then

$$N_s \text{ is also proportional to } N. \tag{4.34}$$

From (4.32), (4.33) and (4.34) $N_s$ must be given by

$$N_s \propto \frac{1}{L^2} \cdot N \cdot L^2 d\Omega = N d\Omega.$$

Hence

$$N_s = \sigma(\theta) N d\Omega, \tag{4.35}$$

where $\sigma(\theta)$ is the constant of proportionality.
The proportionality constant $\sigma(\theta)$ is called *differential cross-section* of the scattering.
The number of incoming particles through the shaded area

$$= N b d\phi \cdot db. \tag{4.36}$$

For elastic scattering, where there is no loss or gain of particles we must have

$$N\sigma(\theta)d\Omega = Nbd\phi db.$$

Therefore,

$$\sigma(\theta) = \frac{bdbd\phi}{d\Omega} = \frac{bdbd\phi}{\sin\theta d\theta d\phi} = \frac{bdb}{\sin\theta d\theta}.$$

Since cross-section must be a positive quantity we write

$$\sigma(\theta) = \left|\frac{bdb}{\sin\theta\ d\theta}\right|. \tag{4.37}$$

In principle $\sigma$ is a function of $\theta$ and $\phi$, but for our case of scattering by a central potential field, the Hamiltonian $= (p^2/2m)+V(r)$. This is rotationally invariant, so angular momentum is conserved and the motion is confined to a plane, as in the case of central orbit. Hence scattering is expressed in terms of the angle $\theta$ only. As $H$ is conservative energy is conserved. We could have represented the cross-section in terms of conserved quantities K.E. and angular momentum. We instead expressed in terms of impact parameter. The expression (4.37) gives angular distribution of the scattered particle. The total cross-section of scattering is obtained by integrating $\sigma(\theta)$, i.e.,

$$\sigma_{\text{total}} = \int \sigma(\theta)d\Omega. \tag{4.38}$$

We now consider a few application of own results.

## 4.8 Scattering of a Particle by Hard Sphere

Consider a beam of particles incident on a *perfectly elastic* sphere of radius $R$ as shown in the fig. 4.5.

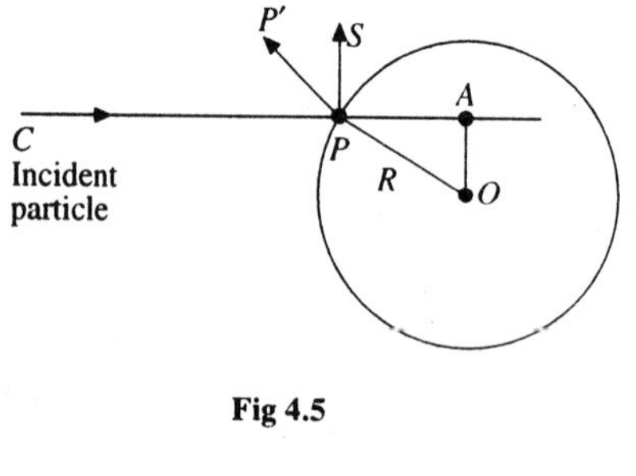


Fig 4.5

Let $O$ be the centre of the sphere and $P$ be a point on its surface, where one incident particle hits the target. $CP$ is the direction of the incident beam. $OP = R$, the radius of the sphere. $OPP'$ is along the radius vector. $OA$ is perpendicular to $CPA$. Then $OA = b =$ impact parameter. Let $\angle CPP' = \alpha$ be the angle of incidence. Since sphere is perfectly elastic, the scattered particle direction is along $PS$, where $\angle P'PS = \alpha$. The angle $\angle SPA = \theta$, is the angle between incident and outgoing direction of the particle. Therefore, $\theta =$ scattering angle.

$$\theta = \pi - 2\alpha$$

$b = OA = OP \sin OPA = R \sin \alpha$, as $\angle CPP' = \angle OPA = \alpha$.

$$b = R \sin\left(\frac{\pi}{2} - \frac{\theta}{2}\right) = R \cos \frac{\theta}{2}.$$

Hence,

$$-db = \frac{1}{2} R \sin \frac{\theta}{2}\, d\theta.$$

Therefore,

$$\sigma(\theta) = \left|\frac{b db}{\sin \theta d\theta}\right| = \left|\frac{R \cos\frac{\theta}{2} \cdot \frac{1}{2} R \sin \frac{\theta}{2} d\theta}{\sin \theta d\theta}\right|; \quad \sigma(\theta) = \frac{1}{4} R^2.$$

This gives differential cross-section. For $\sigma_{\text{tot}}$,

$$\sigma_{\text{tot}} = \int \sigma(\theta) d\Omega = \frac{1}{4} R^2 \int d\Omega = 4\pi \frac{1}{4} R^2 = \pi R^2.$$

Thus, the total cross-section just equals the area of the circular opaque disc of radius $R$.

## 4.9 Rutherford Scattering

This process consists in the scattering of $\alpha$-particles by atomic nucleus by means of the coulomb field. The potential $V$ is taken as $k/r$, where $k$-denotes the strength of the interaction. Here the $\alpha$-particles are moving in the coulomb field with energy $E > 0$. The equation of the path is given by the integral

$$\theta' = -M \int \frac{du}{\sqrt{(E - V - M^2 u^2)}}$$

with $V = ku$. The equation of the path is

$$r = \frac{2M^2/k}{-1 + e \cos \theta'}, \quad e = \sqrt{\left(1 + \frac{4EM^2}{k^2}\right)}.$$

As $E > 0$, the eccentricity, $e$ becomes greater than unity, hence the path is a hyperbola. (See the diagram fig. 4.6).

$ABA'$ is the hyperbola along which the particle is approaching towards the centre, $S$ and $S'$ are the focii. Let $a$ and $b$ are semi-major and minor axes then $SO = ae$, where $O$ is the origin $SC$ is the perpendicular distance $p$, denotes the impact parameter.

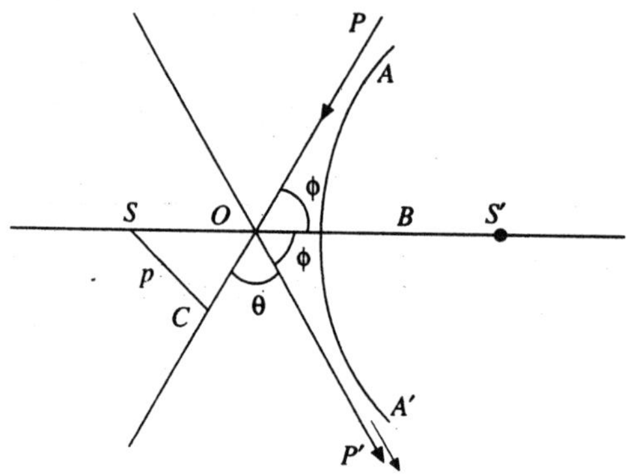


**Fig 4.6**

$PO$ denotes the asymptote of the hyperbola; it gives the direction of the incoming $\alpha$-particle. The point $P$ has the co-ordinates $(r, \theta')$ the polar co-ordinate $r$, being mesured from $S$ to $P$ and $\theta'$ is the angle made by $r$ with the horizontal. When the particle is coming from infinity $PC$ is the asymptote which makes the angle $\phi$ with horizontal. After scattering the $\alpha$-particle follows asymptote $OP'$. Hence, the angle between $PO$ and $OP'$ is the scattering angle $\theta$.
Therefore from the figure 4.6

$$\theta = \pi - 2\phi, \quad \text{or,} \quad \phi = \frac{\pi - \theta}{2}. \tag{4.39}$$

From the same figure we can also find $\angle SOC = \phi$. Therefore,

$$SC = p = SO \cdot \sin\phi = ae\sin\phi, \tag{4.40}$$

where $SO = ae$ is used.
For the asymptote $PO$, we have $r = \infty$. Therefore, from the hyperbola equation

$$r = \frac{\tilde{l}}{(-1 + e\cos\theta')}.$$

We find

$$-1 + e\cos\theta' = 0, \quad \text{i.e.,} \quad \cos\theta' = \frac{1}{e}$$

and also $\theta'$ reduces to $\phi$ when $r \to \infty$. Thus

$$\cos\phi = \frac{1}{e}. \tag{4.41}$$

Hence, from (4.40) and (4.41), the impact parameter $p$ is

$$p = a \cdot \frac{1}{\cos\phi} \sin\phi = a \tan\phi. \tag{4.42}$$

From (4.39), $p$ becomes

$$p = a \tan\left(\frac{\pi}{2} - \frac{\theta}{2}\right) = a \cot\frac{\theta}{2}.$$

Then,

$$dp = -\frac{1}{2} a \operatorname{cosec}^2\frac{\theta}{2} d\theta$$

and $pdp$ becomes

$$pdp = -\frac{1}{2} a^2 \cot\frac{\theta}{2} \cdot \operatorname{cosec}^2\frac{\theta}{2} d\theta.$$

The differential cross-section $\sigma|\theta|$ becomes,

$$\sigma(\theta) = \left|\frac{pdp}{\sin\theta d\theta}\right| = \frac{1}{2} a^2 \cot\frac{\theta}{2} \cdot \operatorname{cosec}^2\frac{\theta}{2} \cdot \frac{1}{2}\frac{1}{\sin\theta} = \frac{1}{4} a^2 \frac{1}{\sin^4(\theta/2)}. \tag{4.43}$$

From

$$r = \frac{(2M^2/k)}{(-1 + e\cos\theta')}$$

we find latus rectum of the hyperbola is $\tilde{l}$, given by

$$\tilde{l} = \frac{2M^2}{k}, \quad \text{i.e.,} \quad a(e^2 - 1) = \frac{2M^2}{k}.$$

Again,

$$e = \sqrt{\left(1 + \frac{4EM^2}{k^2}\right)}.$$

then

$$e^2 - 1 = \frac{4EM^2}{k^2}.$$

Therefore,

$$a = \frac{(2M^2/k)}{(e^2 - 1)} = \frac{2M^2/k}{4EM^2/k^2} = \frac{k}{2E}.$$

Hence, from (4.43),

$$\sigma(\theta) = \frac{k^2}{16E^2} \operatorname{cosec}^4\frac{\theta}{2}; \quad \text{and as } E = \frac{1}{2}mv^2$$

we get for

$$\sigma(\theta) = \frac{k^2}{4m^2v^4} \operatorname{cosec}^4\frac{\theta}{2}. \tag{4.44}$$

Equation (4.44) is the well-known *Rutherford formula*. The total cross-section

$$\sigma_t = \int d\Omega \frac{k^2}{4m^2v^4} \operatorname{cosec}^4 \frac{\theta}{2} = \infty;$$

as can be easily seen from

$$\int_0^\pi \sin\theta d\theta \operatorname{cosec}^4 \frac{\theta}{2},$$

which diverges.

One may regard this result unphysical. But we should remember, the potential $k/r$ has an infinite range. This implies that all particles will get scattered, leading to a divergent result. The angular distribution given by

$$\operatorname{cosec}^4 \frac{\theta}{2}$$

law is extremely well matched by experiment. It is interesting to note that Rutherford formula is also true in quantum theory.

## 4.10 Centre of Mass and Laboratory System of Scattering

Consider the scattering of two particles of masses $m_1$ and $m_2$ with co-ordinates $\mathbf{r}_1$ and $\mathbf{r}_2$ respectively. Let $\mathbf{R}$ and $\mathbf{r}$ be the C.M. and relative co-ordinates which are defined by

$$\mathbf{R} = \frac{(m_1\mathbf{r}_1 + m_2\mathbf{r}_2)}{(m_1 + m_2)}, \quad \mathbf{r} = \mathbf{r}_1 - \mathbf{r}_2. \tag{4.45}$$

From these,

$$\mathbf{r}_1 = \mathbf{R} + \frac{m_2}{M}\mathbf{r} \quad \text{and} \quad \mathbf{r}_2 = \mathbf{R} - \frac{m_1}{M}\mathbf{r}$$

$$M = m_1 + m_2. \tag{4.46}$$

Hence,

$$m_1\dot{\mathbf{r}}_1 = m_1\dot{\mathbf{R}} + \frac{m_1m_2}{M}\dot{\mathbf{r}}, \quad m_2\dot{\mathbf{r}}_2 = m_2\dot{\mathbf{R}} - \frac{m_1m_2}{M}\dot{\mathbf{r}}. \tag{4.47}$$

In the centre of mass system $\dot{\mathbf{R}} = 0$, then,

$$m_1\dot{\mathbf{r}}_1 = \frac{m_1m_2}{M}\dot{\mathbf{r}} = \mathbf{p}^*.$$

Then $m_2\dot{\mathbf{r}}_2 = -\mathbf{p}^*$ as seen from (4.47).

Let us assume that after the scattering the masses have co-ordinate $\mathbf{r}_1'$ and $\mathbf{r}_2'$.

The introducing C.M. and relative co-ordinates

$$\mathbf{R}' = \frac{(m_1\mathbf{r}'_1 + m_2\mathbf{r}'_2)}{(m_1 + m_2)}$$

and $\mathbf{r}' = \mathbf{r}'_1 - \mathbf{r}'_2$. we can write

$$m_1\dot{\mathbf{r}}'_1 = m_1\dot{\mathbf{R}}' + \frac{m_1 m_2}{M}\dot{\mathbf{r}}' \quad \text{and} \quad m_2\dot{\mathbf{r}}'_2 = m_2\dot{\mathbf{R}}' - \frac{m_1 m_2}{M}\dot{\mathbf{r}}'.$$

In the centre of mass system we choose $\dot{\mathbf{R}}' = 0$ so we get

$$m_1\dot{\mathbf{r}}'_1 = \frac{m_1 m_2}{M}\dot{\mathbf{r}}' = \mathbf{q}^*.$$

then $m_2\dot{\mathbf{r}}'_2 = -q^*$. By conservation of energy we get

$$|p^*| = |q^*|. \tag{4.48}$$

In the laboratory system one of the particles is at rest and treated as target while the other particle is incident on it and gets scattered. Before scattering let $\mathbf{p}_1$ be the momentum of $m_1$ then

$$\mathbf{p}_1 = m_1\dot{\mathbf{R}} + \frac{m_1 m_2}{M}\dot{\mathbf{r}}$$

which can be written as

$$\mathbf{p}_1 = m_1\dot{\mathbf{R}} + \mathbf{p}^*.$$

Similarly,

$$\mathbf{p}_2 = m_2\dot{\mathbf{R}} - \frac{m_1 m_2}{M}\dot{\mathbf{r}} = m_2\dot{\mathbf{R}} - p^*.$$

In the laboratory system we take $m_2$ to be at rest. then

$$\mathbf{p}_2 = 0. \tag{4.49}$$

Then. $m_2\dot{\mathbf{R}} = p^*$. then using this we get for $\mathbf{p}_1$.

$$\mathbf{p}_1 = \frac{m_1}{m_2}p^* + p^*. \tag{4.50}$$

After scattering. we write

$$\mathbf{q}_1 = m_1\dot{\mathbf{R}}' + \frac{m_1 m_2}{M}\dot{\mathbf{r}}' \quad \text{and} \quad \mathbf{q}_2 = m_2\dot{\mathbf{R}}' - \frac{m_1 m_2}{M}\dot{\mathbf{r}}'.$$

From momentum conservation. $\dot{\mathbf{R}} = \dot{\mathbf{R}}'$. therefore.

$$\mathbf{q}_1 = \frac{m_1}{m_2}\mathbf{p}^* + \mathbf{q}^* \quad \text{and} \quad \mathbf{q}_2 = \mathbf{p}^* - \mathbf{q}^*. \tag{4.51}$$

As $\mathbf{p}_2 = 0$. we also have

$$\mathbf{p}_1 = \mathbf{q}_1 + \mathbf{q}_2. \tag{4.52}$$

We now utilize (4.50) to (4.51) to obtain the following diagram [Fig. (4.7)]

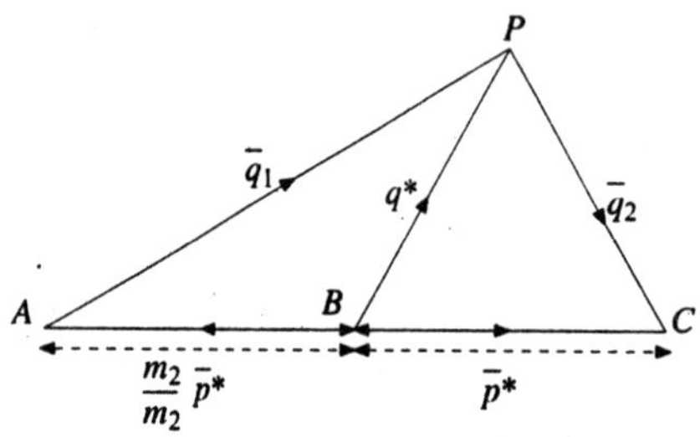


Fig 4.7

$$\overline{AC} = \overline{AB} + \overline{BC}, \quad \overline{AC} = \mathbf{p}_1,$$
$$\overline{AB} = \frac{m_1}{m_2}\mathbf{p}^*, \quad \overline{BC} = \mathbf{p}^*$$
$$\overline{AP} = \mathbf{q}_1; \quad \overline{PC} = \mathbf{q}_2$$
$$\mathbf{p}_1 = \mathbf{q}_1 + \mathbf{q}_2; \quad \mathbf{p}_1 = \frac{m_1}{m_2}p^* + \mathbf{p}^*.$$

We have

$$\mathbf{q}_1 = \frac{m_1}{m_2}\mathbf{p}^* + \mathbf{q}^*$$

as given by the relation: $\overline{AP} = \overline{AB} + \overline{BP}$.
Also,

$$\mathbf{q}_i = \mathbf{p}^* - \mathbf{q}^*, \quad \text{i.e.,} \quad \overline{PC} = \overline{BC} - \overline{BP}.$$

Let $\theta^*$ is the C.M. scattering angle, then $\theta^*$ is the angle between $\overline{BP}$ and $\overline{BC}$. If $\theta$ is the laboratory scattering angle, then $\theta$ is the angle between $\overline{AB}$ and $\overline{AP}$.
From this diagram (4.8) $\tan\theta = PD/AD$, where $PD$ is the perpendicular to $AB$. Then

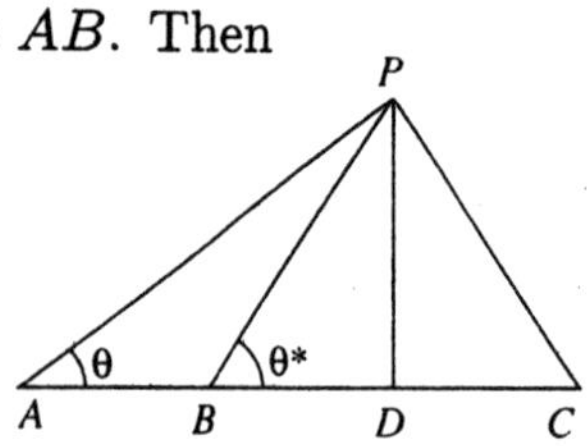


Fig 4.8

$$\tan\theta = \frac{BP\sin\theta^*}{(m_1/m_2)p^* + BP\cos\theta^*}$$
$$= \frac{\sin\theta^*}{(m_1/m_2) + \cos\theta^*},$$

where we have used $p^* = q^*$ in magnitude.

Therefore,

$$\tan\theta_{\text{Lab}} = \frac{\sin\theta_{cm}}{(m_1/m_2) + \cos\theta_{cm}}. \tag{4.53}$$

If $m_2 \gg m_1$, then $\theta_{\text{Lab}} = \theta_{\text{C.M.}}$ For equal masses,

$$\tan\theta_{\text{Lab}} = \tan\frac{\theta_{cm}}{2}, \quad \text{i.e.,} \quad \theta_{\text{Lab}} = \frac{\theta_{cm}}{2}.$$

The equation (4.53) can be solved for $\cos\theta_{cm}$. we have

$$\tan\theta_L = \frac{m_2\sin\theta_{cm}}{(m_1 + m_2\cos\theta_{cm})}.$$

Then,

$$\cos\theta_{cm} = \frac{m_1}{m_2}\sin^2\theta_L \pm \cos\theta_L\left(1 - \frac{m_1^2}{m_2^2}\sin^2\theta_L\right)^{1/2}$$

For real $\theta_{cm}$, and $m_1 > m_2$,

$$\sin \theta_L^{\max} = \frac{m_2}{m_1}.$$

*To find the relation between the C.M. and Lab-system of scattering.*
We note that the solid angle $d\Omega^*$ in C.M. system is

$$d\Omega^* = \sin \theta^* d\theta^* d\phi$$

($\phi$ is not charged as it is not occuring in cross-section). The differential cross-section in C.M. system is $d\sigma/d\Omega^*$. If $d\sigma/d\Omega$ where

$$d\Omega = \sin \theta \, d\theta \, d\phi.$$

is the cross-section in Lab-system then

$$\frac{d\sigma}{d\Omega} = \frac{d\sigma}{d\Omega^*} \cdot \frac{d\Omega^*}{d\Omega} = \frac{d\sigma}{d\Omega^*} \cdot \frac{\sin \theta^* d\theta^* d\phi}{\sin \theta d\theta d\phi}.$$

$$\text{or.} \left(\frac{d\sigma}{d\Omega}\right)_{\text{Lab}} = \left(\frac{d\sigma}{d\Omega^*}\right)_{\text{C.M.}} \cdot \frac{d(\cos \theta^*)}{d(\cos \theta)}.$$

Put $\cos \theta = z$ and $\cos \theta^* = z'$, we have

$$\left(\frac{d\sigma}{d\Omega}\right)_{\text{Lab}} = \left(\frac{d\sigma}{d\Omega^*}\right)_{\text{C.M.}} \frac{dz^*}{dz}. \tag{4.54}$$

The quantity $dz^*/dz$ can be obtained from

$$\tan \theta = \frac{\sin \theta^*}{(m_1/m_2) + \cos \theta^*}.$$

$$\text{i.e.,} \ z^2 = \frac{[(m_1/m_2) + z^*]^2}{1 + 2(m_1/m_2)z^* + (m_1/m_2)^2}. \tag{4.55}$$

## 4.11 A few Simple Applications

We have seen that the Hamiltonian of our system is

$$H = \frac{\mathbf{p}^2}{2m} + V(r): \quad \mathbf{p}^2 = (\text{momentum})^2$$

and $V$ = central potential which is a function of radial distance only. Thus, $H$ is invariant under rotation, as a result the angular momentum vectors **M** is a constant of motion. In addition, the following vector called *Lenz-vector* is also a constant of motion.

**Example 1.** *The vector* $\mathbf{V} \times \mathbf{M} + \alpha \frac{\mathbf{r}}{r}$ *is a constant of motion.*

***Solution:*** Because, total time-derivative of the expression is

$$\dot{\mathbf{V}} \times \mathbf{M} + \mathbf{V} \times \dot{\mathbf{M}} + \alpha \frac{\dot{\mathbf{r}}}{r} + (\alpha \cdot \mathbf{r}) \left( -\frac{1}{r^2} \right) \frac{d}{dt} r.$$

Since $M$ is a constant of motion, $\dot{\mathbf{M}} = 0$ and since

$$r^2 = X^2 + Y^2 + Z^2,$$

we find

$$2r \cdot \dot{r} = 2(\mathbf{X} \cdot \dot{\mathbf{X}} + \mathbf{Y} \cdot \dot{\mathbf{Y}} + \mathbf{Z} \cdot \dot{\mathbf{Z}}).$$

Therefore,

$$\frac{d}{dt}(r) = \frac{\mathbf{r} \cdot \dot{\mathbf{r}}}{r}.$$

Then

$$\frac{d}{dt} \left( \mathbf{V} \times \mathbf{M} + \alpha \frac{\mathbf{r}}{r} \right)$$

becomes

$$\dot{\mathbf{V}} \times \mathbf{M} + \alpha \frac{\mathbf{V}}{r} - (\alpha \mathbf{r}) \frac{(\mathbf{r} \cdot \dot{\mathbf{r}})}{r^3}.$$

Since, $\mathbf{M} = m\mathbf{r} \times \mathbf{V}$, we easily see that the above expression reduces to

$$m(\mathbf{V} \cdot \dot{\mathbf{V}})\mathbf{r} - m\mathbf{V}(\mathbf{r} \cdot \dot{\mathbf{V}}) + \alpha \frac{\mathbf{V}}{r} - \alpha \mathbf{r} \cdot \frac{(\mathbf{r} \cdot \mathbf{V})}{r^3}.$$

Using the relation

$$m\dot{\mathbf{V}} = \frac{\alpha \mathbf{r}}{r^3}$$

in the above expression we find that it vanishes. We easily check the magnitude of the Lenz-vector $= \alpha e$, where $e$ = eccentricity of the orbit.
To conclude this section we consider a few examples in the theory of scattering.

**Example 2.** *From the relation between the scattering angle in laboratory and C.M. system we found that*

$$\theta_L = \frac{1}{2} \theta_{cm},$$

*where masses* $m_1$ *and* $m_2$ *are equal. It is interesting to note that in such a case masses always lie on a circumference of a circle. See the vector-diagram (fig. 4.9).*

***Solution:*** We easily see that when $m_1 = m_2$, $|\overline{AB}| = |\overline{BC}|$ and as

$$|p^*| = |q^*|, \quad |\overline{AB}| = |\overline{CP}|.$$

Further

$$\theta = \frac{1}{2}\theta^*.$$

Thus, if with radius $|p^*|$ we draw a circle with $B$ as centre, then the circle must pass through $A, C$ and $P$. That is the particles lie on a circle. After the collision, the two particles move at right angle to each each other.

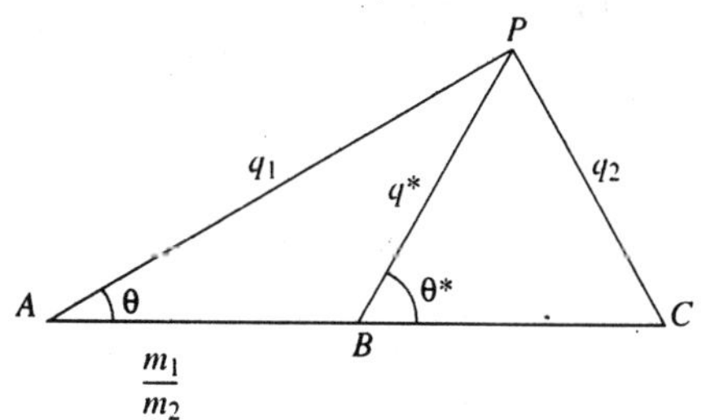


**Fig 4.9**

We have mentioned earlier that there are not many potential functions for which the integral of the orbit equation can be performed in terms of simple functions. We consider here a simple case of scattering.

**Example 3.** *To find the angle of deflection and the corresponding differential cross-section for the scattering of a beam of particle by a field of potential*

$$V = \frac{\alpha}{r^2} = \alpha u^2$$

***Solution:*** The integral to be evaluated for the determination of the orbit for the potential is

$$\theta = -M\int \frac{du}{\sqrt{(E - V - M^2u^2)}},$$

where $V = \alpha u^2$. We rewrite this as

$$\theta = -\frac{M}{M'}\int \frac{M'du}{\sqrt{(E - \alpha u^2 - M^2u^2)}} = -\frac{M}{M'}\int \frac{M'du}{\sqrt{(E - M'^2u^2)}},$$

where $M'^2 = M^2 + \alpha$.

Thus,

$$\alpha_1\theta = -\int \frac{M'du}{\sqrt{(E - M'^2u^2)}},$$

where $\alpha_1 = M'/M$.

This can be easily integrated and the orbit equation is

$$r = \frac{M'}{\sqrt{E}\ \cos\alpha_1\theta}.$$

We note that the path is invariant under $\theta \to -\theta$, so the path is symmetric. The angle the asymptote makes with the horizontal is $\phi$ say: then $\phi = \theta$ evaluated when $r \to \infty$ in the orbit equation. Then $\phi = \pi/2\alpha_1$, as $r \to \infty$, we get $\cos\alpha_1\theta = 0$ giving $\theta = \pi/2\alpha_1$. As in Rutherford scattering angle of deflection defined by $\Theta$ is

$$\Theta = \pi - 2\phi = \pi - \frac{\pi}{\alpha_1}.$$

Therefore,

$$\alpha_1 = \frac{\pi}{(\pi - \Theta)}.$$

But

$$\alpha_1 = \frac{M'}{M}.$$

Hence,

$$\alpha_1^2 = \frac{M'^2}{M^2} = \frac{\pi^2}{(\pi - \Theta)^2}.$$

But

$$M'^2 = M^2 + \alpha.$$

therefore,

$$1 + \frac{\alpha}{M^2} = \frac{\pi^2}{(\pi - \Theta)^2}.$$

Therefore

$$\frac{M^2}{\alpha} = \frac{(\pi - \Theta)^2}{[\pi^2 - (\pi - \Theta)^2]}.$$

But

$$M^2 = \frac{l^2}{2\mu}.$$

where $l$ = angular momentum. If $v$ be the velocity of the particle then $l = \mu v \rho$; where $\mu$ is the mass and $\rho$ is the impact parameter.
Thus,

$$\frac{(\mu^2 v^2 \rho^2)}{2\mu\alpha} = \frac{(\pi - \Theta)^2}{[\pi^2 - (\pi - \Theta)^2]}; \quad \rho^2 = \frac{2\alpha}{\mu v^2} \frac{(\pi - \Theta)^2}{[\pi^2 - (\pi - \Theta)^2]}.$$

From which we can calculate

$$\sigma(\Theta) = \frac{\rho d\rho}{\sin\Theta d\Theta}$$

to be equal to

$$\frac{2\pi^2\alpha}{\mu v^2} \cdot \frac{(\Theta - \pi)}{\Theta^2(\Theta - 2\pi)^2} \cdot \frac{1}{\sin\Theta}.$$

We calculated Rutherfords formula with

$$V = \frac{k}{r}, \quad k > 0.$$

It is interesting to ask the following question.

**Example 4.** *What changes if any, there would be if the sign of k is changed in the Rutherford formula?*

***Solution:*** As the cross-section depends on $k^2$, there would be no change in cross-section. But from the orbit for $k < 0$, we have

$$r = \frac{\tilde{l}}{(1 + e\cos\theta)}$$

rather than

$$\frac{\tilde{l}}{(-1 + e\cos\theta)}$$

which we obtained when $k > 0$. Therefore, the angle made by the asymptote with the $x$-axis is obtained from $r \to \infty$, from

$$r = \frac{\tilde{l}}{(1 + e\cos\theta)},$$

the angle is $\phi$, where

$$\cos\phi = -\frac{1}{e} \quad \text{and} \quad e > 1.$$

We conclude this section by mentioning an important result in Virial theorem.

**Example 5.** *Consider a system in which the total forces acting on particles consist of conservative forces* $\mathbf{F}'_i$ *and frictional forces proportional to velocities, then, for such a system, the virial theorem holds in the following form*

$$\bar{T} = -\frac{1}{2}\sum_i \overline{\mathbf{F}'_i \cdot \mathbf{r}_i}$$

*provided the steady motion is achieved.*

***Solution:*** We consider the kinetic energy of the system, $T$,

$$\begin{aligned} T &= \frac{1}{2}\sum_i m_i \mathbf{V}_i^2 = \frac{1}{2}\sum_i \mathbf{m}_i \mathbf{V}_i \mathbf{V}_i \\ &= \frac{1}{2}\sum_i \mathbf{p}_i \cdot \mathbf{V}_i = \frac{d}{dt}\left(\frac{1}{2}\sum_i \mathbf{p}_i \cdot \mathbf{r}_i\right) - \frac{1}{2}\sum_i \dot{\mathbf{p}}_i \cdot r_i. \end{aligned}$$

Since, the time-average of a quantity which is a total time derivative of a bounded function is zero, therefore,

$$\mathbf{T} = -\frac{1}{2}\sum_i \overline{\dot{\mathbf{p}}_i \cdot \mathbf{r}_i} = -\frac{1}{2}\sum_i \overline{(\mathbf{F}'_i + \mathbf{f}'_i)\cdot r_i} = -\sum_i \overline{\mathbf{F}'_i \cdot \mathbf{r}_i} - \sum_i \overline{\mathbf{f}'_i \cdot \mathbf{r}_i}.$$

where we have used Newtonic equation.

$$\dot{\mathbf{p}}_i = \mathbf{F}'_i + \mathbf{f}'_i.$$

But as $\mathbf{f}'_i = C\mathbf{V}_i$,

$$\sum_i \mathbf{f}'_i \cdot \mathbf{r}_i = C\sum_i \mathbf{V}_i \cdot \mathbf{r}_i.$$

The quantity

$$\sum_i \mathbf{V}_i \cdot \mathbf{r}_i = \frac{d}{dt}\left(\frac{1}{2}\sum_i \mathbf{r}_i^2\right).$$

The average of

$$\frac{d}{dt}\left(\frac{1}{2}\sum_i \mathbf{r}_i^2\right)$$

again is zero.
Hence,

$$\mathbf{T} = -\frac{1}{2}\sum_i \overline{\mathbf{F}'_i \cdot \mathbf{r}_i}.$$

## Exercises

1. Two bodies of reduced mass, $\mu$ and relative separation $\mathbf{r}$ interact with a central force described by a potential $V(r)$. The total energy is $E$ and angular momentum is $l$. Show that the equation of the orbit is

$$t = \int \left[\frac{2}{\mu}E - V - \frac{l^2}{2\mu r^2}\right]^{-1/2} dr.$$

2. Find the effective cross-section for scattering of particle by a spherical potential well: $U = 0$; $r > a$ and $U = -U_0$ for $r < a$.

[*Hints:* Incident beam parallel to horizontal axis hits at $P$ on the well which divides the media with potential $U_1$ and $U_2$, $V_1$ and $V_2$ are the velocities of

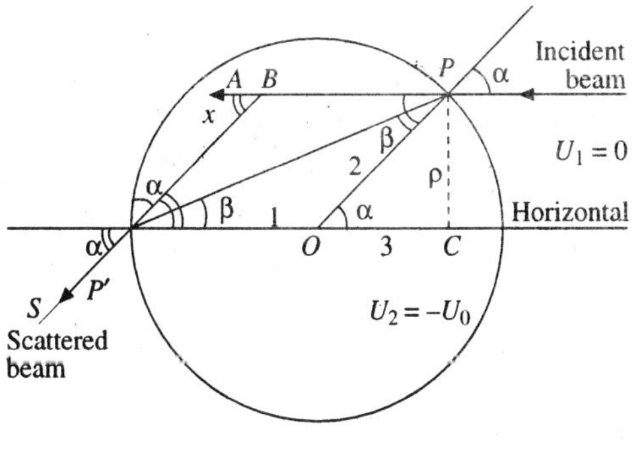


Fig 4.10

particles outside and inside. From conservation of total energy

$$\frac{1}{2}mV_1^2 + U_1 \quad \text{and} \quad \frac{1}{2}mV_2^2 + U_2$$

we get

$$\frac{\sin\alpha}{\sin\beta} = \frac{V_2}{V_1} = n = \sqrt{\left(1 + \frac{2}{mV_1^2}(U_1 - U_2)\right)}.$$

Impact parameter $\rho = CP$. Incidents beam makes an angle $\alpha$ with normal $OP$, then $\angle POC = \alpha$. Hence if $a$ = radius of well then $\rho = a\sin\alpha$. Here $\beta$ = angle of refraction. $\angle PP'O = \beta$ also. $P'S$ is the direction of scattered wave $PB$ is the incident direction. Hence

$$\angle P'BA = \chi$$

is the angle of deflection and

$$\chi = \angle BPP' + \angle BP'P; \quad \angle BPP' = \angle BP'P = \alpha - \beta.$$

Hence

$$\chi = 2(\alpha - \beta), \quad \text{or,} \quad \beta = \alpha - \frac{\chi}{2}.$$

Then

$$\frac{\sin\beta}{\sin\alpha} = \frac{\sin\left(\alpha - \frac{1}{2}\chi\right)}{\sin\alpha} = \frac{1}{n}.$$

That is

$$\cos\frac{1}{2}\chi - \cot\alpha\sin\frac{\chi}{2} = \frac{1}{n}.$$

From this

$$\cot\alpha = \frac{(n\cos\frac{\chi}{2} - 1)}{n\sin\frac{\chi}{2}},$$

$$\text{or, } \sin^2\alpha = (1 + \cot^2\alpha)^{-1} = \frac{(n^2\sin^2\frac{\chi}{2})}{(n^2 + 1 - 2n\cos\frac{\chi}{2})}.$$

Hence

$$\rho^2 = a^2 \sin^2 \alpha = a^2 \frac{n^2 \sin^2 \frac{\lambda}{2}}{1 + n^2 - 2n \cos \frac{\lambda}{2}}.$$

From this the differential cross-section.

$$\frac{d\sigma}{d\Omega} = \frac{\rho d\rho}{\sin \chi \cdot d\chi} = \frac{a^2 n^2}{\cos \frac{\lambda}{2}} \frac{(n \cos \frac{\lambda}{2} - 1)(n - \cos \frac{\lambda}{2})}{\left(n^2 + 1 - 2n \cos \frac{\lambda}{2}\right)^2}.$$

After integration of all angles $\sigma_{Tot} = \pi a^2$.]

3. Show that for any repulsive central potential. $V$. a formal solution of the angle of scattering is

$$\theta = \pi + 2s \int_0^{u_0} \frac{du}{\sqrt{1 - [V(u)/E] - s^2 u^2}}.$$

where $u = 1/r$. $u_c$ corresponds to a turning point of the orbit and $s$ is a constant.

4. A planet is moving with time period $T$ round the sun in a circular orbit the sun being at the centre. Suddenly the motion of the planet is stopped. Show that the time taken by the planet to collide the sun is $(\sqrt{2}/8)T$.

[*Hints:* One particle is sun. Then the other particle is tracing a circular path around it. The radius of the circle is $a$ with sun at its centre. For circular orbit time period

$$T = \frac{2\pi}{\sqrt{\mu}} a^{3/2}.$$

$\mu$ = Potential strength/m.

As the particle is stopped suddenly in its path. its velocity is zero. The particle will then be attracted by the sun. Hence the equation of motion of the particle is

$$\frac{d^2x}{dt^2} = -\frac{\mu}{x^2}.$$

We can solve this. to find

$$\frac{dx}{dt} = \sqrt{\left[2\mu\left(\frac{1}{x} - \frac{1}{a}\right)\right]}.$$

which immediately gives

$$\sqrt{2\mu}t = \frac{\pi}{2} a^{3/2}.$$

Hence

$$\frac{t}{T} = \frac{\sqrt{2}}{8}.]$$

5. Find the law of force if the orbit is given by: (*a*) $r = \frac{2k}{1+\cos\theta}$:(*b*) $r^n = A\cos n\theta + B\sin n\theta$: (*c*) $r = a\sin n\theta$.

   [*Ans.* (a) $F \propto \frac{1}{r^2}$: (b) $F \propto \frac{1}{r^{2n+3}}$: (c) $F \propto \frac{2n^2a^2}{r^5} - \frac{n^2-1}{r^3}$.]

6. A particle of mass unity moves in a resesting media under a central force $P$: show that the orbit is

$$\frac{d^2u}{d\theta^2} + u = \frac{P}{h^2}u^{-2},$$

where

$$h = h_0 \exp\left\{-\int \frac{R}{v}\cdot dt\right\}.$$

$R$ is the resistance.

[*Hints:* The Lagrangian equation is

$$\ddot{r} - r\dot{\theta}^2 = -P - R\cos\phi.$$

where $\phi$ is the angle between the tangent at $P$ of the orbit and the radius vector **r**. The resistance force $R$ is acting along the tangent opposing the motion. Since the rate of change of angular momentum is equal to the moment of $R$ we get

$$\frac{d}{dt}(r^2\dot{\theta}) = -\mathbf{r}\times\mathbf{R} = -Rr\sin\phi = -Rr^2\frac{d\theta}{ds}$$

then

$$\frac{d}{dt}(r^2\dot{\theta}) = -Rr^2\frac{\dot{\theta}}{v}; \quad r^2\dot{\theta} = h = h_0\exp\left\{-\int\frac{R}{v}\cdot dt\right\}.$$

We have

$$\frac{dr}{dt} = \frac{dr}{d\theta}\dot{\theta} = \frac{1}{r^2}\frac{dr}{d\theta}h.$$

Therefore,

$$\frac{d^2r}{dt^2} = \frac{d}{dt}\left(\frac{1}{r^2}\frac{dr}{d\theta}\right)h - \dot{r}\frac{R}{v}.$$

We use this in

$$\ddot{r} - r\dot{\theta}^2 = -P - R\cos\phi = -P - R\frac{dr}{dS} = -P - R\frac{\dot{r}}{v}.$$

Then this reduces to

$$h\frac{d}{dt}\left(\frac{1}{r^2}\frac{dr}{d\theta}\right) - r\dot{\theta}^2 = -P$$

change to $r = 1/u$, then we get

$$\frac{d^2u}{d\theta^2} + u = \frac{P}{h^2u^2}]$$

7. A particle of mass $m$ is projected from infinity towards the origin with any velocity and is acted upon by a force $\mu u^3$ at right angle to the radius vector. Find the orbit.

[*Hints.* We take $m = 1$. The radial acceleration $\ddot{r} - r\dot{\theta}^2 = 0$: and

$$\frac{d}{dt}(r^2\dot{\theta}) = r\mu u^3 = \mu u^2. \quad \text{or,} \quad 2(r^2\dot{\theta})\frac{d}{dt}(r^2\dot{\theta}) = 2\mu\frac{d\theta}{dt}; \quad u = \frac{1}{r}$$

is used or,

$$\frac{d}{dt}(r^2\dot{\theta})^2 = 2\mu u\frac{d\theta}{dt}..$$

Therefore,

$$(r^2\dot{\theta})^2 = 2\mu\theta + A.$$

At $\theta = 0$, $\dot{\theta} = 0$, hence $A = 0$. Then $r^2\dot{\theta} = \sqrt{2\mu\theta}$. We use this to reduce $\ddot{r} - r\dot{\theta}^2$ to

$$2\theta\left(\frac{d^2u}{d\theta^2} + u\right) + \frac{du}{d\theta}.$$

Hence we have

$$2\theta\left(\frac{d^2u}{d\theta^2} + u\right) + \frac{du}{d\theta} = 0.$$

Put $u = \theta^{1/4}f(\theta)$. Then $f(\theta)$ satisfies

$$\frac{d^2f}{d\theta^2} + \frac{1}{\theta}\frac{df}{d\theta} + \left(1 - \frac{1}{16\theta^2}\right)f = 0.$$

This is an equation for the Bessel function of order $\frac{1}{4}$. Thus the orbit is

$$u = \theta^{1/4}\cdot J_{1/4}(\theta).$$

$J_{1/4}(\theta)$ stands for the Bessel function.]

Chapter V

# Small Oscillation

## 5.1 Introduction

In this chapter, we want to discuss another important application of the Lagrangian equations of motion. This concerns us with the small oscillatory motion developed in a conservative system when it is slightly displaced from its stable equilibrium position. Since, the system is conservative, the forces are derivable from time-independent potential function. Let $q_1, \cdots, q_n$ be the set of generalised co-ordinates which represent the system after removing the time-independent holonomic constraints. As the system is in equilibrium, the forces acting on the system is zero. As,

$$F_i = -\frac{\partial V(q_1, \cdots, q_n)}{\partial q_i}.$$

we must have for equilibrium

$$\frac{\partial V}{\partial q_i} = 0, \quad i = 1, \cdots, n. \tag{5.1}$$

Equation (5.1) then implies that the potential of the stable equilibrium state, is an *extremum*. Let the equilibrium position be given by

$$q_i = q_{i0} \;\; (i = 1, 2, \cdots, n),$$

then one can show that at $q_i = q_{i0}$, the potential is minimum. Because, if we slightly displace the system from $q_{i0}$, i.e., away from the equilibrium position, the potential energy increases. But as the system is conservative the energy is a constant of motion. Therefore, away from the minimum position of the potential, the kinetic energy decreases and as a result no motion can develop in the system. So, the system remains in a stable equilibrium at the position, where the potential has the minimum value. Therefore, we must have

$$\frac{\partial^2 V}{\partial q_i \partial q_j} > 0. \tag{5.2}$$

Let

$$\eta_i \ (i = 1, \cdots, n)$$

be the small displacement given to the system around the equilibrium points $q_i = q_{i0}$. Then the potential at $q_{i0} + \eta_i$ becomes

$$V(q_{i0} + \eta_i).$$

Expanding $V$ in powers of $\eta_i$'s, we get

$$V(q_{i0} + \eta_i) = V(q_{i0}) + \sum_i \eta_i \left(\frac{\partial V}{\partial q_i}\right)_{q_i=q_{i0}} + \sum_i \sum_j \frac{\eta_i \eta_j}{2} \left(\frac{\partial^2 V}{\partial q_i \partial q_j}\right)_{\substack{q_i=q_{i0}\\ q_j=q_{j0}}} + \cdots .$$

But

$$\frac{\partial V}{\partial q_i} = 0,$$

so, the potential function $V$ as *measured from the minimum value* $V(q_{i0})$ is (keeping the second term)

$$V = \frac{1}{2} \sum_i \sum_j \left(\frac{\partial^2 V}{\partial q_i \partial q_j}\right)_{\substack{q_i=q_{i0}\\ q_j=q_{j0}}} \eta_i \eta_j > 0. \tag{5.3}$$

as $\partial^2 V > 0$ for minimum. We write

$$V_{ij}(q_{i0}) \text{ for } \left(\frac{\partial^2 V}{\partial q_i \partial q_j}\right)$$

evaluated at $q_i = q_{i0}$, etc. Therefore, $V$ is (dropping higher order terms)

$$V = \sum_{ij} \frac{1}{2} V_{ij} \eta_i \eta_j \tag{5.4}$$

where

$$V_{ij} = V_{ji}. \tag{5.5}$$

For a time-independent holonomic system with generalised co-ordinates $q_1, \cdots, q_n$, the kinetic energy $T$ is given by

$$T = \frac{1}{2} \sum_{i,k} a_{ik} \dot{q}_i \dot{q}_k,$$

where

$$a_{ik} = a_{ki}. \tag{5.6}$$

Hence, at points

$$q_i = q_{i0} + \eta_i \quad (i = 1, \cdots, n).$$

(5.5) becomes

$$T = \frac{1}{2}\sum_{i,k} a_{ik}(q_{i0} + \eta_i)\dot{\eta}_i\dot{\eta}_k$$

where $q_{i0}$'s are constant points. Expanding $a_{ik}$ around $q_{i0}$, we get

$$a_{ik}(q_{i0} + \eta_i) = a_{ik}(q_{i0}) + \sum_i \eta_i \left(\frac{\partial a_{ik}}{\partial q_i}\right)_{q_i = q_{i0}} + \cdots.$$

We only keep the first term in the expansion of $T$.

$$T = \frac{1}{2}\sum_{i,k} a_{ik}(q_{i0}\cdots)\dot{\eta}_i\dot{\eta}_k. \qquad (5.7)$$

Then $T$ and $V$ are given for the small displacement of the system by equation (5.4) and (5.7) respectively.

## 5.2 Lagrangian equation for system with many degrees of freedom with small displacements

In the last section, we have obtained the expressions for the kinetic and potential energy of the system with many holonomic degrees of freedom, when the system is displaced around its stable equilibrium position. Thus, from (5.4) and (5.7) we get,

$$L = T - V = \frac{1}{2}\sum_{i,k}(a_{ik}\dot{\eta}_i\dot{\eta}_k - V_{ik}\eta_i\eta_k). \qquad (5.8)$$

where $\eta_i$'s are the generalised position co-ordinates.
From (5.8), the Lagrangian equation becomes

$$\sum_k \frac{d}{dt}(a_{ik}\dot{\eta}_k) + \sum V_{ik}\eta_k = 0. \qquad (5.9)$$

$$\text{i.e., } \sum_k (a_{ik}\ddot{\eta}_k + V_{ik}\eta_k) = 0. \qquad (5.10)$$

In deriving (5.9), we have used

$$a_{ik} = a_{ki}; \quad V_{ik} = V_{ki}$$

and they are constants. We now seek a solution of (5.10) of the following form

$$\text{i.e., } \eta_k = A_k \exp(i\omega t). \qquad (5.11)$$

Putting (5.11) in (5.10), we find

$$\sum_k (V_{ik} - \omega^2 a_{ik}) A_k = 0 \quad (i = 1, \cdots, n). \tag{5.12}$$

The condition that $A_k$ has non-trivial solution, we must have

$$\det(V_{ik} - \omega^2 a_{ik}) = 0. \tag{5.13}$$

(5.13) is the characteristic equation, giving $n$-values of $\omega^2$ as the eigen-frequencies. Since, $V_{ik}$ and $a_{ik}$ are symmetric, the eigen-frequencies of $\omega^2$ are *real*. Further from (5.12), we see

$$\omega^2 = \frac{V_{ik} x_i x_k}{a_{ik} x_i x_k} > 0.$$

We conclude that eigen-frequencies $\omega_r$'s are real positive or negative. Let

$$\omega_1, \cdots, \omega_n$$

are the eigen-frequencies. If *all the* $\omega_r$*'s are different*, then the co-efficients $A_k$ are proportional to the minors of the determinant given by (5.12) with $\omega = \omega_r$. Let these minors be $\Delta_{kr}$. Then, if we write equation (5.12) as

$$\sum_k b_{ik} A_k = 0.$$

we find that for the minors

$$\sum_k b_{ik} \Delta_{k\alpha} = 0.$$

Then $A_k$'s are proportional to $\Delta_{k\alpha}$'s. A particular solution of the differential equation (5.10) is therefore,

$$\eta_k = \Delta_{kr} C_r \exp(\pm i\omega_r t).$$

where $C_r$'s are complex constants. The general solution is the sum of all these particular solutions.

$$\eta_k = \sum_{r=1}^{n} \Delta_{kr} C_r \exp(\pm i\omega_r t) \tag{5.14}$$

$$= \sum_r \Delta_{kr} \theta_r, \tag{5.15}$$

where

$$\theta_r = C_r \exp(\pm i\omega_r t). \tag{5.16}$$

Thus the time-variation of each co-ordinate is a super-position of simple periodic oscillations $\theta_1, \theta_2, \cdots, \theta_n$ (where $\ddot{\theta}_r + \omega_r^2 \theta_r = 0$) with arbitrary amplitudes $C_1, C_2, \cdots, C_n$ and phases, but with definite frequency.

We note that from (5.15), we can find $\theta_r$'s in term of $\eta_1, \eta_2, \cdots, \eta_n$ by solving the matrix equation

$$\eta_k = \sum_r \Delta_{kr}\theta_r. \tag{5.17}$$

The $\theta_r$'s execute simple harmonic oscillation and $\theta_r$'s are obtained as a linear combination of the generalised co-ordinates $\eta_1, \eta_2, \cdots, \eta_n$. These $\theta_r$'s are now the generalised co-ordinate; each of them executes simple harmonic motion. These are called the *normal co-ordinates.* They are completely independent, each having definite frequency of small oscillation.

When the eigen-frequencies are not distinct, one needs little alterations in the arguments, however, the general form of the solution remains the same.

## 5.3 Molecular Vibration: $CO_2$-molecule

In this section, we apply the method discussed in the last section to obtain the frequency and normal co-ordinate of vibration of a linear tri-atomic molecule. In particular, we consider here the $CO_2$-molecule. The configuration of the molecule is given by the following diagram (Fig. 5.1).

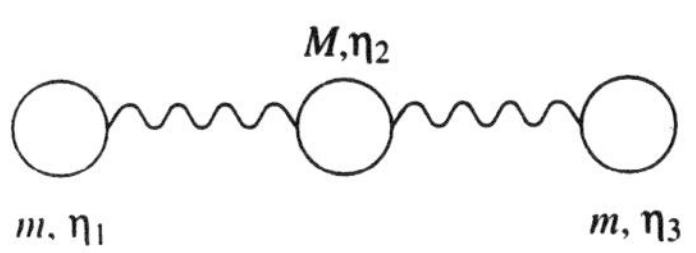


**Fig 5.1**

The oxygen atoms are connected to the carbon atom by springs of a given spring constant $\mu$. Let $\eta_1$, $\eta_2$ and $\eta_3$ are the displacements of Oxygen. Carbon and Oxygen atoms from their equilibrium positions. Let $m$ be the mass of oxygen, and $M$ is the mass of carbon atoms. Hence, the Kinetic energy of the $CO_2$ is

$$T = \frac{1}{2}m(\dot{\eta}_1^2 + \dot{\eta}_3^2) + \frac{1}{2}M\dot{\eta}_2^2. \tag{5.18}$$

We are assuming that the atoms interact harmonically. Then the potential energy $V$ is

$$V = \frac{1}{2}\mu(\eta_2 - \eta_1)^2 + \frac{1}{2}\mu(\eta_3 - \eta_2)^2. \tag{5.19}$$

The Lagrangian $L$ is $T - V$, where $T$ and $V$ are given by (5.18) and (5.19). We can easily derive the Lagrangian equation for $\eta_1, \eta_2$ and $\eta_3$ from

$$\frac{\partial L}{\partial \eta_i} - \frac{d}{dt}\left(\frac{\partial L}{\partial \dot{\eta}_i}\right); \quad i = 1, 2, 3.$$

The equations are

$$\left.\begin{aligned} m\ddot{\eta}_1 &= \mu(\eta_2 - \eta_1) \\ M\ddot{\eta}_2 &= \mu(\eta_3 - \eta_2) - \mu(\eta_2 - \eta_1) \\ m\ddot{\eta}_3 &= -\mu(\eta_3 - \eta_2). \end{aligned}\right\} \tag{5.20}$$

We now seek a solution of the form $\eta_i = a_i e^{\pm i\omega t}$. Then

$$\left.\begin{aligned} -m\omega^2 a_1 &= \mu(a_2 - a_1) \\ -M\omega^2 a_2 &= \mu(a_3 - a_2) - \mu(a_2 - a_1) \\ -m\omega^2 a_3 &= -\mu(a_3 - a_2). \end{aligned}\right\} \quad (5.21)$$

The characteristic determinant to determine the frequency is

$$\begin{vmatrix} (m\omega^2 - \mu) & \mu & 0 \\ \mu & (M\omega^2 - 2\mu) & \mu \\ 0 & \mu & (m\omega^2 - \mu) \end{vmatrix} = 0. \quad (5.22)$$

$$\text{i.e.,} \quad (m\omega^2 - \mu)[(M\omega^2 - 2\mu)(m\omega^2 - \mu) - 2\mu^2] = 0.$$

which gives from the first factor

$$\omega = \pm\sqrt{\left(\frac{\mu}{m}\right)} \quad (5.23)$$

and the second factor yields:

$$\omega^2 = \sqrt{\left(\frac{\mu(M + 2m)}{Mm}\right)} \quad (5.24)$$

$$\text{and} \quad \omega^2 = 0. \quad (5.25)$$

From (5.20) and

$$\eta_i = a_i \exp(\pm i\omega t)$$

we have

$$\left.\begin{aligned} (m\omega^2 - \mu)a_1 + \mu\cdot a_2 + 0\cdot a_3 &= 0 \\ \mu a_1 + (M\omega^2 - 2\mu)a_2 + \mu a_3 &= 0 \\ 0\cdot a_1 + \mu a_2 + (m\omega^2 - \mu)a_3 &= 0. \end{aligned}\right\}. \quad (5.26)$$

To find the eigen vector corresponding to $\omega = 0$, we put $\omega = 0$ in (5.26) and solve for $a_1, a_2$ and $a_3$. We find

$$a_1 = a_2 = a_3 = \rho \text{ (a constant, say)}.$$

Normalising we get,

$$m(a_1^2 + a_3^2) + Ma_2^2 = 1.$$

Then

$$\rho = \frac{1}{\sqrt{(2m + M)}}.$$

For $\omega^2 = \mu/m$, we put this in (5.26) and solve for $a_1$, $a_2$ and $a_3$. We get

$$a_2 = 0 \quad \text{and} \quad a_1 = -a_3 = \rho.$$

Normalisation yields

$$m(a_1^2 + a_3^2) = 1.$$

Then

$$\rho = \frac{1}{\sqrt{2m}}.$$

Hence, the eigen vector is

$$\left(\frac{1}{\sqrt{2m}}, 0, -\frac{1}{\sqrt{2m}}\right).$$

For

$$\omega^2 = \frac{(M\mu + 2m\mu)}{Mm},$$

we find from (5.26), the following values of $a_1, a_2$ and $a_3$

$$a_1 = a_3 = \rho \text{ (say)} \quad \text{and} \quad 2\mu\rho + \frac{M\mu}{m}a_2 = 0, \quad \text{i.e.,} \quad a_2 = -\frac{2m}{M}\rho.$$

Using the normalisation

$$m(a_1^2 + a_3^2) + Ma_2^2 = 1,$$

we find

$$\rho = \sqrt{\left(\frac{M}{2m(M+2m)}\right)}.$$

Hence, the eigen vector is

$$\left(\sqrt{\frac{M}{2m(M+2m)}}; -\frac{2m}{M}\sqrt{\frac{M}{2m(M+2m)}}; \sqrt{\frac{M}{2m(M+2m)}}\right).$$

## 5.4 Some Simple Examples

**Example 1.** *Find the eigen frequencies and normal co-ordinates of a vibrating system characterised by a Lagrangian for three degrees of freedom, namely*

$$L = \frac{1}{2}(\dot\eta_1^2 + \dot\eta_2^2 + \dot\eta_3^2) - \alpha^2(\eta_1^2 + \eta_2^2 + \eta_3^2 - \eta_1\eta_3).$$

***Solution:*** The Lagrangian equations of motion are

$$\ddot\eta_1 = -2\alpha^2\eta_1 + \alpha^2\eta_3 \tag{5.27}$$

$$\ddot\eta_2 = -2\alpha^2\eta_2 \tag{5.28}$$

$$\ddot\eta_3 = -2\alpha^2\eta_3 + \alpha^2\eta_1. \tag{5.29}$$

From (5.28) we see $\eta_2$ is a normal co-ordinate with frequency $\omega^2 = 2\alpha^2$.

From (5.27) and (5.29) above we see, that the coupled system is easily separated by introducing the normal co-ordinate

$$Q_1 = \eta_1 + \eta_2$$
$$\text{and} \quad Q_3 = \eta_1 - \eta_2.$$

Then (5.27) and (5.28) give:

$$\ddot{Q}_1 = -\alpha^2 Q_1$$
$$\ddot{Q}_3 = -3\alpha^2 Q_3.$$

Hence

$$\omega_1^2 = \alpha^2 \quad \text{and} \quad \omega_2^2 = 3\alpha^2$$

are the frequencies of normal co-ordinate $Q_1$ and $Q_3$. Then $(Q_1, \eta_2, Q_3)$ are the normal co-ordinates.

**Example 2.** *Find normal co-ordinates and frequency modes for a system with Lagrangian*

$$L = \frac{1}{2}(\dot{x}^2 + \dot{y}^2) - \frac{1}{2}\omega_0^2(x^2 + y^2) + \alpha xy; \quad \alpha > 0.$$

***Solution:*** The pair of equations of motion is

$$\ddot{x} = -\omega_0^2 x + \alpha y$$
$$\ddot{y} = -\omega_0^2 y + \alpha x.$$

Therefore,

$$(\ddot{x} + \ddot{y}) = -\omega_0^2(x + y) + \alpha(x + y)$$
$$(\ddot{x} - \ddot{y}) = -\omega_0^2(x - y) - \alpha(x - y).$$

Normal co-ordinates $Q_1 = x + y$ and $Q_2 = x - y$, obey

$$\ddot{Q}_1 + (\omega_0^2 - \alpha)Q_1 = 0$$
$$\ddot{Q}_2 + (\omega_0^2 + \alpha)Q_2 = 0.$$

For $\omega_0^2 > \alpha$, $Q_1 = Ae^{\pm i\omega t}$, where $\omega^2 = \omega_0^2 - \alpha > 0$ and $Q_2 = Be^{\pm i\omega' t}$, where $\omega'^2 = \omega_0^2 + \alpha$.

**Example 3.** *Oscillation of particle on a string. Consider a light string of length $(n+1)l$ with $n$ equal masses $m$ spaced along it at regular interval $l$. The string is stretched to a tension $F$. We shall consider transverse oscillations of the particles. We use as the generalised co-ordinates the displacements $y_1, y_2, \cdots, y_n$. The kinetic energy $T$ is*

$$T = \frac{1}{2}m(\dot{y}_1^2 + \dot{y}_2^2 + \cdots + \dot{y}_n^2).$$

***Solution:*** The equilibrium length between $j$th and $(j+1)$th particles is $l$. But when the particles are displaced the length is $l + \delta l$.

The value of $l + \delta l$ is

$$l + \delta l = \left[l^2 + (y_{j+1} - y_j)^2\right]^{1/2} = l\left[1 + \frac{(y_{j+1} - y_j)^2}{2l^2}\right]$$

approximately, thus

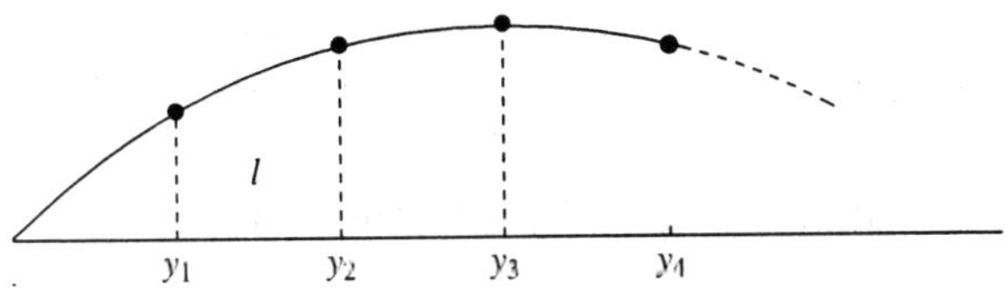

Fig 5.2

$$\delta l = \frac{(y_{j+1} - y_i)^2}{2l}.$$

If $F$ is the tension, the potential $V = F \cdot \delta l$, i.e.,

$$V = \frac{F}{2l}\left[y_1^2 + (y_2 - y_1)^2 + \cdots + (y_n - y_{n-1})^2 + y_n^2\right].$$

We have used

$$y_0 = y_{n+1} = 0.$$

The Lagrangian $L = T - V$. The equations of motion are

$$\begin{aligned}
\ddot{y}_1 &= \frac{F}{ml}[-2y_1 + y_2] \\
\ddot{y}_2 &= \frac{F}{ml}[y_1 - 2y_2 + y_3] \\
\vdots \quad & \quad \vdots \quad \vdots \quad \vdots \\
\ddot{y}_n &= \frac{F}{ml}[y_{n-1} - 2y_n].
\end{aligned}$$

Put

$$\frac{F}{ml} = \omega_0^2$$

and seek solutions:

$$y_n = A_n \exp(i\omega t).$$

Then

$$\begin{aligned}
&(2\omega_0^2 - \omega^2)A_1 - \omega_0^2 A_2 = 0 \\
&-\omega_0^2 A_1 + (2\omega_0^2 - \omega^2)A_2 - \omega_0^2 A_3 = 0 \\
&\vdots \quad \vdots \quad \vdots \quad \vdots \quad \vdots \quad \vdots \\
&-\omega_0^2 A_{n-1} + (2\omega_0^2 - \omega^2)A_n = 0.
\end{aligned}$$

The condition that $A_n$'s have solutions is

$$D_n = \begin{vmatrix} (2\omega_0^2 - \omega^2) & -\omega_0^2 & 0 & \cdots & \cdots & \cdots & 0 \\ -\omega_0^2 & (2\omega_0^2 - \omega^2) & -\omega_0^2 & \cdots & 0 & \cdots & \cdots \\ \cdots & \cdots & \cdots & \cdots & \cdots & \cdots & 0 \\ 0 & \cdots & \cdots & \cdots & -\omega_0^2 & (2\omega_0^2 - \omega^2) & -\omega_0^2 \\ 0 & \cdots & \cdots & \cdots & \cdots & -\omega_0^2 & (2\omega_0^2 - \omega^2) \end{vmatrix} = 0.$$

Expanding the determinant with respect to the last row we get

$$D_n = (2\omega_0^2 - \omega^2)D_{n-1} - \omega_0^4 D_{n-2}.$$

Put

$$D_n = \omega_0^{2n} D_n^1$$

then.

$$D_n^1 = \left(2 - \frac{\omega^2}{\omega_0^2}\right) D_{n-1}^1 - D_{n-2}^1.$$

To solve this difference equation. put

$$D_n^1 = \sin(n+1)\theta.$$

then

$$\sin(n+1)\theta + \sin(n-1)\theta = \left(2 - \frac{\omega^2}{\omega_0^2}\right) \sin n\theta.$$

This is satisfied if

$$2\cos\theta = \left(2 - \frac{\omega^2}{\omega_0^2}\right). \quad \omega^2 = 4\omega_0^2 \sin^2\frac{\theta}{2}. \quad \text{i.e.,} \quad \omega = \pm 2\omega_0 \sin\frac{\theta}{2}.$$

The zeros of $D_n$ are given by

$$\sin(n+1)\theta = 0. \quad \text{i.e.,} \quad \theta = \frac{m\pi}{n+1}. \quad m = 1, 2, \cdots, n.$$

Hence,

$$\omega_m = \pm 2\omega_0 \sin\frac{m\pi}{2(n+1)}$$

are the eigen frequencies.

## 5.5 Anharmonic Oscillator

The entire discussion that we have presented about small oscillation was concerned with keeping only second order terms in the expansion of $T$ and $V$. As a consequence, we had linear harmonic oscillator type equation. One may always take higher-order terms in the expansion of $T$ and $V$. For example in the expansion of $V$, we can include terms of the type

$$\sum_{ijk} V_{ijk}\eta_i\eta_j\eta_k$$

$$\text{and} \sum_{ijkl} V_{ijkl}\eta_i\eta_j\eta_k\eta_l \text{ etc.}$$

Similarly in the expansion of $T$.
Like

$$\sum_{ijk} \eta_{ijk}\dot{\eta}_i\dot{\eta}_j\eta_k, \quad \sum_{ijkl} m_{ijkl}\dot{\eta}_i\dot{\eta}_j\eta_k\eta_l \text{ etc.}$$

Keeping up to term in third order, one would arrive at equations like

$$(\ddot{\eta}_i + \omega_i^2\eta_i) = f_i(\eta_i, \dot{\eta}_i, \ddot{\eta}_i).$$

For instance, keeping the third and fourth order terms in $V$ only and keeping

$$T = \frac{1}{2}\sum_{ij} a_{ik}\dot{\eta}_i\dot{\eta}_k$$

only, we would come to the following equation.

$$\sum_j (a_{ij}\ddot{\eta}_j + c_{ij}\eta_j) + \sum_{j,k} V_{ijk}\eta_i\eta_k + \sum_{jkl} V_{ijkl}\eta_j\eta_k\eta_l = 0.$$

The presence of the last two terms introduce non-linear terms in the usual oscillator equation. These are known as *anharmonic terms* and the oscillator is called *anharmonic oscillator*. One can set up successive approximation techniques to solve the non-linear equation. We are not going to develop this any further here.

## 5.6 Small Oscillation Under Constraint

Suppose that there is an additional constraint on the co-ordinates of the system whose $K.E$ and $P.E.$ are given by

$$2T = \sum_{ij} a_{ij}\dot{q}_i\dot{q}_j, \quad 2V = \sum b_{ij}q_iq_j.$$

Let the constraint equation be given by

$$f(q_1, q_2, \cdots, q_n) = 0. \tag{5.30}$$

Since the relation is to be satisfied in the position of equilibrium at which we assume the co-ordinates vanish, it may be expanded in the form

$$f = a_1q_1 + a_2q_2 + \cdots + a_nq_n = 0, \tag{5.31}$$

where we have neglected higher powers in $q_i$'s.

We transform the system to normal co-ordinate, $\eta_1, \eta_2, \cdots, \eta_n$, then

$$2T = \dot{\eta}_1^2 + \dot{\eta}_2^2 + \cdots \tag{5.32}$$

$$2V = \lambda_1\eta_1^2 + \lambda_2\eta_2^2 + \cdots \tag{5.33}$$

$$\text{and} \quad f = b_1\eta_1 + b_2\eta_2 + \cdots. \tag{5.34}$$

We now use Lagranges method of indeterminate multiplier and write for the Lagrangian of the system as

$$L = T - V - \mu f,$$

where $\mu$ is Lagrange's indeterminate multiplier.

The Lagrange's equation then gives us

$$\frac{d}{dt}\frac{\partial T}{\partial \dot{\eta}_r} + \frac{\partial V}{\partial \eta_r} + \mu\frac{\partial f}{\partial \eta_r} = 0 \quad (r = 1, \cdots, n) \tag{5.35}$$

$$\text{or,} \quad \ddot{\eta}_r = -\lambda_r\eta_r - \mu b_r. \tag{5.36}$$

We put

$$\left.\begin{aligned} \eta_r &= \alpha_r \sin(pt + \varepsilon) \quad (r = 1, \cdots, n) \\ \mu &= \nu \sin(pt + \varepsilon) \end{aligned}\right\} \tag{5.37}$$

in equation (5.36) and (5.34). We get

$$\alpha_r(\lambda_r - p^2) + \nu b_r = 0 \quad (r = 1, \cdots, n) \tag{5.38}$$

$$\text{and} \quad b_1\alpha_1 + b_2\alpha_2 + \cdots + b_n\alpha_n = 0. \tag{5.39}$$

Eliminating the $\alpha$'s from (5.38) we get from (5.39)

$$\frac{b_1^2}{\lambda_1 - p^2} + \frac{b_2^2}{\lambda_2 - p^2} + \cdots + \frac{b_n^2}{\lambda_n^2 - p^2} = 0. \tag{5.40}$$

The equation (5.40) gives a polynomial of degree $n - 1$ in $p^2$.

For instances

$$F(p^2) = b_1^2(\lambda_2 - p^2)\cdots(\lambda_n - p^2) + \cdots = 0.$$

We see that if $\lambda_1 < \lambda_2 < \lambda_3 \cdots$, then $F(\lambda_1) > 0$ and $F(\lambda_2)$ is $< 0$ and so on, so that the roots in $p^2$ separate $\lambda_1, \lambda_2, \cdots, \lambda_n$. That is the periods under constraint separate the free periods. If there is a second constraint there would be $(n - 2)$ oscillations and they will separate the $(n - 1)$ oscillations obtained by imposing one constraint on the original $n$-oscillations.

## 5.7 Equal Roots of the Determinantal Equation

Let $D$ stand for the determinant whose zeros are the normal modes of oscillations. If two roots of $D$ are identical then

$$\frac{dD}{d(\omega^2)} = 0.$$

In the solution of linear differential equation with constant co-efficients when two roots are equal, then the form of the solution is of the form

$$(A + Bt)\sin\omega t.$$

But we then cannot get stable oscillation with such forms of solution and the equality of roots is not to be interpreted as instability but an indeterminacy of the co-efficients $x_r$'s in

$$q_r = x_r \exp(i\omega t).$$

In fact, equality of two roots means that two normal modes have the same frequency and in such a case there is an indeterminacy in the normal co-ordinates themselves. Consider a system with

$$\begin{aligned} 2T &= \dot{\xi}_1^2 + \dot{\xi}_2^2 + \cdots + \dot{\xi}_n^2 \\ \text{and} \quad 2V &= \lambda_1\xi_1^2 + \lambda_2\xi_2^2 + \cdots + \lambda_n\xi_n^2. \end{aligned}$$

$\xi$'s are normal co-ordinates. If in this we put $\lambda_1 = \lambda_2$, then it is obvious that if we put

$$\begin{aligned} \xi_1 &= \eta_1\cos\alpha - \eta_2\sin\alpha \\ \text{and} \quad \xi_2 &= \eta_1\sin\alpha + \eta_2\cos\alpha. \end{aligned}$$

then $\dot{\xi}_1^2 + \dot{\xi}_2^2$ transforms into

$$\begin{aligned} &\dot{\eta}_1^2 + \dot{\eta}_2^2 \\ \text{and} \quad &\xi_1^2 + \xi_2^2 \text{ to } \eta_1^2 + \eta_2^2 \end{aligned}$$

for any $\alpha$. That is $\eta_1$ and $\eta_2$ are also normal co-ordinates for whatever value of $\alpha$.
In other words $\xi_1$ and $\xi_2$ are degenerate modes: the degeneracy is lifted by slightly changing the values of the constants.

## 5.8 Small Oscillation about Steady Motion

A particle of mass $m$ is placed inside a hollow-sphere and is under the influence of gravity. The sphere is rotating around its vertical axis with constant angular velocity, $\omega$. Find the small oscillation of the particle around its steady motion.

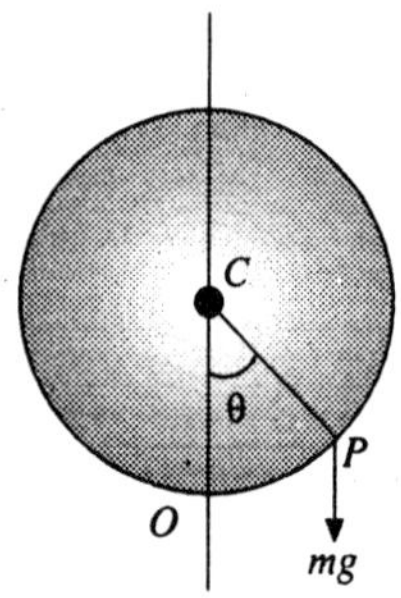


Fig 5.3

$C$ is centre of sphere, $P$ is position of the particle of mass $= m$. $CO$ is the vertical axis. $CP = r =$ radius of the sphere.

The co-ordinate of $P$ in polar variable $(r, \theta, \phi)$ of which $r =$ constant, $\phi = \omega t$, $\omega =$ angular velocity of rotation around $OC$. The K.E. is therefore,

$$\begin{aligned} T &= \frac{1}{2}m(\dot{r}^2 + r^2\dot{\theta}^2 + r^2\sin^2\theta\dot{\phi}^2) \\ &= \frac{1}{2}m(r^2\dot{\theta}^2 + r^2\omega^2\sin^2\theta). \end{aligned}$$

The potential energy

$$V = mgr(1 - \cos\theta).$$

The Lagrangian

$$L = T - V \quad \text{and} \quad \frac{d}{dt}\left(\frac{\partial L}{\partial\dot{\theta}}\right) = \frac{\partial L}{\partial\theta}$$

gives

$$r^2\ddot{\theta} = r^2\omega^2\sin\theta\cos\theta - gr\sin\theta. \tag{5.41}$$

If the motion of the particle is steady around $\theta = \alpha$, thus the condition of steady motion gives $\ddot{\theta} = 0$ and then from (5.41)

$$r^2\omega^2\sin\theta\cos\theta - gr\sin\theta = 0$$

at $\theta = \alpha$, so,

$$g = r\omega\cos\alpha. \tag{5.42}$$

To find the small oscillation around $\alpha$, we put $\theta = \alpha + \chi$ in (5.41) then for small $\chi$

$$r\ddot{\chi} = \frac{1}{2}a\omega^2\sin(2\alpha + 2\chi) - g\sin(\alpha + \chi)$$

reduces to

$$a\ddot{\chi} = -g\chi\frac{\sin^2\alpha}{\cos\alpha},$$

where we used (5.42).

Hence the period of oscillation is

$$2\pi\sqrt{\frac{a\cos\alpha}{g\sin^2\alpha}}.$$

## 5.9 Forced and Damped Oscillation

So long we described only free oscillatory motion. We now complete our discussion by considering forced and damped oscillation. We consider the motion of a one-dimensional oscillator in presence of a time dependent external field of force. We assume that the field is weak enough so that the displacement may not be large. In addition to the harmonic potential $\frac{1}{2}kx^2$, we also have an additional time-dependent potential $U(x,t)$. Writing $U(x,t)$ as

$$U(0,t) + x\frac{\partial U(0,t)}{\partial x} + \cdots$$

we see that the first term does not contribute to the equation of motion. We identify $-\frac{\partial U(0,t)}{\partial x}$ as an external force $F(t)$. Thus, the Lagrangian $L$ for the present case becomes

$$L = \frac{1}{2}m\dot{x}^2 - \frac{1}{2}kx^2 + xF(t).$$

Hence the equation of motion is

$$m\ddot{x} + kx = F(t).$$

This is the equation for a forced oscillation. We first discuss the solution by taking $F = f\cos(\gamma t + \delta)$ where $f$, $\gamma$ and $\delta$ are constants. Then our equation becomes

$$\ddot{x} + \omega^2 x = \left(\frac{f}{m}\right)\cos(\gamma t + \delta); \quad w^2 = \left(\frac{k}{m}\right).$$

The free part has the solution $a\cos(wt + \beta)$ where $a$ and $\beta$ are constants. If we put for the operator $\frac{d}{dt}$, the symbol $D$, then the complete solution is

$$x = a\cos(wt + \beta) + \frac{(f/m)}{D^2 + w^2}\cos(\gamma t + \delta).$$

By writing

$$(D^2 + w^2)^{-1} = A_0 + A_2D^2 + A_4D^4 + \cdots,$$

we can easily find

$$x = a\cos(wt + \beta) + \frac{f\cos(\gamma t + \delta)}{m(w^2 - \gamma^2)}.$$

We see from this solution that when the free oscillation frequency, $w$ becomes equal to the frequency of the external field, the displacement becomes infinite. This phenomenon is known as *resonants*.

To see the behaviour of $x$ near resonance $w = \gamma$ we express the expression for $x$ in the following way by adjusting the constants.

$$x = a\cos(wt + \beta) + \frac{f}{m(w^2 - \gamma^2)} \cdot (\cos(\gamma t + \delta) - \cos(wt + \delta)).$$

Taking limit $w \to \gamma$ we have

$$x = a\cos(wt + \beta) + \frac{ft}{2mw}\sin(wt + \delta).$$

Expressing sine and cosine in terms of exponential, we have

$$x = A\exp(iwt) + B\exp(it(w + \varepsilon))$$

where $\varepsilon$ is small so that $\varepsilon^2$ etc. are neglected.
Thus

$$x = (A + B\exp(i\varepsilon t))\exp(iwt).$$

Hence the amplitude depends on time. Put

$$A = A_0 e^{i\alpha_0}, \quad B = B_0 e^{i\beta_0}.$$

The square of the amplitude then becomes

$$C^2 = |\text{amp}|^2 = A_0^2 + B_0^2 + 2A_0B_0\cos(\varepsilon t + \beta_0 - \alpha_0).$$

Therefore $C$ lies between

$$|A_0 - B_0| \text{ and } |A_0 + B_0|.$$

This phenomenon is known as *beats*.
Since the Lagrangian of the forced oscillator is time-dependent the energy is not conserved. The source term transfers energy. To find the total energy transferred we consider the following:
The energy

$$E = \frac{m}{2}(\dot{x}^2 + w^2x^2) = \frac{1}{2}|y|^2,$$

where $y = \dot{x} + iwx$.
Then from the forced oscillator equation we find

$$\frac{dy}{dt} = iwy + \frac{F(t)}{m}.$$

This is also equal to

$$\frac{d}{dt}(e^{-iwt}y) = e^{-iwt}\frac{F(t)}{m}.$$

Hence

$$y = \exp(iwt)\int_{-\infty}^{t} F(t)e^{-iwt}\frac{dt}{m},$$

where we assumed $y = 0$ at $t = -\infty$. The value of $E$ is now equal to

$$E = \frac{1}{2m}\left|\int_{-\infty}^{t} F(t)e^{-iwt}dt\right|^2.$$

Hence the total energy transfered is

$$E = \frac{1}{2m}\left|\int_{-\infty}^{\infty} F(t)e^{-iwt}dt\right|^2 .$$

If the external force acts for a short time such that $wt \ll 1$. then the quantity $\int F dt$ is the momentum and

$$E = \frac{1}{2m}\left|\int F dt\right|^2 .$$

This shows that a force of short duration gives a momentum

$$\int F dt$$

to the system without bringing any noticable change in displacement.

## Damped Oscillation

We consider the motion of an one-dimensional oscillator under the influence of a dissipative force. The Lagrangian of the oscillator is as usual equal to

$$\frac{1}{2}m\dot{x}^2 - \frac{1}{2}kx^2$$

and the dissipative force is given by the Rayleigh dissipative function $R = \lambda\dot{x}^2$. $\lambda$ is a constant: $\lambda > 0$. The Lagrangian equation is now

$$\frac{d}{dt}\left(\frac{\partial L}{\partial \dot{x}}\right) - \frac{\partial L}{\partial x} = -\frac{\partial R}{\partial \dot{x}}.$$

Therefore, the relevant equation is

$$m\ddot{x} + kx + 2\lambda\dot{x} = 0.$$

Put

$$\frac{k}{m} = \omega_0^2 \quad \text{and} \quad \frac{\lambda}{m} = \lambda_1,$$

then the equation becomes

$$\ddot{x} + 2\lambda_1\dot{x} + w_0^2 x = 0$$

we can have a solution of the form $e^{mt}$ where

$$m = -\lambda_1 \pm \sqrt{(\lambda_1^2 - w_0^2)}:$$

$$x = e^{-\lambda_1 t}\cdot A\cos(wt + \alpha)$$

$$\text{and} \quad w = \sqrt{(w_0^2 - \lambda_1^2)}.$$

Suppose $\lambda_1 \ll w_0$: in this case

$$x = Ae^{-\lambda_1 t}\cos(w_0 t + \alpha)$$

We now calculate average values of $x^2$ and $\dot{x}$ over a period of a cycle. As $\lambda_1$ is small we assume that the amplitude of the damped oscillator will not change during the period. Therefore, the mean energy $E$ of the damped oscillator is $E_0 e^{-2\lambda_1 t}$, where $E_0$ is the mean energy of the free-oscillator. When $\lambda_1 = w_0$ the solution takes the form

$$(A + Bt)e^{-\lambda_1 t}.$$

The solution is not periodic.
To find the energy of the damped oscillator we use the definition of $E$ given by

$$E = \dot{x}\frac{\partial L}{\partial \dot{x}} - L.$$

Using equation of motion for the damped oscillator under a Rayleigh dissipative function, we get

$$\frac{dE}{dt} = -\dot{x}\frac{\partial R}{\partial \dot{x}} = -2R.$$

As $R > 0$, so there is always a loss of energy in a dissipative system. The rate of loss of energy is twice the damping force.

**Example 1.** *The point of suspension of plane simple pendulum has a horizontal motion $x = a\cos wt$. Show that simple pendulum executes forced oscillation.*

***Solution:*** The $x$-coordinate of the bob of the pendulum is

$$x = b\sin\theta + a\cos wt.$$

The $y$-coordinate is

$$y = b\cos\theta;$$

$b$ = length of pendulum.
The Lagrangion $L$ is equal to

$$\frac{1}{2}(\dot{x}^2 + \dot{y}^2) - gl(1 - \cos\theta).$$

where mass $m = 1$ and $\theta$ = angle at time $t$, made by the string to the vertical taken as $y$-axis.

$$L = \frac{1}{2}(b^2\dot{\theta}^2 + a^2w^2\sin^2 wt - 2aw\sin wt \cdot b\cos\theta \cdot \dot{\theta}) - gl(1 - \cos\theta).$$

Then

$$\frac{d}{dt}\left(\frac{\partial L}{\partial \dot{\theta}}\right) = \frac{\partial L}{\partial \theta}$$

gives:

$$\ddot{\theta} - \frac{aw^2}{b} \cos wt \cos \theta + \frac{g}{b} \sin \theta = 0.$$

For small $\theta$, we get

$$\ddot{\theta} + \frac{g}{b}\theta = \frac{aw^2}{b} \cos wt.$$

The solution is

$$\dot{\theta} = A \cos kt + B \sin kt + \frac{aw^2}{b} \cdot \frac{\cos wt}{k^2 - w^2}$$

where $k^2 = g/b$.
This solution was discussed in the text.

**Example 2.** *The motion of a system is described by a single coordinate: its energy at any instant is $(1/2)m\dot{x}^2 + (1/2)cx^2$ and the time-rate of frictional damping of its energy is $(1/2)k\dot{x}^2$. Find the motion.*

***Solution:*** From the given problem we have

$$\frac{d}{dt}\left(\frac{1}{2}m\dot{x}^2 + \frac{1}{2}cx^2\right) = -\frac{1}{2}k\dot{x}^2.$$

This yields:

$$m\ddot{x} \cdot \dot{x} + cx \cdot \dot{x} = -\frac{1}{2}k\dot{x}^2, \quad \text{or,} \quad m\ddot{x} + \frac{1}{2}k\dot{x} + cx = 0.$$

The solution

$$x = Ae^{-(kt/2m)} \cos\left[t\sqrt{\frac{c}{m} - \frac{k^2}{16m^2}} + B\right].$$

Suppose, for instance, the particle was initially at rest at a certain distance from the origin. The constants $A$ and $B$ are found and they are finite. The factor $\exp(-kt/2m)$ gradually decreases with time and ultimately the vibration vanishes after some length of time.

**Example 3.** *If the motion described in the last example is sustained by an external force $A \cos pt$ then the forced oscillation is maximum where*

$$p^2 = \frac{c}{m} - \frac{k^2}{8m^2}.$$

*Find an expression for the value of the maximum amplitude and the phase lag.*

***Solution:*** The relevant equation is

$$\left(m\frac{d^2}{dt^2}+\frac{1}{2}k\frac{d}{dt}+e\right)x = A\cos pt.$$

The particular integral gives the amplitude

$$x = A\frac{1}{-mp^2+e+\frac{1}{2}kD}\cos pt; \quad D=\frac{d}{dt}.$$

We can easily write down the solution

$$\begin{aligned}x &= A\frac{(e-mp^2)-\frac{1}{2}kD}{(e-mp^2)^2-\frac{1}{4}k^2D^2}\cos pt\\ &= A\frac{(e-mp^2)\cos pt+\frac{1}{2}kp\sin pt}{(e-mp^2)^2+\frac{1}{4}k^2p^2}.\end{aligned}$$

This we can put as

$$x = A\cdot\frac{\cos(pt-\varepsilon)}{\left\{e-mp^2)^2+\frac{1}{4}k^2p^2\right\}^{1/2}}; \quad \tan\varepsilon=\frac{kp}{2(e-mp^2)}.$$

The amplitude $x$ is large where the dinominator

$$\left\{(e-mp^2)^2+\frac{1}{4}k^2p^2\right\}^{1/2}$$

is minimum and the minimum occurs when

$$\frac{d}{dp}\sqrt{\left\{(e-mp^2)^2+\frac{1}{4}k^2p^2\right\}}=0;$$

this occurs when

$$p^2=\frac{e}{m}-\frac{k^2}{8m}.$$

At this value of $p$, the value of the amplitude becomes:

$$\frac{k}{2}\sqrt{\left(\frac{e}{m}-\frac{k^2}{16m^2}\right)}$$

$$\text{and } \tan\varepsilon=\frac{kp}{2(e-mp^2)}=\frac{4mp}{k^2}.$$

Therefore the phase

$$\varepsilon=\tan^{-1}\left(\frac{4mp}{k^2}\right).$$

## Exercises

1. Find the normal modes of vibration of a system characterised by the Lagrangian

$$L = \frac{1}{2}(\dot{x}^2 + \dot{y}^2) - \frac{1}{2}\omega_0^2(x^2 + y^2) + \alpha\dot{x}\dot{y}.$$

[*Hints:* One easily finds that with $X$ and $Y$ as normal co-ordinate, where

$$X = x + y \quad \text{and} \quad Y = x - y,$$

the equations are

$$(\alpha + 1)\ddot{X} = -\omega_0^2 X$$
$$\text{and} \quad (1 - \alpha)\ddot{Y} = -\omega_0^2 Y.$$

We can now find the eigen frequency modes.]

2. The Lagrangian of a system is given by

$$L = \frac{1}{2}(\dot{x}_1^2 + \dot{x}_2^2 + \dot{x}_3^2) - \frac{1}{2}\left[x_1^2 + (x_2 - x_1)^2 + (x_3 - x_2)^2 + x_3^2\right].$$

Find the eigen frequencies and eigen vectors of the system.

3. Find the normal modes of vibration for a system given by the Lagrangian

$$L = \dot{q}_1^2 + \dot{q}_2^2 + \dot{q}_1\dot{q}_2 - q_1^2 - q_2^2.$$

4. A particle of mass $m$ is constrained to move on a smooth plane curve under the influence of gravity the equation of the curve is given by

$$s = 4a\sin\phi.$$

where $s$ = arc length along the curve, $a$ is a constant, $\phi$ being the angle made by the tangent to the curve with the $x$-axis. Find the period of small oscillation.

5. A mass $m_1$ hangs from a fixed point at the end of very long string of length $l_1$. To $m_1$ is suspended a mass $m_2$ by a string whose length is $l_2$ and is very small compared to $l_1$. Prove that the time period of small oscillation of $m_2$ is

$$2\pi\sqrt{\frac{m_1 l_2}{(m_1 + m_2)g}}.$$

[*Hints:* $\phi_1$ **and** $\phi_2$ **are the** angles made by $l_1$ and $l_2$ with vertical. ($\phi_1 \ll 1$, $\phi_2 \ll 1$) **the Lagrangian** $\boldsymbol{L}$ is

$$\begin{aligned} L &= \frac{1}{2}(m_1 + m_2)l_1^2\dot{\phi}_1^2 + \frac{1}{2}m_2 l_2^2\dot{\phi}^2 + m_2 l_1 l_2\dot{\phi}_1\dot{\phi}_2 \\ &\quad - g\frac{1}{2}(m_1 + m_2)l_1\phi_1^2 - \frac{1}{2}g m_2 l_2\phi_2^2. \end{aligned}$$

The equations of motion are

$$(m_1+m_2)l_1\ddot{\phi}_1+m_2l_2\ddot{\phi}_2+m_1m_2g\phi_1 = 0$$
$$l_1\ddot{\phi}_1+l_2\ddot{\phi}_2+g\phi_2 = 0.$$

Put $\phi_2 = A_2\exp(i\omega t)$ and $\phi_1 = A_1e^{i\omega t}$, then

$$\left.\begin{aligned} A_1(m_1+m_2)(g-l_1\omega^2)-A_2\omega^2m_2l_2 &= 0\\ -A_1l_1\omega^2+A_2(g-l_2\omega^2) &= 0\end{aligned}\right\}.$$

The roots of the characteristic equation are

$$\omega_{1,2} = \frac{g}{2m_1l_1l_2}\Big[(m_1+m_2)(l_1+l_2)\pm\sqrt{(m_1+m_2)}\sqrt{(m_1+m_2)(l_1+l_2)^2-4m_1l_1l_2}\Big].$$

where $l_1 \gg l_2$, $\omega_1$ becomes

$$\omega_1 = \frac{g}{m_1l_2}(m_1+m_2);\ \omega_2 = 0.$$

Hence the time period is

$$2\pi\sqrt{\frac{m_1l_2}{(m_1+m_2)g}}.]$$

6. Find the eigenfrequency of the following two-degrees of freedom characterised by the pain of equations:

$$\left.\begin{aligned}\left(\tfrac{16}{3}D^2+\tfrac{3g}{a}\right)\theta+2D^2\phi &= 0\\ 2D^2\theta+\left(\tfrac{1}{3}D^2+\tfrac{g}{a}\right)\phi &= 0\end{aligned}\right\}:\quad D=\frac{d}{dt}.$$

[*Hints:* Put $\theta = \theta_0e^{iwt}$, and $\phi = \phi_0e^{iwt}$.

Then

$$\left(-\frac{16}{3}w^2+\frac{3g}{a}\right)\theta_0+2(-w^2)\phi_0 = 0$$

$$\text{and}\quad\left(-\frac{4}{3}\omega^2+\frac{g}{a}\right)\phi_0+2(-w^2)\theta_0 = 0.$$

The eigenfrequencies are obtained from the zeros of the determinant.

$$\begin{vmatrix} -\frac{16}{3}w^2+\frac{3g}{a}, & -2w^2\\ -2w^2 & -\frac{1}{3}w^2+\frac{g}{a}\end{vmatrix} = 0;$$

$$\text{or, } w^4-\frac{3g}{a}w^2+\frac{27}{28}\frac{g^2}{a^2} = 0;\ w_1^2 = \frac{3g}{14a}(7+2\sqrt{7}),\ w_2^2 = \frac{3g}{14a}(7-2\sqrt{7}).$$

7. A mass $m$ is attached to a fixed point by a light spring and its vertical time period of oscillation is $2\pi/p_1$. If a second mass $m'$ is suspended from $m$ by another light spring and the period of vertical oscillation of $m'$ about $m$ be $2\pi/p_2$ when $m$ is held fixed, show that when both the masses are free the periods $2\pi/n$ of the normal modes of vibration vertically of the system is given by

$$n^4 - n^2\left[p_1^2 + \left(1 + \frac{m'}{m}\right)p_2^2\right] + p_1^2p_2^2 = 0.$$

[*Hints:* Let $x$ and $y$ are the vertical heights from a fixed origin of the two masses $m$ and $m'$ at a given instant of time. Let $k$ and $k'$ are spring constants of the two strings one which suspends $m$ and the other $m'$. Then

$$p_1^2 = \frac{k}{m} \text{ and } p_2^2 = \frac{k'}{m'}.$$

The K.E. is

$$\begin{aligned} T &= \frac{1}{2}m\dot{x}^2 + \frac{1}{2}m'\dot{y}^2 \\ \text{and } V &= \frac{1}{2}kx^2 + \frac{1}{2}k'(y-x)^2 - mgx - mgy. \end{aligned}$$

The Lagrangian equation of motions are from $L = T - V$.

$$\begin{aligned} m\ddot{x} &= -kx - k'(x-y) + mg \\ \text{and } m'\ddot{y} &= -k'(y-x) + m'g. \end{aligned}$$

If $x_0$ and $y_0$ are the equilibrium positions then.

$$\begin{aligned} kx_0 + k'(x_0 - y_0) &= mg \\ \text{and } k'(y_0 - x_0) &= m'g. \end{aligned}$$

We eliminate $m$ and $m'$ from the equation by using these two values of $mg$ and $m'g$.

Put $x - x_0 = \eta_1$ and $y - y_0 = \eta_2$, then

$$\begin{aligned} m\ddot{\eta}_1 + k\eta_1 + k'(\eta_1 - \eta_2) &= 0 \\ \text{and } m'\ddot{\eta}_2 + k'(\eta_2 - \eta_1) &= 0. \end{aligned}$$

we now put $\eta_i = \eta_{i0}e^{int}$ then the determinantal equation is

$$\begin{vmatrix} -n^2 + p_1^2 + \frac{m'}{m}p_2^2 & -\frac{m'}{m}p_2^2 \\ -p_2^2 & -n^2 + p_2^2 \end{vmatrix} = 0.$$

This gives

$$n^4 - n^2\left[p_1^2 + \left(1 + \frac{m'}{m}\right)p_2^2\right] + p_1^2p_2^2 = 0.$$

Chapter VI

# Rotating Frame

## 6.1 Angular Velocity; Rate of change of a Vector

So far we have used inertial frame of reference, in which Newton's equation of motion takes simple form. But there are many a problem which can easily be solved in non-inertial frames of reference. For instance, while discussing the motion of particles near earth's surface it is suitable to use frame rigidly attached to the earth and rotates along with it. We would like to discuss the Lagrangian equation in such a rotating frame.

Consider a solid body rotating with constant angular velocity $\Omega$ about a fixed axis. The direction of the vector $\bar{\Omega}$ is along the fixed axis and its direction is fixed by the right-hand rule. For example, for the earth the direction of $\bar{\Omega}$ is along the polar axis pointing towards the north pole. For simplicity, we take the origin to lie on the axis of rotation. We have already noticed that if a vector $\mathbf{r}$ is rotated around the $z$-axis, the axis of rotation, through an angle $\delta\varphi$, then the vector $\mathbf{r}$ changes to $\mathbf{r}'$, where,

$$\mathbf{r}' = \mathbf{r} + \delta\bar{\phi} \times \mathbf{r}$$

when $\delta\bar{\phi}$ is the vector along the positive $z$-axis.

Thus, the change in the vector $\mathbf{r}$ is $\delta\mathbf{r}$, where

$$\delta\mathbf{r} = (\delta\bar{\phi} \times \mathbf{r}). \tag{6.1}$$

Let $\mathbf{i}$, $\mathbf{j}$ and $\mathbf{k}$ are the unit vectors along $x$, $y$ and $z$-axes of the rotating frame, fixed in the body then under rotation through $\delta\varphi$,

$$\mathbf{i} \to \mathbf{i} + (\delta\bar{\phi} \times i). \tag{6.2}$$

i.e., $\delta\mathbf{i} = \delta\bar{\phi} \times \mathbf{i}$. Likewise

$$\delta\mathbf{j} = \delta\bar{\phi} \times \mathbf{j} \quad \text{and} \quad \delta\mathbf{k} = (\delta\bar{\phi} \times \mathbf{k}).$$

Hence, the rate of change of $\mathbf{i}$, $\mathbf{j}$ and $\mathbf{k}$ with time due to the rotational change is given by

$$\frac{d\mathbf{i}}{dt} = \frac{d\bar{\phi}}{dt} \times \mathbf{i}, \quad \text{i.e.,} \quad \frac{d}{dt}\mathbf{i} = (\bar{\Omega} \times \mathbf{i}). \tag{6.3}$$

where

$$\bar{\Omega} = \frac{d\bar{\omega}}{dt}.$$

Thus,

$$\frac{d}{dt}\mathbf{j} = \bar{\Omega} \times \mathbf{j} \quad \text{and} \quad \frac{d}{dt}\mathbf{k} = \bar{\Omega} \times \mathbf{k}.$$

Let us now consider a vector **a** with components $a_x$, $a_y$, $a_z$ in the rotating frame with **i**, **j**, **k** as unit vectors along $x$, $y$, $z$ axes, then

$$\mathbf{a} = a_x\mathbf{i} + a_y\mathbf{j} + a_z\mathbf{k}. \tag{6.4}$$

We want to find the rate of change of **a**, with time, when the frame is rotating, we will denote by $\dot{\mathbf{a}}$ as the rate of change of **a** with time.
When the vector **a** is changing we must note that the components $a_x$, $a_y$ and $a_z$ are moving with time. We denote these change by $\frac{d}{dt}a_x$, etc. Again, due to rotation, the fundamental unit vectors **i**, **j**, **k** are also changing with time. Hence, rate of change of **a** with time is

$$\dot{\mathbf{a}} = \frac{d}{dt}a_x \cdot \mathbf{i} + \frac{d}{dt}a_y \cdot \mathbf{j} + \frac{d}{dt}a_z \cdot \mathbf{k} + a_x\frac{d\mathbf{i}}{dt} + a_y\frac{d\mathbf{j}}{dt} + a_z\frac{d\mathbf{k}}{dt} \tag{6.5}$$

$$= \frac{d}{dt}a_x\mathbf{i} + \frac{d}{dt}a_y\mathbf{j} + \frac{d}{dt}a_z \cdot \mathbf{k} + a_x\bar{\Omega} \times \mathbf{i} + a_y\bar{\Omega} \times \mathbf{j} + a_z\bar{\Omega} \times \mathbf{k}. \tag{6.6}$$

We now take for **a**, the position vector **r**, where now

$$a_x = x, \quad a_y = y, \quad a_z = z.$$

Thus, for this case we get from (6.6)

$$\dot{\mathbf{r}} = \mathbf{v} + \bar{\Omega} \times \mathbf{r}. \tag{6.7}$$

Where we have used

$$\frac{d}{dt}x = \dot{x} = v_x, \text{ etc.},$$

the usual velocities of the position co-ordinate. We now utilize (6.7) to write down the Lagrangian.

## 6.2 Lagrangian equation of motion; Coriolis Force

Since the Lagrangian is a scalar quantity we still use the same form of $L$ in this frame also, i.e.,

$$L = T - V. \tag{6.8}$$

The Kinetic energy

$$T = \frac{1}{2}m\dot{\mathbf{r}}^2 \quad \text{and} \quad V = V(r).$$

for a single particle. Therefore,

$$T = \frac{1}{2}m(\mathbf{v} + \bar{\Omega} \times \mathbf{r})^2 = \frac{1}{2}m\mathbf{v}^2 + \frac{1}{2}m(\mathbf{r} \times \bar{\Omega})^2 + m\mathbf{v}(\bar{\Omega} \times \mathbf{r})$$
$$V = V(r). \tag{6.9}$$

We use (6.9) in (6.8) to obtain the Lagrangian equation

$$\frac{d}{dt}\left(\frac{\partial L}{\partial \mathbf{v}}\right) = \frac{\partial L}{\partial \mathbf{r}}.$$

Thus,

$$\delta L = m\mathbf{v}\cdot\delta\mathbf{v} + m(\bar{\Omega} \times \mathbf{r})(\bar{\Omega} \times \delta\mathbf{r}) + m\delta\mathbf{v}(\bar{\Omega} \times \mathbf{r}) + m\mathbf{v}(\bar{\Omega} \times \delta\mathbf{r}) - \delta V(r).$$

From this we find

$$\frac{\partial L}{\partial \mathbf{v}} = m\mathbf{v} + m(\bar{\Omega} \times \mathbf{r}). \tag{6.10}$$

Equation (6.10) defines the conjugate moments $\mathbf{p}_r$ corresponding to the co-ordinate $\mathbf{r}$, i.e.,

$$\mathbf{p}_r = m\mathbf{v} + m(\bar{\Omega} \times \mathbf{r}). \tag{6.11}$$

Again

$$\frac{\partial L}{\partial \mathbf{r}} = m\mathbf{v} \times \bar{\Omega} + m(\bar{\Omega} \times \mathbf{r}) \times \bar{\Omega} - \frac{\partial V}{\partial r}.$$

Hence, the equation of motion becomes:

$$m\dot{\mathbf{v}} = 2m(\mathbf{v} \times \bar{\Omega}) + m(\bar{\Omega} \times \mathbf{r}) \times \bar{\Omega} - \frac{\partial V}{\partial \mathbf{r}}. \tag{6.12}$$

The first term on the right-hand side of (6.12) is called *Coriolis Force* and the second term on the right-hand is known as *centrifugal force*.
The Hamiltonian of the system $H$ is

$$H = \mathbf{p}_r \cdot \dot{\mathbf{r}} - L = \frac{1}{2}m\mathbf{v}^2 - \frac{1}{2}m(\bar{\Omega} \times \mathbf{r})^2 + V(r). \tag{6.13}$$

## 6.3 Applications

**Example 1.** *Deflection of a freely falling particle in a rotating media. To find the deflection of a freely falling body from the vertical caused by earth's rotation assuming the angular velocity of earth's rotation is small.*

***Solution:*** $CPZ$ is along the $z$-axis which is in the vertically upward direction. The particle is falling along the $z$-axis freely under gravity. $C$ is the centre of the earth and $P$ is the origin of the $xyz$-axes system (Fig 6.1).

$CA$ is the polar axis of rotation of the earth with angular velocity $\omega$. $CPZ$ makes an angle $\theta$ with $CA$. The line $PN$ is parallel to $CA$. Then $\angle NPZ = \theta$.

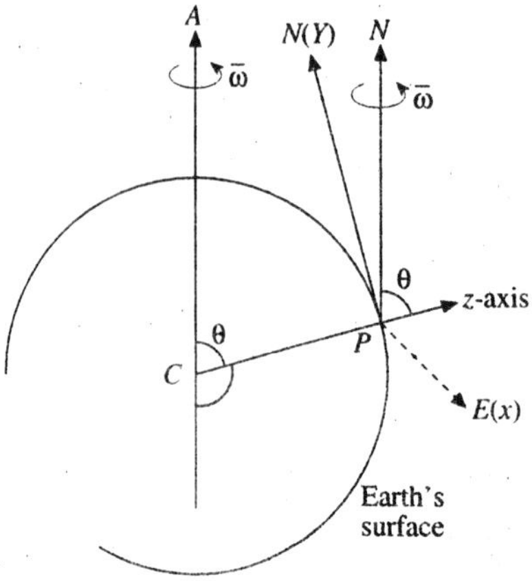


Fig 6.1

$PN(Y)$ is the tangent at $P$ and is along the north direction denoting $y$-axis. The line $PE(x)$ is along $x$-axis towards eastern direction. The angular velocity $\bar{\omega}$ is along $PN$.

Thus, $\bar{\omega}$ has no component along $x$-axis. The component of $\bar{\omega}$ along $y$-axis is $\omega \sin\theta$, while the component of $\bar{\omega}$ along $z$-axis is $\omega\cos\theta$. Thus,

$$\bar{\omega} = (0, \omega\sin\theta, \omega\cos\theta).$$

The velocity $\mathbf{v}$ of the particle is denoted by $\mathbf{v} = (\dot{x}, \dot{y}, \dot{z})$.
The equation of motion when the earth is rotating with constant angular velocity $\bar{\omega}$ is

$$m\ddot{\mathbf{r}} = 2m(\mathbf{v}\times\bar{\omega}) - \frac{\partial V}{\partial \mathbf{r}}.$$

As the particle is freely falling under gravity, $V = mgz$.
Therefore,

$$-\frac{\partial V}{\partial z} = -mg.$$

We now write the equations of motion for the three components,

$$\left.\begin{aligned} m\ddot{x} &= 2m(\dot{y}\omega\cos\theta - \dot{z}\omega\sin\theta) \\ m\ddot{y} &= -2m(\dot{x}\omega\cos\theta) \\ m\ddot{z} &= 2m\dot{x}\omega\sin\theta - mg. \end{aligned}\right\} \tag{6.14}$$

To solve these equations we put in the right-hand side the values of $\dot{x}$, $\dot{y}$ and $\dot{z}$, when $\bar{\omega}$ is absent. In that case

$$\ddot{x} = 0, \quad \ddot{y} = 0, \quad \ddot{z} = -g$$

which gives

$$\dot{x} = 0, \quad \dot{y} = 0, \quad \dot{z} = -gt$$

as the particle is falling along $z$-axis. With these values of $\dot{x}$, $\dot{y}$ and $\dot{z}$, we find for (4.14)

$$\ddot{x} = 2gt\omega\sin\theta; \quad \ddot{y} = 0; \quad \ddot{z} = -g.$$

The $y$ and $z$ equations are same as in the case when $\vec{\omega} = 0$.
The $x$-component equation changes. Integrating $x$-component equation we get

$$x = \frac{1}{3} g t^3 \omega \sin\theta.$$

But from $\ddot{z} = -g$ we find

$$t = \sqrt{\frac{2h}{g}}.$$

where $h$ is the height from which the particles is falling. Thus,

$$x = \frac{1}{3}\omega \sin\theta \cdot g \left(\frac{2h}{g}\right)^{3/2}.$$

Thus after a time $t$, the particle will be found in the eastern direction at a distance of

$$\frac{1}{3}\omega \sin\theta \cdot g \left(\frac{2h}{g}\right)^{3/2}.$$

Normally it will fall at the origin.

**Example 2.** ***(Projectile motion):*** *A particle is thrown with an initial velocity* **v** *in the $yz$-plane with components $v_y$ and $v_z$ at $t = 0$. To find the deflection in the path.*

***Solution:*** The equations of motion in rotating frame are (we neglect $\omega^2$-dependent term as in the last example).

$$\left.\begin{aligned} m\ddot{x} &= 2m(\dot{y}\omega\cos\theta - \dot{z}\omega\sin\theta) \\ m\ddot{y} &= -2m\dot{x}\omega\cos\theta \\ m\ddot{z} &= 2m\dot{x}\omega\sin\theta - mg. \end{aligned}\right\} \tag{6.15}$$

In the $\vec{\omega} = 0$ case, $\ddot{x} = 0$, $\dot{x} = 0$ at $t = 0$.

$$\ddot{y} = 0, \quad \dot{y} = v_y \quad \text{and} \quad \ddot{z} = -g; \quad \dot{z} = -gt + v_z.$$

We use these values of $\dot{x}$, $\dot{y}$ and $\dot{z}$ on the right-hand side of (6.15). Then,

$$\begin{aligned} &\ddot{x} = 2[v_y\omega\cos\theta - (-gt + v_z)\omega\sin\theta] \\ &\left.\begin{aligned} \ddot{y} &= 0 \\ \ddot{z} &= -g \end{aligned}\right\}. \end{aligned}$$

We note that with this approximation, the last two equation are the same as the ordinary projectile equation in $yz$-plane. But $\ddot{x} \neq 0$ because of $\omega \neq 0$. Hence the presence of $\vec{\omega}$ will take the projectile away from the $yz$-plane on to a point in the $x$-axis given by the solution of $\ddot{x}$ equation. This equation can be easily solved to give for $x$.

$$x = 2v_y\omega\cos\theta\cdot\frac{t^2}{2} - 2\left(-\frac{1}{6}gt^3 + v_z\frac{t^2}{2}\right)\omega\sin\theta. \tag{6.16}$$

From $\dot{z} = -gt + v_z$ we find

$$z = -g\frac{t^2}{2} + v_z t.$$

Hence, the time of flight

$$t = 2\frac{v_z}{g}.$$

If we put this value of $t$ in (6.16) we get the point on the $x$-axis, where the projectile will fall after the time of flight given by.

$$x = \frac{4v_z^2}{g^2}\left(v_y\omega\cos\theta - \frac{1}{3}v_z\omega\sin\theta\right) \quad \text{in the eastern direction.}$$

**Example 3.** ***(Foucault Pendulum):*** *To find the effect of earth's rotation on the oscillation of a simple pendulum.*

***Solution:*** We neglect the motion of the vertical component, then the oscillation is in the $xy$-plane. The equations are

$$m\ddot{x} = 2mv_y\omega\cos\theta - \frac{\partial V}{\partial x}$$

$$\text{and} \quad m\ddot{y} = -2mv_x\omega\cos\theta - \frac{\partial V}{\partial y}.$$

Choose

$$V = \frac{1}{2}k^2m(x^2 + y^2).$$

thus

$$\ddot{x} = 2\omega\cos\theta\dot{y} - k^2x \quad \text{and} \quad \ddot{y} = -2\omega\cos\theta\dot{x} - k^2y.$$

Write $\xi = x + iy$, then

$$\ddot{\xi} + 2i\omega\cos\theta\dot{\xi} + k^2\xi = 0.$$

This is a damped harmonic oscillator in one dimension, the oscillation of which is well-known. When $\bar{\omega} = 0$, this equation is a simple harmonic equation. Hence the presence of earth's rotation can be tested from the damped oscillator.

In discussing all the above problems we have neglected the centrifugal term which is of $O(\Omega^2)$ and can be neglected as $\Omega$ is small. Writing $\omega_R$ for $\omega\cos\theta$, we have

$$\ddot{x} = -k^2x - 2\omega_R y$$

$$\text{and} \quad \ddot{y} = -k^2y - 2\omega_R x.$$

From these we derive by multiplying the first by $y$ and the second by $x$, and then subtracting,

$$\frac{d}{dt}(y\dot{x} - x\dot{y}) = -\frac{d}{dt}(y^2 + x^2)\omega_R.$$

Integrating and changing to

$$x = r\cos\theta,$$
$$y = r\sin\theta$$

variable, we find

$$r^2\dot{\theta} = -\omega_R r^2 \quad \text{or,} \quad \dot{\theta} = -\omega_R.$$

Hence the plane of oscillation rotates with angular velocity $\omega_R$.

As a different kind of effect in the use of rotating frame of co-ordinates we discuss *Larmor Effect.*

Consider the effect of a magnetic field on a particle of charge $q$ in an orbit round a fixed point charge $q'$. The equation of motion including the magnetic and electric force is

$$m\ddot{\mathbf{r}} = -\frac{qq'}{r^2}[\mathbf{e}_r] + \frac{q}{c}(\dot{\mathbf{r}} \times \mathbf{B}).$$

where $\mathbf{B}$ is the magnetic field. Let us rewrite this using the expression

$$\dot{\mathbf{r}} = \mathbf{V} + (\vec{\omega} \times \mathbf{r})$$
$$\text{and} \quad \ddot{\mathbf{r}} = \dot{\mathbf{V}} + 2\vec{\omega} \times \mathbf{V} + \vec{\omega} \times (\vec{\omega} \times \mathbf{r}).$$

Then the equation in the presence of electric and magnetic force becomes

$$\dot{\mathbf{V}} + 2(\vec{\omega} \times \mathbf{V}) + \vec{\omega} \times (\vec{\omega} \times \mathbf{r}) = -\frac{qq'}{mr^2}[\mathbf{e}_r] + \frac{q}{mc}(\mathbf{V} + \vec{\omega} \times \mathbf{r}) \times \mathbf{B}).$$

We now choose

$$\vec{\omega} = -\frac{q}{2mc}\mathbf{B}.$$

Then, the term involving $\vec{\omega} \times \mathbf{V}$ drops out and after simplification we get

$$\dot{\mathbf{V}} = -\frac{qq'}{mr^2}\mathbf{e}_r + \left(\frac{q}{2mc}\right)\mathbf{B} \times (\mathbf{B} \times \mathbf{r}).$$

We now assume that the magnetic field is weak such that

$$\left(q\frac{B}{2mc}\right)^2 = \omega^2 \ll \frac{qq'}{mr^2} = \omega_0^2.$$

Then the equation of motion becomes

$$m\dot{\mathbf{V}} \equiv m\ddot{\mathbf{r}} = -\frac{qq'}{r^2}\mathbf{e}_r.$$

But this means the motion of the particle in an inverse square-law field. Consequently in the rotating frame, the orbit would be a precessing ellipse with an angular velocity $\omega$, since $\omega \ll \omega_0^2$, the axes of the ellipse will precess only through a small angle in each revolution. This effect is known as *Larmor effect* and the angular velocity of precession is

$$\omega_L = \frac{qB}{(2mc)};$$

$\omega_L$ is called the *Larmor frequency.*

This effect can lead to observable changes in the atomic spectra emitted in the presence of magnetic field (Zeeman effect).

## 6.4 Larmor Frequency

In this section we consider the effect of a magnetic field on a particle of charge $q$ moving in an orbit round a fixed particles whose charge is $q'$. The equation of motion in presence of the magnetic field. $\mathbf{B}$ is

$$m\frac{d\mathbf{v}}{dt} = -\frac{qq'}{r^2}\hat{e}_r + \frac{q}{c}(\mathbf{v} \times \mathbf{B}),$$

where $\hat{e}_r$ = unit vector along $\mathbf{r}$ and $m$ is the mass; the velocity $\mathbf{v}$ = absolute time-derivative of the position vector $\mathbf{r}$. That is

$$\mathbf{v} = \dot{\mathbf{r}} + \bar{\omega} \times \mathbf{r},$$

where $\bar{\omega}$ is the angular velocity with which the frame is rotating. Therefore

$$\frac{d\mathbf{v}}{dt} = \ddot{\mathbf{r}} + (\bar{\omega} \times \dot{\mathbf{r}}) + (\bar{\omega} \times \dot{\mathbf{r}}) + \bar{\omega} \times (\bar{\omega} \times \mathbf{r}).$$

Hence the equation of motion becomes

$$\ddot{\mathbf{r}} + 2(\bar{\omega} \times \dot{\mathbf{r}}) + \bar{\omega} \times (\bar{\omega} \times \mathbf{r}) = -\frac{qq'}{mcr^2}\hat{e}_r + \frac{q}{mc}[\dot{\mathbf{r}} + (\bar{\omega} \times \mathbf{r}) \times \mathbf{B}].$$

If we now make the ausatz,

$$\bar{\omega} = -\frac{q}{2mc}\mathbf{B},$$

then the equation of motion reduces to

$$\ddot{\mathbf{r}} = -\frac{qq'}{mr^2}\hat{e}_r + \left(\frac{q}{2mc}\right)^2 \mathbf{B} \times (\mathbf{B} \times \mathbf{r}).$$

In the weak field approximation

$$\left(\frac{qB}{2mc}\right)^2 = \omega^2 \ll \frac{qq'}{mr^2}.$$

The equation of motion becomes

$$\ddot{\mathbf{r}} = -\frac{qq'}{mr^2}\hat{e}_r.$$

This is just the equation of motion of a particle in an inverse square-law field. Hence the orbit is an ellipse in the rotating frame of reference. Hence, in the original non-rotating frame it is a precessing ellipse precessing with angular velocity, $\omega$. This effect is known as *Larmor effect* and the angular velocity of precession $\omega_L$ is

$$\omega_L = \frac{qB}{2mc}$$

is called the *Larmor frequency*. This effect results in an observable shift in the spectral lines of atom in presence of a magnetic field, known as *Zeeman effect*.

If $\mathbf{J}$ is angular momentum, then

$$\frac{d\mathbf{J}}{dt} = \mathbf{r} \times \mathbf{F} = \frac{q}{c}\mathbf{r} \times (\mathbf{v} \times \mathbf{B}).$$

We choose $\mathbf{B} = \mathbf{k}B$ and the orbit to be circular. In the weak field approximation, $d\mathbf{J}/dt$ becomes $-(qB/2mc)(\mathbf{k} \times \mathbf{J})$. That is $\mathbf{J}$ is rotating around $\mathbf{B}$ with Larmour frequency; the orbital plane is thus precessing around $\mathbf{B}$.

## 6.5 Centrifugal and Coriolis forces

At the begining of this chapter we obtained the Lagrangian equation of motion of a particle moving on the surface of the earth, taking into consideration the effect of earth's rotation around its axis. If '$\mathbf{g}$' is the acceleration due to gravity and $\mathbf{F}$ is the external force acting on the particle of mass $m$, the equation of motion is

$$m\ddot{\mathbf{r}} = \mathbf{F} + m\mathbf{g} - m(\mathbf{w} \times (\mathbf{w} \times \mathbf{r})) - 2m\mathbf{w} \times \mathbf{v}.$$

The third term is known as centrifugal force and the last term is the Coriolis force. We have discussed the effect of Coriolis force on the motion of a particle moving on the surface of the earth. In particular we have determined the deflection from coplanarity of the path of a particle thrown from the earth's surface with an initial velocity. Though this deflection is small, nevertheless it plays an important role in guided missiles. Another terrestrial application of the Coriolis force concerns the formation of cyclones. When a low pressure region occurs, all rushes to this region from all directions. In the northern hemisphere it will be deflected to the right due to Coriolis force. Thus in a cyclone the wind whirls in the anti-clockwise direction in Northern hemisphere and wind will rotate in the clockwise direction in southern hemisphere.

We write the equation of motion as

$$m\ddot{\mathbf{r}} = \mathbf{F} + m[\mathbf{g} - \mathbf{w} \times (\mathbf{w} \times \mathbf{r})] - 2m\mathbf{w} \times \mathbf{v}.$$

The second term on the right hand side is called the effective acceleration due to gravity.

$$\mathbf{g}^* = \mathbf{g} - \mathbf{w} \times (\mathbf{w} \times \mathbf{r}).$$

Since at the pole $\mathbf{w}$ is parallel to radius vector $\mathbf{r}$ the $\mathbf{g}^*$ equals $g$ while $\mathbf{w} \times (\mathbf{w} \times \mathbf{r})$ is maximum at the equator. Taking the angular velocity of the earth's rotation. $w$ to be equal to $7.29 \times 10^{-5}$ radians/sec. and radius of the earth equal to $6.4 \times 10^6 m$ we find the magnitude of the centrifugal acceleration to be $3.4 \times 10^{-2}$ m/sec$^2$.

This is only 0.3% of the value of $g$.

**Example 1.** *In latitude 45° N a gun is fired due north at an object distant 20 km. this being the maximum range of the gun. Show that if earth's rotation is not taken into account. then shell will fall 44 meters towards east from the mark and that if the gun is fired due south then the shell will fall 88 meters towards west of the mark.*

***Solution:*** Using the same reference frame as in example 1 we write the equations of motion as

$$\begin{aligned} \ddot{x} &= -2w(\sin\theta\dot{z} - \cos\theta\dot{y}); \\ \ddot{y} &= -2w\cos\theta.\dot{x} \\ \text{and} \quad \ddot{z} &= +2w\sin\theta\cdot\dot{x} - g. \end{aligned}$$

We use for $\dot{y}$ and $\dot{z}$ the values where $w = 0$; then

$$\begin{aligned} \dot{y} &= V\cos\alpha \\ \text{and} \quad \dot{z} &= V\sin\alpha - gt; \end{aligned}$$

$V$ = velocity of projection of the shell and $\alpha$ = angle of elevation towards noth. $\alpha = 45°$ with positive $y$-axis.

The angle $\theta = 45°$ as the colattitude. the equation for $\ddot{x}$ is then

$$\ddot{x} = 2wV\cos(\alpha + \theta) + 2w\sin\theta\cdot gt$$

Hence

$$x = wV\cos(\alpha + \theta)t^2 + \frac{1}{3}wg\sin\theta\cdot t^3.$$

If the time of flight

$$t = 2V\frac{\sin\alpha}{g}$$

is used then

$$x = 4V\frac{w}{g^2}\sin^2\alpha\left(\frac{2}{3}\sin\alpha\sin\theta + \cos(\alpha+\theta)\right)$$

when the shells are fired due north. $\alpha = 45^\circ$. $\theta = 45^\circ$ and the range $= \frac{V^2}{g} =$ 20 km. In that case for $g \simeq 10$ m/sec$^2$, the value of $x = 44$ m approx. When the shell are fired due south, then $\alpha = 135^\circ$ and $\theta = 45^\circ$. In this case using the same value of the range and the angular velocity

$$w = \frac{2\pi}{(24\times 60\times 60)} \text{ radian/sec,}$$

we get $x = -88$ meter. This is west ward, 88 meters awat from this mark.

**Example 2.** *A train in northern hemisphere is travelling southward along a meridian of the earth with a velocity $V$: show that in latitude $\lambda$, it processes around western railway with a force equal to $\left(\frac{2Vw}{g}\right)\sin\lambda$ times its own weight: $w$ = angular velocity of the earth around its axis.*

***Solution:*** From the last example we have

$$\ddot{x} = 2Vw\cos(\alpha+\theta)$$

where $\alpha$ = angule of projection. $\theta$ = colatitude $90^\circ - \lambda$: $V$ = velocity of the object: $\alpha = 180^\circ$ as the train moves southward. Then

$$\ddot{x} = -2Vw\cos\theta, \quad \text{or,} \quad \ddot{x} = -2Vw\cos\left(\frac{\pi}{2}-\lambda\right) = -2Vw\sin\lambda.$$

Hence the force is westwardly $= F = m\ddot{x} = -m2Vw\sin\lambda$ or the force $= \frac{2Vw}{g}\sin\lambda$ times its own weight.

## Exercises

1. Find the equation of motion for a particle moving in a rotating frame with variable angular velocity. Show that it has an extra term $-m\dot{\omega}\times\mathbf{r}$. Interpret this term physically.

2. The angular velocity of electron in lowest Bohr orbit in hydrogen atom is $\simeq 2\times 10^{16}\ S^{-1}$. What is the largest magnetic field which may be regarded as small in this case. Determine Larmor frequency in a field of 20 KG.

3. The orbit of an electron of charge $(-e)$ around a proton $(+e)$ is a circle of radius $a$, in a plane perpendicular to the magnetic field $\mathbf{B}$ (uniform field). By writing the equation of motion in a frame, rotating with the electron, show that the angular velocity $\omega$ satisfies

$$m\omega^2 - \frac{eB\omega}{c} - \frac{Ze^2}{a^3} = 0.$$

4. Derive $d\mathbf{J}/dt = -(qB/2mc)(\mathbf{k} \times \mathbf{J})$.

*Hints:* We choose $\mathbf{B} = \mathbf{k}B$ and the particle is moving in a circular orbit with radius vector $\mathbf{r}$ in a plane which makes an angle $\alpha$ with the $xy$-plane. We have $\mathbf{r}\cdot\mathbf{v} = 0$. For simplicity choose unit vectors $\mathbf{n}$, $\mathbf{a}$ and $\mathbf{b}$ such that $\mathbf{n}$ is normal to orbital plane, $\mathbf{a} = \mathbf{k} \times \mathbf{n}$ and $\mathbf{b} = \mathbf{n} \times \mathbf{a}$. Then $\mathbf{J} = mrv\mathbf{n}$. We have

$$\begin{aligned} \mathbf{r} &= (r\cos\psi, r\sin\psi, 0); \\ \mathbf{v} &= (-v\sin\psi, v\cos\psi, 0) \\ \text{and} \quad \mathbf{B} &= B(0, \sin\alpha, \cos\alpha). \end{aligned}$$

$\psi$ is the angle between $\mathbf{a}$ and $\mathbf{r}$. Then

$$\frac{d\mathbf{J}}{dt} = \frac{qBrv}{c}\sin\alpha(-\sin^2\psi, \sin\psi\cos\psi, 0).$$

In the weak field approximation we replace the right hand side by its average value over a complete cycle.

$$\frac{d\mathbf{J}}{dt} = -\frac{qvBr}{2c}\sin\alpha\cdot\mathbf{a}.$$

Again $\mathbf{k} \times \mathbf{J} = mrv\sin\alpha\cdot\mathbf{a}$. Hence

$$\frac{d\mathbf{J}}{dt} = -\left(\frac{qB}{2mc}\right)(\mathbf{k} \times \mathbf{J}).$$

Chapter VII

# Dynamics of a Rigid Body

## 7.1 Kinetic Energy of a Rigid Body

A *rigid body* is considered to be a set of discrete mass points whose orientation does not change under any external force. The position of a rigid body is fixed as soon as three non-colinear points in the body are fixed. Any arbitrary point in the body can be localized with reference to these three basic points. To fix 3-points $A$, $B$ and $C$ in a body we need to know 9 co-ordinates. But as $|AB|$, $|BC|$ and $|CA|$ are fixed, we have three constraints. Thus, we have 6-independent degrees of freedom. These degrees of freedom consist of 3-degrees of translation and another three degrees of rotation of the body. We thus have both translatory kinetic energy of the body and also rotational kinetic energy of the system. To find an expression for the kinetic energy of the system we consider the following co-ordinate system.

### 7.1.1 Expression for kinetic energy of a rigid body

Let $B$ be a rigid body, $Oxyz$ is a body fixed axes system. The body is free

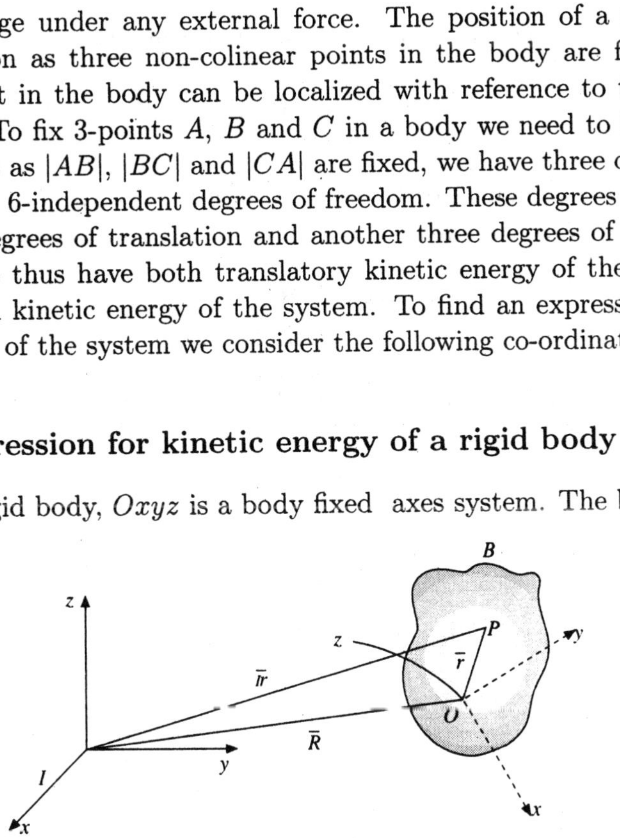


Fig 7.1

to rotate around any axis through $O$. Let $P$ be any point in the body, where $\overline{OP} = \mathbf{r}$, and $Ixyz$ be a fixed inertial frame and suppose (Fig 7.1)

$$\overline{IP} = \mathbf{Ir} \quad \text{and} \quad \overline{IO} = \mathbf{R}.$$

Therefore,

$$\overline{IP} = \overline{IO} + \overline{OP}.$$

We have,

$$\mathbf{IP} = \mathbf{R} + \mathbf{r}. \tag{7.1}$$

Let the body-fixed axes system $Oxyz$ rotate through an angle $\delta\bar{\omega}$, then the vector $\overline{OP}$, i.e., $\mathbf{r}$ will change as

$$\mathbf{r} \rightarrow \mathbf{r} + \delta\mathbf{r}$$

where

$$\delta\mathbf{r} = \delta\bar{\omega} \times \mathbf{r}. \tag{7.2}$$

Hence, the rate of change of $\mathbf{r}$ with respect to time, is

$$\frac{d\mathbf{r}}{dt} = \bar{\Omega} \times \mathbf{r}. \tag{7.3}$$

where $\bar{\Omega}$ is the angular velocity of rotation of the body.
From equation (7.1) we have

$$\frac{d\mathbf{IP}}{dt} = \frac{d\mathbf{R}}{dt} + \frac{d\mathbf{r}}{dt}. \tag{7.4}$$

If $\mathbf{v}$ is the velocity of the point $P$ relative to the inertial frame and $\mathbf{V}$ is the velocity of translation of the body relative to inertial frame, then from (7.3) and (7.4),

$$\mathbf{v} = \mathbf{V} + \bar{\Omega} \times \mathbf{r}. \tag{7.5}$$

The kinetic energy of $B$ relative the inertial frame is then

$$T = \text{K.E.} = \sum_i \frac{1}{2} m_i (\mathbf{v}_i)^2.$$

where $m_i$ is the mass of $i$th discrete point and $\mathbf{v}_i$ is the velocity of the point relative to the inertial frame, $Ixyz$.
Using (7.5)

$$T = \text{K.E.} = \frac{1}{2} \sum_i m_i (\mathbf{V}_i + \bar{\Omega} \times \mathbf{r}_i)^2. \tag{7.6}$$

Since

$$\mathbf{V} = \frac{d\mathbf{R}}{dt} \quad \text{and} \quad \mathbf{R}$$

is the same for all the mass points $m_i$ we have $\mathbf{V}_i = \mathbf{V}$ with components $(V_x, V_y, V_z)$. The components of $\mathbf{v}_i$ is $(v_x, v_y, v_z)_i$ and $\mathbf{r}_i(x, y, z)_i$; where the suffix $i$ stands for the particle index $i$. In the following calculation we will drop the suffix for the particle symbol (i.e., the suffix $i$) and write the expression for the kinetic energy, $T$, as,

$$T = \frac{1}{2}\sum m(\mathbf{V} + \Omega \times \mathbf{r})^2.$$

We now write this in terms of components and use

$$\Omega = (\Omega_x, \Omega_y, \Omega_z).$$

Then

$$T = \frac{1}{2}\sum m\left[(V_x + \Omega_y z - \Omega_z y)^2 + (V_y + \Omega_z x - \Omega_x z)^2 + (V_z + \Omega_x y - \Omega_y x)^2\right].$$

We can easily put this expression in much simpler form, if we assume that the centre of mass of the body $B$ is at $O$, which is the origin of the body-fixed axes system. In that case terms like

$$\sum mx, \quad \sum my \quad \text{and} \quad \sum mz$$

are all zero. The expression for $T$ reduces to the following:

$$T = \frac{1}{2}\sum m\mathbf{V}^2 + \frac{1}{2}\sum m\Big[\Omega_x^2(y^2 + z^2) + \Omega_y^2(z^2 + x^2) + \Omega_z^2(x^2 + y^2)$$
$$-2\Omega_x\Omega_y xy - 2\Omega_y\Omega_z yz - 2\Omega_z\Omega_x xz\Big].$$

If we use

$$I_{xx} = \sum m(y^2 + z^2) \text{ etc.}$$
$$\text{and } I_{xy} = \sum mxy \text{ etc.} \quad \text{and} \quad \sum m = M.$$

then

$$T = \frac{1}{2}M\mathbf{V}^2 + \frac{1}{2}I_{xx}\Omega_x^2 + \frac{1}{2}I_{yy}\Omega_y^2 + \frac{1}{2}I_{zz}\Omega_z^2$$
$$-I_{xy}\Omega_x\Omega_y - I_{yz}\Omega_y\Omega_z - I_{zx}\Omega_z\Omega_x. \tag{7.7}$$

In deriving equation (7.7) we have used

$$I_{xy} = I_{yx},$$
$$I_{yz} = I_{zy}$$
$$\text{and } I_{zx} = I_{xz}.$$

We want to express the equation (7.7) for $T$ in terms of tensorial notation. We associate $(x_1, x_2, x_3)$ for $(x, y, z)$ and use

$$I_{11} = I_{xx},$$
$$I_{22} = I_{yy},$$
$$I_{33} = I_{zz}$$
$$\text{and } I_{12} = I_{xy} \quad \text{etc.}$$

We define by $I_{ik}$ the following

$$I_{ik} = \sum m \left[ (x_1^2 + x_2^2 + x_3^2)\delta_{ik} - x_i x_k \right]. \tag{7.8}$$

This immediately yields

$$I_{xx} = I_{11} = \sum m(y^2 + z^2) \text{ etc.}$$
$$\text{and} \quad I_{12} = I_{xy} = \sum -mxy.$$

Using (7.8), the expression for $T$ is

$$T = \frac{1}{2} M\mathbf{V}^2 + \frac{1}{2} \sum_{i,k} I_{ik}\Omega_i\Omega_k. \tag{7.9}$$

If the centre of mass $O$ coincide with $I$, then the translation degree is frozen and $T$ becomes rotational K.E. $= T_{rot}$ and

$$T_{rot} = \frac{1}{2} \sum_{i,k} I_{ik}\Omega_i\Omega_k. \tag{7.10}$$

The tensor $I_{ik}$ is called *Inertia Tensor*. $I_{ik} = I_{ki}$. $I_{ik}$ is a symmetric second-rank tensor. We can put this as a $3 \times 3$ matrix:

$$I_{ik} = \begin{vmatrix} \sum m(y^2 + z^2) & -\sum mxy & -\sum mxz \\ -\sum myx & \sum m(z^2 + x^2) & -\sum myz \\ -\sum mzx & -\sum mzy & \sum m(x^2 + y^2) \end{vmatrix}.$$

Using a principal axes transformation we can diagonalize this matrix and can put $T_{rot}$ as

$$T_{rot} = \frac{1}{2} \sum_i I_i\Omega_i^2. \tag{7.11}$$

When $I_i$ are the 3 moments of inertia along $x$, $y$ and $z$-directions. If

$$I_1 = I_2 \neq I_3.$$

We call the body a *symmetric top*. When

$$I_1 = I_2 = I_3.$$

then the body is known as *spherical top*: when $I_1 = I_2$ but $I_3 = 0$, it is a rotator. It has two degrees of rotational symmetry. For

$$I_1 \neq I_2 \neq I_3$$

the body is called an *asymmetric top*.

## 7.2 Angular Momentum of a Rigid Body

We define the angular momentum of a rigid body with respect to the inertial frame. (Fig. 7.2). For this we considers a mass point $m_i$ in the body with co-ordinate $\mathbf{r}_i$ with respect to body-fixed $Oxyz$ system. If $\mathbf{IO} = \mathbf{R}$ and $\overline{IP} = \mathbb{r}_i$ as we have defined in the earlier section, then

$$\mathbb{r}_i = \mathbf{R} + \mathbf{r}_i.$$

Taking the time-derivative of both sides we would get

$$\mathbf{v}_i = \mathbf{V} + \bar{\Omega} \times \mathbf{r}_i.$$

Thus, the angular momentum $\mathbf{M}_i$ for the $i$th mass point of the rigid body defined with respect to inertial frame is

$$\mathbf{M}_i = \mathbb{r}_i \times \mathbf{p}_i,$$

where $\mathbf{p}_i = m_i \mathbf{v}_i$, i.e.,

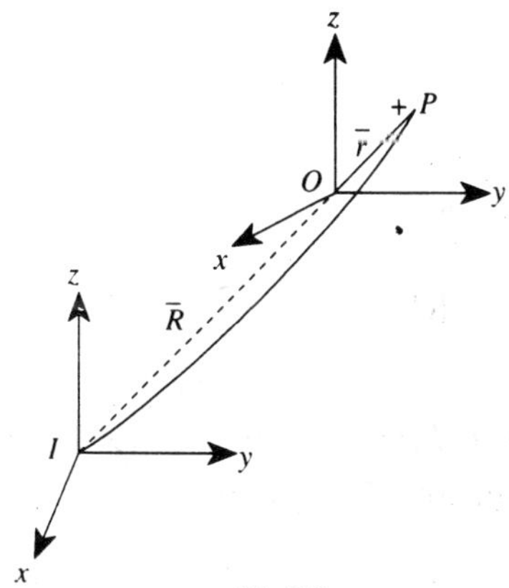


Fig 7.2

$$\mathbf{M}_i = \mathbb{r}_i \times (\mathbf{V} + \bar{\Omega} \times \mathbf{r}_i) m_i.$$

Hence, the angular momentum of the whole rigid body,

$$\mathbf{M} = \sum_i M_i.$$

In the following we drop the suffix, $i$, for our convenience

$$\mathbf{M} = \sum m(\mathbf{V} + \Omega \times r) \times \mathbb{r}, \tag{7.12}$$

i.e.,

$$\begin{aligned} M &= \sum m \left[(\mathbf{R} + \mathbf{r}) \times (\mathbf{V} + \bar{\Omega} \times \mathbf{r})\right] \\ &= \sum m(\mathbf{R} \times \mathbf{V}) + \sum m[\mathbf{r} \times (\bar{\Omega} \times \mathbf{r})]. \end{aligned}$$

In deriving the last expression we have dropped terms like

$$\sum m\mathbf{r}.$$

Writing $\mathbf{M}_0$ for the angular momentum of the centre of mass we get

$$\mathbf{M} = \mathbf{M}_0 + \sum m \left[\bar{\Omega} r^2 - \mathbf{r}(\bar{\Omega} \cdot \mathbf{r})\right].$$

Dropping $\mathbf{M}_0$ term and writing in component form, we have

$$\mathbf{M}_i = \sum m \left[\Omega_i(x_1^2 + x_2^2 + x_3^2) - x_i\left(\sum_k x_k\Omega_k\right)\right]$$
$$= \sum_k \left[\sum m(x_1^2 + x_2^2 + x_3^2)\delta_{ik} - x_i x_k\right]\Omega_k. \tag{7.13}$$

The expression in the third bracket is $I_{ik}$. Hence

$$M_i = \sum_k I_{ik}\Omega_k.$$

Equation (7.13) can be put as

$$\mathbf{M}_i = \sum_k \left[\sum m(x_1^2 + x_2^2 + x_3^2)\delta_{ik} - x_i x_k\right]\bar{\Omega}_k = \sum_k I_{ik}\Omega_k. \tag{7.14}$$

where $I_{ik}$ is the symmetric inertia tensor.
Equation (7.14) can be written symmetrically in a matrix form

$$\mathbf{M} = I \cdot \bar{\Omega}. \tag{7.15}$$

## 7.3 Lagrangian Equation: Conservation of Angular Momentum

The Lagrangian for the rigid body is as usual

$$L = T - V$$

then,

$$L = \frac{1}{2}M\mathbf{V}^2 + \frac{1}{2}\sum_{i,k} I_{ik}\Omega_i\Omega_k - U(\text{Ir}). \tag{7.16}$$

where the first term denotes translational kinetic energy, the second term is the rotational part, while $U$ is the potential function depending on the distance from $I$. Recall $\Omega_i = \dot{\phi}_i$, the angular velocity. $\phi_i$ is a co-ordinate here. Therefore,

$$\frac{\partial L}{\partial \dot{\phi}_i} = \sum_l I_{il}\dot{\phi}_l = \sum_l I_{il}\Omega_l = M_i, \tag{7.17}$$

where $M_i$ = $i$th component of angular momentum.
We now find $\partial L/\partial \phi$. Since

$$\delta U = -\sum_i \mathbf{f}_i \delta \text{Ir}.$$

We get

$$\delta U = -\sum_i \mathbf{f}_i \delta(\mathbf{R} + \mathbf{r}_l) = -\sum \mathbf{f}_i \cdot (\delta\bar{\phi} \times \mathbf{r}_l).$$

Hence

$$\frac{\partial U}{\partial \bar{\phi}} = -\sum_i (\mathbf{r}_i \times \mathbf{f}_i)$$

As $L$ depends on $\phi_l$ only through $U(\Pi)$, we then get,

$$\frac{d}{dt}\left(\frac{\partial L}{\partial \dot{\bar{\phi}}}\right) = -\frac{\partial U}{\partial \bar{\phi}} = \sum_l (\mathbf{r}_l \times f_i).$$

Hence from equation (7.17) we get

$$\frac{d}{dt}\mathbf{M} = \sum_i (\mathbf{r}_i \times \mathbf{f}_i). \tag{7.18}$$

The right-hand side of equation (7.18) denotes the moment of the derived force $\mathbf{f}_i$, i.e., the torque

$$\sum_i (\mathbf{r}_i \times \mathbf{f}_i).$$

Hence, it follows from the Lagrangian equation that the rate of change of angular momentum is equal to the moment of the force, i.e., *torque*. Hence, if the moment vanishes, then angular momentum is a constant of motion.

## 7.4 Eulerian Angles

We have seen from our Lagrangian equation that the equation of motion of the rigid body depends on its centre of mass co-ordinate with respect to the inertial system and the three angles $\phi_1, \phi_2$ and $\phi_3$, where the vector $\bar{\phi}$ specifies the direction of the axis of rotation of the moving rigid body. If we freeze the translatory motion of the body then the co-ordinate of the system are entirely $\phi_{1,2,3}$, i.e., $\bar{\phi}$ denoting the axis of rotation. The centre of mass $O$ is now at $I$.

Instead of using the co-ordinate $\phi_{1,2,3}$ to take account of the rotational degrees of freedom, we introduce another set of three angles called *Eulerian angles* to characterize a 3-dimensional rotation. To introduce this we first consider a two dimensional rotation of a vector $(x, y)$ in the $xy$-plane through angle $\phi$. Let the new-axis system is given by $(x', y')$. Then the relationship of $(x', y')$ with $(x, y)$ is given by

$$\begin{pmatrix} x' \\ y' \end{pmatrix} = \begin{pmatrix} \cos\phi & \sin\phi \\ -\sin\phi & \cos\phi \end{pmatrix} \begin{pmatrix} x \\ y \end{pmatrix}. \tag{7.19}$$

Consider now that this vector is rotating in three dimension space around the $z$-axis through the angle $\phi$. Then $(x, y, z)$ changes to $(x', y', z')$, where of course $z = z'$ and the relationship of $(x', y')$ with $(x, y)$ is as given in (7.19). Therefore, the relationship of $(x', y', z')$ with $(x, y, z)$, where $(xy)$ vector rotates around $z$-axis through $\phi$, is

$$\begin{pmatrix} x' \\ y' \\ z' \end{pmatrix} = \begin{pmatrix} \cos\phi & \sin\phi & 0 \\ -\sin\phi & \cos\phi & 0 \\ 0 & 0 & 1 \end{pmatrix} \begin{pmatrix} x \\ y \\ z \end{pmatrix}. \tag{7.20}$$

The matrix in front of $(x, y, z)$ vector rotates the vector to $(x', y', z')$ through an angle $\phi$ around the $z$-axis.

Consider now the following 3-successive rotations which carry the moving frame axes $(x, y, z)$ to $(x', y', z')$.

We apply first a rotation through an angle $\phi$ with angular velocity $\omega_\phi$ around $z$-axis by which the $(x, y, z)$ is changed to $(\xi, \eta, \zeta)$ axes. Then using equation (7.20), we get

$$\begin{pmatrix} \xi \\ \eta \\ \zeta \end{pmatrix} = \begin{pmatrix} \cos\phi & \sin\phi & c \\ -\sin\phi & \cos\phi & 0 \\ 0 & 0 & 1 \end{pmatrix} \begin{pmatrix} x \\ y \\ z \end{pmatrix}. \tag{7.21}$$

We call this $\bar{\xi} = A\mathbf{X}$, where

$$A = \begin{pmatrix} \cos\phi & \sin\phi & 0 \\ -\sin\phi & \cos\phi & 0 \\ 0 & 0 & 1 \end{pmatrix}. \tag{7.22}$$

We now apply a second rotation to $(\xi, \eta, \zeta)$ through an angle $\theta$ with angular velocity $\omega_\theta$ around the $\xi$-axis then

$$(\xi, \eta, \zeta) \to (\xi', \eta', \zeta') \text{ axes.}$$

The corresponding transformation matrix is

$$\begin{pmatrix} \xi' \\ \eta' \\ \zeta' \end{pmatrix} = \begin{pmatrix} 1 & 0 & 0 \\ 0 & \cos\theta & \sin\theta \\ 0 & -\sin\theta & \cos\theta \end{pmatrix} \begin{pmatrix} \xi \\ \eta \\ \zeta \end{pmatrix}. \quad \text{or,} \quad \bar{\xi}' = B\bar{\xi}.$$

where

$$B = \begin{pmatrix} 1 & 0 & 0 \\ 0 & \cos\theta & \sin\theta \\ 0 & -\sin\theta & \cos\theta \end{pmatrix}. \tag{7.23}$$

We now give the third rotation around $\zeta'$-axis through an angle $\psi$ with angular velocity $\omega_\psi$ which change $(\xi', \eta', \zeta')$ to $(x', y', z')$, the final position. Thus,

$$\begin{pmatrix} x' \\ y' \\ z' \end{pmatrix} = \begin{pmatrix} \cos\psi & \sin\psi & 0 \\ -\sin\psi & \cos\psi & 0 \\ 0 & 0 & 1 \end{pmatrix} \begin{pmatrix} \xi' \\ \eta' \\ \zeta' \end{pmatrix}.$$

Hence $X' = C\xi'$. where

$$C = \begin{pmatrix} \cos\psi & \sin\psi & 0 \\ -\sin\psi & \cos\phi & 0 \\ 0 & 0 & 1 \end{pmatrix}. \tag{7.24}$$

We thus have.

$$(i)\ \begin{pmatrix} x' \\ y' \\ z' \end{pmatrix} = CBA \begin{pmatrix} x \\ y \\ z \end{pmatrix}; \quad (ii)\ \begin{pmatrix} x' \\ y' \\ z' \end{pmatrix} = CB \begin{pmatrix} \xi \\ \eta \\ \zeta \end{pmatrix} \text{ and}$$

$$(iii)\ \begin{pmatrix} x' \\ y' \\ z' \end{pmatrix} = C \begin{pmatrix} \xi' \\ \eta' \\ \zeta' \end{pmatrix}.$$

Hence. the components of the angular velocity vector $(0.0.\omega_\phi)$ on to $(x'.y'.z')$ is given by

$$CBA \begin{pmatrix} 0 \\ 0 \\ \omega_\phi \end{pmatrix}.$$

Likewise components of $(\omega_\theta.0.0)$—vector on $(x'.y'.z')$ is given by

$$CB \begin{pmatrix} \omega_\theta \\ 0 \\ 0 \end{pmatrix}$$

and the components of $(0.0.\omega_\psi)$ on $(x'.y'.z')$ is

$$C \begin{pmatrix} 0 \\ 0 \\ \omega_\psi \end{pmatrix}.$$

Adding all the component along $x'$, along $y'$ and along $z'$ we get the angular velocity along $x'.y'$ and $z'$ by

$$\begin{pmatrix} \Omega_{x'} \\ \Omega_{y'} \\ \Omega_{z'} \end{pmatrix} = CBA \begin{pmatrix} 0 \\ 0 \\ \omega_\phi \end{pmatrix} + CB \begin{pmatrix} \omega_\theta \\ 0 \\ 0 \end{pmatrix} + C \begin{pmatrix} 0 \\ 0 \\ \omega_\psi \end{pmatrix}.$$

The angles $\theta.\phi.\psi$ are known as *Eulerian angles.*
Expanding the matrix product $CBA.CB$ and $C$ we get the angular velocities with moving-frame in terms of Eulerian angles as follows:

$$\begin{aligned} \Omega_{x'} &= \sin\psi\sin\theta\dot\phi + \cos\psi\dot\theta \\ \Omega_{y'} &= \cos\psi\sin\theta\dot\phi - \sin\psi\dot\theta \\ \Omega_{z'} &= \cos\theta\dot\phi + \dot\psi. \end{aligned} \tag{7.25}$$

where we have written $\dot\phi.\dot\theta$ and $\dot\psi$ for $\omega_\phi.\ \omega_\theta$ and $\omega_\psi$ respectively.

Hence,

$$T_{rot} = \frac{1}{2}\left(I_1'\Omega_{x'}^2 + I_2'\Omega_{y'}^2 + I_3'\Omega_{z'}^2\right), \tag{7.26}$$

where $I_1, I_2$ and $I_3$ are the principal inertia tensor.
The Lagrangian is then $T_{rot} - U$, where $U$ is the potential. We will utilize this Lagrangian to discuss the motion of a symmetric top in the next section.

## 7.5 Symmetric Top

We have already mentioned that for a symmetric top, the principal moment of inertia $I_1' = I_2'$. Then the Lagrangian of the top from equations (7.25) and (7.26)

$$L = \frac{1}{2}I_1'(\dot\theta^2 + \sin^2\theta\dot\phi^2) + \frac{1}{2}I_3'(\cos\theta\cdot\dot\phi + \dot\psi)^2 - Mgl\cos\theta, \tag{7.27}$$

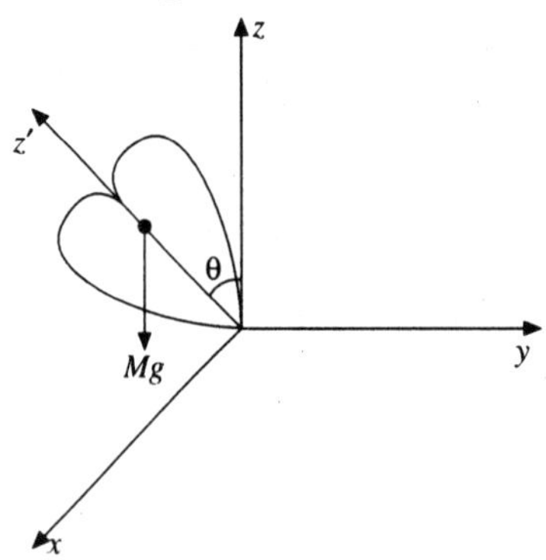


Fig 7.3

where $Mgl\cos\theta$ represents the potential energy of the top as shown in this figure. (we assume the $Ixyz$ and $Ox'y'z'$ axis have the common origin). The moment of inertia $I_1'$ is the sum of principal moment of inertia along the $x$-axis together with $Ml^2$ where $M$ is the mass and $l$ is the distance of C.G. of the top from $O$. (Fig. 7.3).

The Hamiltonian of the symmetric top is $H = T + V$ and $H$ does not contain time explicitly, $H = E =$ Energy and it is conserved.
Thus,

$$E = \frac{1}{2}I_1'(\dot\theta^2 + \sin^2\theta\cdot\dot\phi^2) + I_3'(\cos\theta\dot\phi + \dot\psi)^2 + Mgl\cos\theta. \tag{7.28}$$

Since $L$ does not depend on $\psi$ and also $\phi$, we have $p_\psi$ and $p_\phi$ as constants of motion. Then

$$p_\psi = \frac{\partial L}{\partial\dot\psi} = I_3'(\dot\psi + \cos\theta\dot\phi) = M_3 \text{ (a constant)} \tag{7.29}$$

and $$p_\phi = \frac{\partial L}{\partial\dot\phi} = I_1'\sin^2\theta\dot\phi + I_3'(\cos\theta\dot\phi + \dot\psi)\cos\theta = M_z \text{ (a constant)}. \tag{7.30}$$

Equation (7.30) can be written as

$$p_\phi = I_1'\sin^2\theta\dot\phi + M_3\cos\theta = M_z. \tag{7.31}$$

Then

$$\dot{\phi} = \frac{(M_z - M_3 \cos\theta)}{(I_1' \sin^2\theta)} \tag{7.32}$$

and $\dot{\psi}$ can be written as

$$\dot{\psi} = \frac{M_3}{I_3'} - I_1' \frac{\cos\theta}{\sin^2\theta}(M_z - M_3 \cos\theta). \tag{7.33}$$

Using equation (7.32) and (7.33) the energy $E$ becomes

$$E = \frac{1}{2} I_1' \dot{\theta}^2 + U(\theta). \tag{7.34}$$

where $U(\theta)$ is given by

$$U(\theta) = \frac{1}{2I_1' \sin^2\theta}(M_z - M_3 \cos\theta)^2 + \frac{1}{2I_3'} M_3^2 + Mgl\cos\theta.$$

From (7.34) we get

$$\frac{d\theta}{dt} = \{2[E - U(\theta)]/I_1'\}^{1/2}.$$

We can integrate to obtain

$$t = \int \frac{d\theta}{[2(E - U(\theta)/I_1']^{1/2}}. \tag{7.35}$$

This integral expresses $\theta$ as a function of time. We use this value of $\theta$ to find $\psi$ and $\varphi$. Thus Eulerian co-ordinates are completely known. The range of values of $\theta$ can be found from the condition that $E \geq U(\theta)$.
The function $U(\theta)$ tends to infinity ($M_3 \neq M_z$) when $\theta \to 0$ or $\pi$ and $U(\theta)$ has a minimum between these two values. Hence, the equation

$$E = U(\theta)$$

has two roots, which determine the limiting values $\theta_1$ and $\theta_2$ of the inclination of the axis of the top. (at these value of $\theta$, $\dot{\theta} = 0$).
When $\theta$ varies from $\theta_1$ to $\theta_2$ the derivative $\dot{\phi}$, given by

$$\dot{\phi} = \frac{M_z - M_3 \cos\theta}{I_3'}$$

may or may not change sign depending on the sign of

$$M_z - M_3 \cos\theta.$$

For non-vanishing $\dot{\phi}$ we must have

$$M_z - M_3 \cos\theta \neq 0.$$

We distinguish three different cases:

**Case I:** Let $\theta$ vary between $\theta_1$ and $\theta_2$, such that $\dot{\phi} \neq 0$ and always has the same sign, say $\dot{\phi} > 0$.
In this case the axis of the top *precesses* monotonically about the vertical and at the same time oscillates up and down as shown in fig. 7.4a. This oscillation is called *nutation.* The curve between $\theta_1$ and $\theta_2$ shows the track of the axis on the surface of the sphere whose centre is at the fixed point of the top.

**Case II:** When $\theta$ varies between $\theta_1$ and $\theta_2$ and $\dot{\phi} < 0$ at $\theta = \theta_2$ and $\dot{\phi} > 0$ at $\theta = \theta_1$. The curve traced by the axis in this case is given by Fig. 7.4b.

**Case III:** When $\theta$ varies such that $\dot{\phi} = 0$ at $\theta = \theta_2$ and $\dot{\phi} > 0$ at $\theta = \theta_1$. The resulting track is given in Fig. 7.4c.

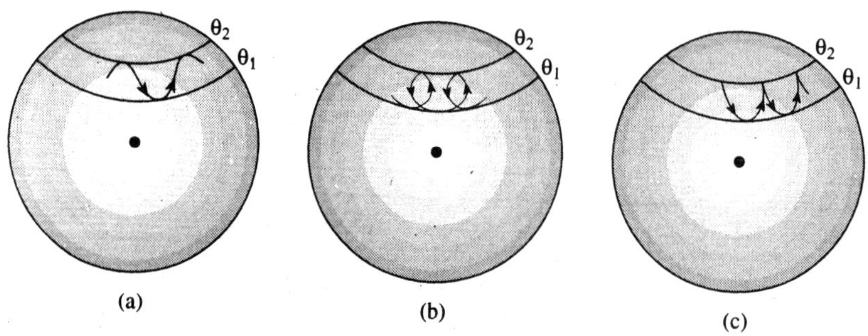


Fig 7.4

## 7.6 Stability of a Vertical Top

If the axis of the top is to be vertical then $\theta = 0$ must hold. But the effective potential $U(\theta)$ at $\theta = 0$ to be finite we must have

$$M_3 = M_z, \quad \text{i.e.,} \quad p_\psi = p_\phi,$$

where

$$p_\psi = I_3'(\dot{\psi} + \cos\theta\dot{\phi}) = I_3'\Omega_3.$$

When $M_z = M_3$

$$U(\theta) = \left[\frac{(I_3'\Omega_3^2)}{2I_1'}\right]\tan^2\frac{\theta}{2} + \frac{1}{2}I_3'\Omega_3^2 + Mgl\cos\theta,$$

where we used

$$\frac{1-\cos\theta}{\sin\theta} = \tan\frac{\theta}{2}.$$

For small values of $\theta$, we expand $U(\theta)$ and retain only terms of order $\theta^2$. Then

$$U(\theta) = \left(\frac{1}{2}I_3'\Omega_3^2 + Mgl\right) + \frac{1}{2}\left(\frac{I_3'^2\Omega_3^2}{4I_1'} - Mgl\right)\theta^2.$$

Since $U(\theta)$ has no term involving $\theta$, then $\theta = 0$ is always a position of equilibrium. It is a stable equilibrium if $U(\theta)$ has a minimum at $\theta = 0$, i.e., if the coefficient of $\theta^2$ is positive. Thus, there is a minimum value of $\Omega_3$ for which a vertical top is *stable*, i.e.,

$$\Omega_3 > \frac{(4I_1'Mgl)}{I_3'^2}. \tag{7.36}$$

Equation (7.36) gives the condition for the stability of a vertical top.

## 7.7 Euler's Equation

We may recall that when we had obtained Lagrange's equation for a rigid body we used the reference frame as the fixed inertial system. As a result the equation of motion was

$$\frac{d}{dt}\mathbf{M} = \mathbf{K}, \tag{7.37}$$

where $\mathbf{M}$ is the angular momentum vector and $\mathbf{K}$ is the moment of the external force $\mathbf{F}$ given by $\mathbf{r} \times \mathbf{F}$. But if we want to involve in our equation of motion (equation 7.37), the angular velocities of the system we must use moving frame of reference fixed in the rigid body itself.
In discussing equation of motion in a rotating frame when the frame is rotating with a constant angular velocity $\bar{\Omega}$ we found that the rate of change of any vector with time denoted by $\dot{\mathbf{a}}$ is given by

$$\dot{\mathbf{a}} = \frac{d}{dt}\mathbf{a} + \bar{\Omega} \times \mathbf{a}.$$

where $\frac{d}{dt}\mathbf{a}$ denotes the time-rate of change of $\mathbf{a}$ in a fixed system. Thus,

$$\dot{\mathbf{M}} = \frac{d\mathbf{M}}{dt} + (\bar{\Omega} \times \mathbf{M}).$$

If any external force is present, then

$$\dot{\mathbf{M}} = \mathbf{K}, \tag{7.38}$$

where $\mathbf{K}$ is the torque of the force.
Hence, from (7.37) and (7.38)

$$\frac{d\mathbf{M}}{dt} + \bar{\Omega} \times \mathbf{M} = \mathbf{K}. \tag{7.39}$$

Equation (7.39) is the correct equation of motion referred to body-fixed rotating frame. $d\mathbf{M}/dt$ is however. calculated by using inertial system of axes. Writing (7.39) in component form we get the following Euler's equation.

$$\frac{dM_1}{dt} + \Omega_2 M_3 - \Omega_3 M_2 = K_1$$

$$\frac{dM_2}{dt} + \Omega_3 M_1 - \Omega_1 M_3 = K_2$$

and
$$\frac{dM_3}{dt} + \Omega_2 M_1 - \Omega_1 M_2 = K_3.$$

We now use.

$$\mathbf{M} = I \cdot \bar{\Omega}.$$

$$\text{or.}\quad M_1 = I_1\Omega_1, \quad M_2 = I_2\Omega_2, \quad M_3 = I_3\Omega_3$$

and assume that there is no external force acting. then

$$I_1\dot{\Omega}_1 + \Omega_2\Omega_3 I_3 - \Omega_3\Omega_2 I_2 = 0$$

$$I_2\dot{\Omega}_2 + \Omega_3\Omega_1 I_1 - \Omega_1\Omega_3 I_3 = 0$$

$$I_3\dot{\Omega}_3 + \Omega_2\Omega_1 I_1 - \Omega_1\Omega_2 I_2 = 0.$$

Thus. $\dot{\Omega}_3 = 0$ for the symmetrical top for which $I_1 = I_2$. That is $\Omega_3 =$ constant. Using

$$I_1 = I_2 \quad \text{and} \quad \Omega_3 = \text{constant}$$

we get from the first two equations

$$\dot{\Omega}_1 + \Omega_2\omega = 0;$$

$$\dot{\Omega}_2 - \omega\Omega_1 = 0.$$

where

$$\omega = \Omega_3\frac{(I_3 - I_1)}{I}, \quad I_1 = I_2 = I.$$

Therefore

$$\Omega_1 = A\sin\omega t$$

$$\text{and}\quad \Omega_2 = A\cos\omega t$$

are the solutions of the $\Omega_1$ and $\Omega_2$ differential equations. This result shows that the components of angular velocity perpendicular to the axis of the top rotates with angular velocity. $\omega$ always remaining with constant magnitude $A$. where

$$A = \sqrt{(\Omega_1^2 + \Omega_2^2)}.$$

Since the component $\Omega_3$ along the axis of the top is also constant we conclude that the vector $\bar{\Omega}$ rotates uniformly with constant magnitude about the axis of the top. As

$$M_1 = I_1\Omega_1.$$
$$M_2 = I_2\Omega_2.$$
$$M_3 = I_3\Omega_3.$$

we find angular momentum **M** also executes similar motion around the axis of the top.

We can easily check that the solution we have obtained from Euler's equation is the same as we have found in our discussion on symmetrical top in section (7.5). Because the value of $\dot{\psi}$ we found was,

$$\dot{\psi} = \frac{M_3}{I_3} - \dot{\phi}\cos\theta = \frac{M\cos\theta}{I_3} - \dot{\phi}\cos\theta = M\cos\theta\left(\frac{1}{I_3} - \frac{1}{I_1}\right)$$
$$= M_3\left(\frac{1}{I_3} - \frac{1}{I_1}\right) = \Omega_3\frac{(I_1 - I_3)}{I_1} = -\omega.$$

This is in agreement with the present result. We used $\Omega_1 = 0$. $\Omega_2 = \dot{\phi}\sin\theta$. $\Omega_3 = \dot{\phi}\cos\theta + \dot{\psi}$ and $M_1 = 0$. $M_2 = M\sin\theta$ $M_3 = M\cos\theta$.

## 7.8 Stability of Rotation about a Principal Axis

In the last section we discussed the case when the principal moments of inertia. $I_1$, $I_2$ and $I_3$ satisfy the relation $I_1 = I_2 \neq I_3$. In the present case we discuss the rotation of a body for which $I_1 \neq I_2 \neq I_3$. Since the principal axes rotate with the rigid body we have

$$\frac{d\mathbf{M}}{dt} + \bar{\Omega} \times \mathbf{M} = \mathbf{K}.$$

where **K** is the torque. Thus in the absence of any external forces we have

$$I_3\dot{\Omega}_3 + (I_2 - I_1)\Omega_1\Omega_2 = 0$$
$$I_1\dot{\Omega}_1 + (I_3 - I_2)\Omega_2\Omega_3 = 0$$
$$\text{and} \quad I_2\dot{\Omega}_2 + (I_1 - I_3)\Omega_1\Omega_3 = 0.$$

If the body is initially rotating with angular velocity $\Omega_3$ around the principal axis $e_3$ (i.e., $z$-axis) then $\Omega_1 = \Omega_2 = 0$, then the first of the above three equations predict that $\dot{\Omega}_3 = 0$, i.e., $\Omega_3$ is a constant.

We now wish to discuss the stability of the motion. That is we now ask what happens if the body is given a small displacement so that its axis of rotation no longer precisely coincides with the $z$-axis. Under such situation

$\Omega_1$ and $\Omega_2$ will no longer be zero, but is small. Then the product $\Omega_1.\Omega_2$ is still small, so that the first of the above three equations will continue to give $\Omega_3$ as a constant.

We can then solve the other two equations for $\Omega_1$ and $\Omega_2$ by substituting

$$\Omega_1 = a_1 e^{pt}$$
$$\Omega_2 = a_2 e^{pt}.$$

where $a_1$ and $a_2$ are constants. If we now eliminate the constants $a_1$ and $a_2$ then we find for $p$, the following

$$p^2 = \frac{(I_3 - I_2)(I_1 - I_3)}{I_1 I_2}\Omega_3^2.$$

We easily see that if

$$I_3 > I_2 \quad \text{and} \quad I_3 > I_1,$$
$$\text{or,} \quad I_3 < I_2 \quad \text{and} \quad I_3 < I_1,$$

roots of $p$ are pure imaginary as $\Omega_3^2$ and the denominator are positive. In this case we have oscillatory solution for the quantities $\Omega_1$ and $\Omega_2$.

If, however,

$$I_1 > I_3 > I_2,$$
$$\text{or,} \quad I_1 < I_3 < I_2$$

the roots are real and values of $\Omega_1$ and $\Omega_2$ may increase with time. We may therefore, conclude that the rotation is stable if $I_3$ is the largest or the least of the three principal moment of inertia.

Consider the rotation of an ellipsoid.

$$\frac{x^2}{a^2} + \frac{y^2}{b^2} + \frac{z^2}{c^2} = 1$$

around $z$-axis.

In this case

$$I_3 = \frac{M(a^2 + b^2)}{5},$$
$$I_2 = \frac{M(c^2 + a^2)}{5}$$
$$\text{and} \quad I_1 = \frac{M(b^2 + c^2)}{5}.$$

Therefore,

$$p^2 = \frac{(b^2 - c^2)(c^2 - a^2)}{(b^2 + c^2)(c^2 + a^2)}\Omega_3^2.$$

We thus see that if

$$b^2 > c^2 \quad \text{and} \quad a^2 > c^2,$$
$$\text{or,} \quad b^2 < c^2 \quad \text{and} \quad a^2 < c^2,$$

we will have stable motion, otherwise the motion will not be a stable one. For a spheriod,

$$\frac{x^2 + y^2}{a^2} + \frac{z^2}{c^2} = 1,$$

we will have

$$p^2 = -\frac{(a^2 - c^2)^2}{(a^2 + c^2)^2}\Omega_3^2,$$

the motion is a stable one.

**Example 1.** *A circular cylinder of radius a surrounds a smaller circular cylinder of radius $b(b < a)$. The former rolls on the latter which is fixed with its generator horizontal.*
*Show that the plane through the axes oscillates like a simple pendulum of equivalent length*

$$(a - b)\left(1 + \frac{k^2}{a^2}\right),$$

*where k is the radius of gyration of the moving cylinder about its axis.*

***Solution:*** $O$ is the centre of the fixed cylinder of radius $b$.
$G$ is the centre of the moving cylinder.
$P$ is point of contact; $OG_1$ is the vertical where $G_1$ is the centre of moving cylinder when it was in contact with fixed cylinder at $P_1$ when $G_1P_1$ is vertical. $\theta = \angle G_1OG$. The potential energy of the moving cylinder

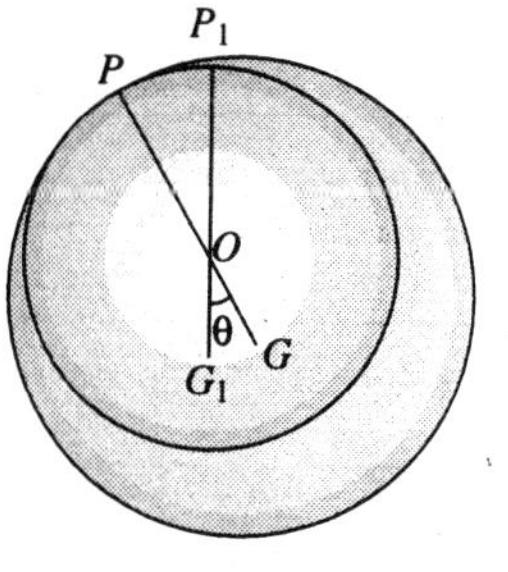

Fig 7.5

$$\begin{aligned} V &= mg \times \text{height of } G \text{ over } G_1 \\ &= mg(a - b)(1 - \cos\theta). \end{aligned}$$

The kinetic energy

$$\text{K.E.} = \frac{1}{2}I\omega^2,$$

where $\omega$ = angular velocity of the cylinder about its centre and

$$I = m(k^2 + a^2),$$

where $k$ = radius of gyration about its axis. The point of contact $P$ has instantaneous velocity zero. Since $P$ has the tangential velocity $a\omega$ and $G$ has $(a - b)\dot{\theta}$ in the opposite direction, hence

$$(a - b)\dot{\theta} = a\omega.$$

Thus,

$$\text{K.E.} = \frac{1}{2}m(k^2 + a^2)\frac{(b-a)^2}{a^2}\dot{\theta}^2 = T$$

$$\text{and} \quad V \simeq mg(b-a)\frac{1}{2}\theta^2.$$

Then from $L = T - V$, we immediately get

$$(b-a)\left(1 + \frac{k^2}{a^2}\right)\ddot{\theta} + g\theta = 0.$$

This is a simple pendulum of length

$$(b-a)\left(1 + \frac{k^2}{a^2}\right).$$

**Example 2.** *A small solid sphere of radius b is placed inside hollow fixed sphere of radius $a (a > b)$ on a horizontal floor. The small sphere is initially placed at the bottom of the bigger one and then slightly disturbed and it rolls; find the period of small oscillation.*

***Solution:*** $O$ is Centre of fixed sphere.

$OP = a$, $G$ is the centre of small moving sphere, $P$ = instantaneous point of contact. $P'$ is the initial point of contact, $OP'$ is vertical and $\angle POP' = \theta$.

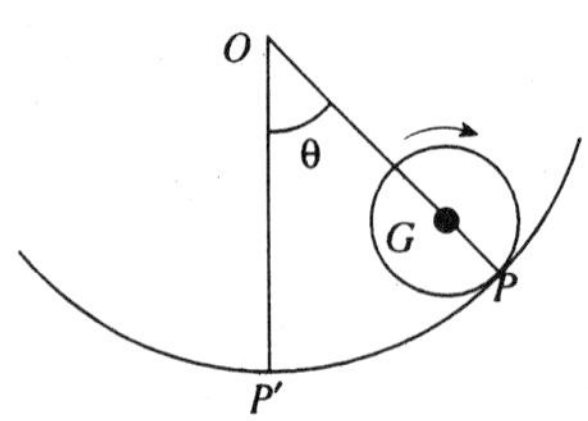


Fig 7.6

The potential energy $= mg(a-b)(1 - \cos\theta)$.

Let the small sphere rotates with angular velocity $\omega$, then $P$ has velocity $b\omega$. Again $G$ has opposite velocity $(a-b)\dot{\theta}$. Hence, as $P$ is at rest instantaneously,

$$b\omega = (a-b)\dot{\theta}.$$

Hence,

$$T = \text{K.E.} = \frac{1}{2}m(k^2 + b^2)\omega^2 = \frac{1}{2}m(k^2 + b^2)\frac{(a-b)^2}{b^2}\dot{\theta}^2.$$

Thus for $T$ and

$$V \simeq \frac{1}{2}m(a-b)\theta^2$$

we get as in the last example the following equivalent simple pendulum.

$$(a-b)\left(1 + \frac{k^2}{b^2}\right)\ddot{\theta} + g\theta = 0.$$

For solid sphere

$$k^2 = \frac{2}{5}b^2.$$

For cylinder

$$k^2 = \frac{1}{2}b^2 \text{ etc.}$$

Hence the period of small oscillation for the small sphere is

$$2\pi\sqrt{\left[\frac{7(a-b)}{5g}\right]}.$$

**Example 3.** *A right solid cone symmetric about its axis, is rolling on a smooth horizontal plane so that at any instant it is in contact with the plane along a generator. If n is the angular velocity of the cone along its axis OC (see figure), ω, the angular velocity of the plane containing the axis of the cone and the vertical ON′ and Ω, the angular velocity of the cone about the instantaneous axis of rotation, OP, then show that*

$$n = \Omega\cos\alpha = -\omega\cos\alpha\cdot\cot\alpha,$$

*where α is the semi-vertical angle of the cone. Show that when there are no external force n and ω are constants.*

***Solution:*** Let $G$ be the C.G. of the cone and let $OP$, be the generator in contact with the plane; $GM$ and $GN$ are drawn perpendiculars from $G$ on $OP$ and $ON'$. Then with the angular velocities in the sense as indicated in the figure, then the velocity of $G$ is $\Omega GM$, or, $-\omega GN$. Hence

$$\Omega = -\omega\cot\alpha.$$

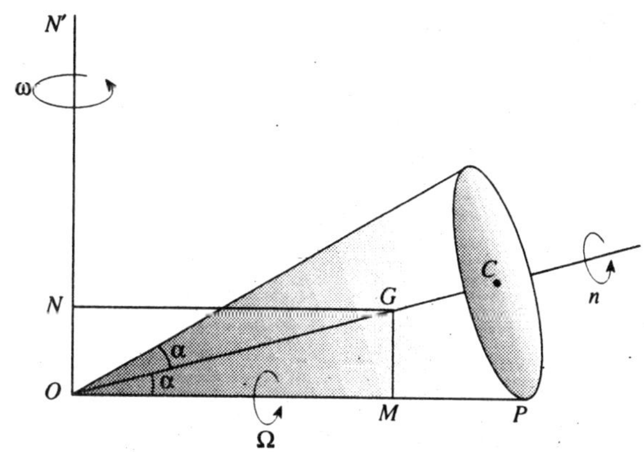


Fig 7.7

The point of contact $P$ has zero velocity. The point $P$ has the velocity relative to $C$ plus the velocity of $C$, therefore

$$n\cdot PC + \omega\cdot OC\cos\alpha = 0, \quad \text{or,} \quad n = -\omega\cot\alpha\cos\alpha.$$

Let $I$ be the moment of inertia of the cone about $OP$. The angular momentum is the about $OP$ is $I\Omega$, and this must be a constant as there is no external force. Thus $\Omega$ is constant. So we get $n$ and $\omega$ also constant. If the angular velocity is large it is possible for the cone to roll on the plane.

**Example 4.** *Find the principal moment of inertia of an ellipsoid $\frac{x^2}{a^2}+\frac{y^2}{b^2}+\frac{z^2}{c^2}=1$. Show that for a homogeneous sphere of radius $R$ it is $\frac{2}{5}MR^2$, where $M$ = mass of the sphere.*

***Solution:*** We apply the transformation

$$\begin{aligned} x &= ax_1; \\ y &= bx_2 \\ \text{and} \quad z &= cx_3. \end{aligned}$$

The ellipsoid reduces to a sphere

$$x_1^2+x_2^2+x_3^2=1$$

of radius unity. Principal moment of inertia along the $z$-axis is

$$I_3=\left(\frac{M}{4\frac{\pi}{3}abc}\right)\iiint(x^2+y^2)dx\,dy\,dz.$$

After the transformation $I_3$ becomes

$$I_3=\left(\frac{M}{4\frac{\pi}{3}abc}\right)\iiint(a^2x_1^2+b^2x_2^2)abc\cdot dx_1dx_2dx_3;$$

we now use spherical polar coordinates. Then

$$\begin{aligned} x_1 &= r\sin\theta\cos\phi \\ x_2 &= r\sin\theta\sin\phi, \\ z &= r\cos\theta. \end{aligned}$$

where $0\le r\le 1$, $0\le\theta\le\pi$, $0\le\varphi\le 2\pi$ and $dx_1dx_2dx_3=r^2dr\sin\theta d\theta\,d\phi$. Then $I_3$ reduces to

$$I_3=\frac{M}{5}(a^2+b^2);$$

similarly,

$$I_1=\frac{M}{5}(b^2+c^2)\quad\text{and}\quad I_2=\frac{M}{5}(a^2+c^2).$$

For a sphere put $a=b=c=R$, then $I=\frac{2}{5}MR^2$, along any diameter.

**Example 5.** *Find the moment of inertia of a solid right circular cone of height $h$, and base radius $R$. The vertex is at the origin and the axis of the cone is along $x$, $y$ and $z$-axis.*

***Solution:*** If $I_1'$ is the moment of inertia of the cone along $x$-axis, then

$$I_1'=\mu\iiint(y^2+z^2)dx\,dy\,dz;$$

$\mu$ = density. To integrate we use cylindrical coordinates $x = \rho\cos\phi$, $y = \rho\sin\phi$ and $z = z$, $dx\,dy\,dz = r\,dr\,d\phi\,dz$, where $0 \le z \le h$ and $0 \le \phi \le 2\pi$; $0 \le \rho \le z\tan\theta$, $\tan\theta = \frac{R}{h}$. Hence $I_1'$ becomes

$$I_1' = \mu\int_0^h dz \int_0^{z\tan\theta} \rho d\rho \int_\theta^{2\pi} d\phi(\rho^2\sin^2\phi + z^2) = \frac{3}{5}M\left(\frac{1}{4}R^2 + h^2\right);$$

$$I_2' = I_1';$$

$$I_3' = \mu\int_0^h dz \int_0^{z\tan\theta} \rho d\rho \int_0^{2\pi} d\phi\,\rho^2 = \frac{3}{10}MR^2.$$

**Example 6.** *A solid homogeneous right circular cone—height h and vertical angle $2\alpha$, oscillates about a horizontal axis passing through the origin; show that the length of the equivalent simple pendulum is $\frac{h}{5}(4 + \tan^2\alpha)$.*

***Solution:*** Let the axis of the cone makes an angle $\theta$ with the vertical ($z$-direction). The vertex is at the origin and $x$-axis is the horizontal axis about which the cone is oscillating. 0 be the vertex and $G$ be the C.G. of the cone lying on its axis. Let $OG = d$. Then the P.E. $= mgd(1 - \cos\theta)$, where $m$ is mass of the solid cone. The K.E. of rotation is $\frac{1}{2}I_1'w^2$; $w = \dot\theta$; $I_1'$ = moment of inertia of the cone about the vertex along the $x$-axis. From energy conservation

$$\frac{1}{2}I_1'\dot\theta^2 + mgd(1 - \cos\theta) = \text{const.}$$

Hence differentiating

$$I_1'\ddot\theta + mgd\sin\theta = 0$$

$$\text{or, } mk^2\ddot\theta + mgd\theta = 0;$$

the equivalent simple pendulum length is $\frac{k^2}{d}$, where $k$ = radium of gyration. Here

$$k^2 = \frac{3}{5}\left(\frac{1}{4}R^2 + h^2\right),$$

where $R$ = base radius:

$$\tan\alpha = \frac{R}{h}; \quad d = \frac{3}{4}h.$$

Hence the pendulum length is

$$\frac{3}{5}\left(\frac{1}{4}\tan^2\alpha\cdot h^2 + h^2\right)\Big/\frac{3}{4}h = \frac{1}{5}h(\tan^2\alpha + 4).$$

**Example 7.** *Show that if $w_1, w_2$, and $w_3$ are the angular velocities of the top along $OA, OB$ and $OC$ denoting the $x, y$ and $z$-axis of the moving top and $\psi, \theta, \phi$ are the Euler's angles then the relationship between $w_1, w_2$ and $w_3$ and $\dot\psi, \dot\theta$ and $\dot\phi$ can be obtained purely from a geometrical consideration (see diagram).*

***Solution:*** Let $OC$ be the axis of the top. $OZ$ is the vertical. $ZOX$ is the vertical plane in which the axis $OC$ was at initial time. $OC$ makes angle $\theta$ with $OZ$ at time $t$.

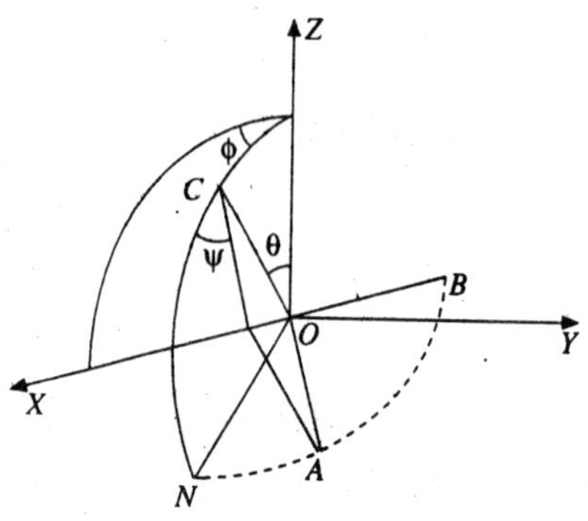


Fig 7.8

Let the plane $ZOC$ turned through an angle $\phi$ from its initial position $ZOX$. $OXYZ$ forms the fixed axes system. Let $OA$ and $OB$ are two perpendicular lines each perpendicular to $OC$. At an instant of time $t$, let $w_1$ and $w_2$, and $w_3$ be the angular velocities of the top along $OA, OB$ and $OC$.

Now consider the motion of $C$ and $A$. Assume $OC =$ unity.

Then

$$\begin{aligned} \dot{\theta} &= \text{velocity of } C \text{ along the arc } ZC \\ &= w_1 \cos\psi - w_2 \sin\psi \qquad \text{(i)} \\ \dot{\phi}\sin\theta &= \dot{\phi} \times \text{perpendicular from } C \text{ on } OZ \\ &= \text{velocity of } C \text{ perpendicular to plane } ZOC \\ &= w_1 \sin\psi + w_2 \cos\psi. \qquad \text{(ii)} \end{aligned}$$

Also

$$\begin{aligned} w_3 &= \text{velocity of } A \text{ along } AB \\ &= \dot{\psi} + \dot{\phi} \times \text{perpendicular from } N \text{ on } OZ \\ &= \dot{\psi} + \dot{\phi}\cos\theta \qquad \text{(iii)} \end{aligned}$$

From (i), (ii) and (iii) we get

$$\begin{aligned} w_1 &= \dot{\phi}\sin\theta\sin\psi + \dot{\theta}\cos\psi \\ w_2 &= \dot{\phi}\sin\theta\cos\psi - \dot{\theta}\sin\psi \\ w_3 &= \dot{\phi}\cos\theta + \dot{\phi}. \end{aligned}$$

**Example 8.** *Find the condition for the steady motion of a symmetrical top.*
***Solution:*** If $I_1'$ is the moment of inertia of the top along $OA$ then for a symmetric top $I_1'$ is also the moment of inertia along $OB$. Let $I_3'$ be that along $OC$. Then K.E. is $T$ equals $\frac{1}{2}I_1'(w_1^2+w_2^2)+\frac{1}{2}I_3'w_3^2$. (see example 7) and P.E. $= mgh(1-\cos\theta)$, where $h =$ distance of $C.\ G$ from 0. The Lagrange's equation is

$$\frac{d}{dt}(I_1'\dot{\theta}^2) - I_1'\dot{\phi}^2\sin\theta\cos\theta + I_3'(\dot{\psi}+\dot{\phi}\cos\theta)\dot{\phi}\cos\theta = mgh\sin\theta$$

and $\dot{\psi} + \dot{\phi}\cos\theta = \text{constant} = n$.

For steady motion $\theta = \alpha$, $\dot{\theta} = 0$, $\ddot{\theta} = 0$, and $\dot{\phi} = \text{const} = w$. Hence we get from the first two equations

$$I_1' w^2 \cos\alpha - I_3' n w + mgh = 0.$$

Since $w$ is real, we get

$$I_3'^2 n^2 > 4 I_1' mgh \cos\alpha.$$

**Example 9.** *Show that a symmetric top executes small oscillation around its steady motion.*

***Solution:*** We have from the Lagrangian, the equation of motion for $\theta$:

$$I_1'\ddot{\theta} - I_1'\dot{\phi}^2 \sin\theta\cos\theta + I_3' n\dot{\phi}\sin\theta = mgh\sin\theta$$

and from $\phi$ equation we get

$$I_1'\dot{\phi}\sin^2\theta + I_3' n\cos\theta = \text{constant} = A.$$

The constant $A$ is given by

$$I_1' w \sin^2\alpha + I_3' n\cos\alpha = A$$

We find $\dot{\phi}$ from the last two equations and put the value in the first equation (in the last equation we used steady state values from example 8). We get

$$I_1'^2\ddot{\theta} - \frac{\cos\theta}{\sin^2\theta}\left[I_1' w\sin^2\alpha + I_3' n\cos\alpha - I_3' n\cos\theta\right]^2 + \frac{I_3' n}{\sin\theta} \times \left[I_1' w\sin^2\alpha + I_3' n\cos\alpha - I_3'\cos\theta\right] = mghI_1'\sin\theta.$$

Put $\theta = \theta_1 + \alpha$ and $\theta_1$ is small and using the steady state result $I_1' w^2\cos\alpha - I_3' n w + mgh = 0$, we have

$$\ddot{\theta}_1 + \Omega\theta_1 = 0,$$

where

$$\Omega_1 = \left[I_1'^2 w^2 - 2I_1' mghw^2\cos\alpha + m^2g^2h^2\right]/I_1'^2 w^2.$$

**Example 10.** *A rod $OA$ of mass $m_1$ and length $2a$, is freely hinged at 0 to a fixed point; at its other end is freely attached another rod $AB$, of mass $m_2$ and length $2b$; the system moves under gravity. Find the Lagrange's equation of motion.*

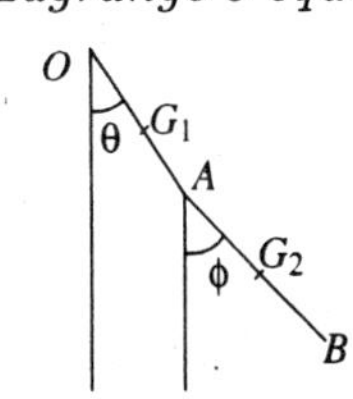

Fig 7.9

***Solution:*** Let $G_1$ and $G_2$ are the C.M. of $OA$ and $AB$ and $\theta$ and $\phi$ are the angles made by them with the vertical. The K.E. of $OA$

$$= \frac{1}{2} m_1 \cdot \frac{4a^2}{3}\dot{\theta}^2.$$

The C.M. $G_2$ is turning around $A$ with velocity $b\dot{\phi}$ while $A$ is turning around $O$ with velocity $2a\dot{\theta}$. Hence the squares of velocity is

$$(2a\dot{\theta}\cos\theta+b\dot{\phi}\cos\phi)^2+(2a\dot{\theta}\sin\theta+b\dot{\phi}\sin\phi)^2 = 4a^2\dot{\theta}^2+b^2\dot{\phi}^2+4ab\dot{\theta}\dot{\phi}\cos(\theta-\phi).$$

Also the K.E. of the rod $AB$ about $G_2$

$$= \frac{1}{2}m_2\frac{b^2}{3}\dot{\phi}^2.$$

Hence the total

$$\text{K.E.} = T = \frac{1}{2}m_1\frac{4a^2}{3}\dot{\theta}^2+\frac{1}{2}m_2\frac{b^2}{3}\dot{\phi}^2+\frac{1}{2}m_2\left[4a^2\dot{\theta}^2+b^2\dot{\phi}^2+4ab\dot{\theta}\dot{\phi}\cos(\theta-\phi)\right].$$

The Potential energy

$$-V = m_1ga\cos\theta + m_2g(2a\cos\theta + b\cos\phi).$$

The Lagrengian $L = T - V$. Hence from

$$\frac{d}{dt}\left(\frac{\partial L}{\partial\dot{\theta}}\right) = \frac{\partial L}{\partial\theta} \quad \text{and} \quad \frac{d}{dt}\left(\frac{\partial L}{\partial\dot{\phi}}\right) = \frac{\partial L}{\partial\phi}$$

we obtain

$$\begin{aligned}&\left(\frac{m_1}{8}+m_2\right)4a\ddot{\theta} + 2m_2b\left[\ddot{\phi}\cos(\theta-\phi) - \dot{\phi}^2\sin(\phi-\theta)\right]\\ = \; & -g(m_1+2m_2)\sin\theta\\ \text{and} \; & \frac{4b}{3}\ddot{\phi} + 2a\ddot{\theta}\cos(\theta-\phi) + 2a\dot{\theta}^2\sin(\phi-\theta) = -g\sin\phi.\end{aligned}$$

**Example 11.** *In example 10, put $a = b$ and assume $\theta = \phi \simeq 0$ and further $\dot{\theta}^2 = \dot{\phi}^2 = 0$. Determine the equations satisfied by $\theta$ and $\phi$ and find the normal modes of small oscillations ($m_1 = m_2$).*

***Solution:*** See example 6 of the exercises at the end of Chapter V.

## Exercises

1. Verify that the product $CB$ is the matrix

$$\begin{pmatrix} \cos\psi & \sin\psi\cos\theta & \sin\phi\sin\theta \\ -\sin\psi & \cos\psi\cos\theta & \cos\psi\sin\theta \\ 0 & -\sin\theta & \cos\theta \end{pmatrix}.$$

2. Show the three elements of the last column in the product $CBA$ are $\sin\psi\sin\theta$, $\cos\psi\sin\theta$ and $\cos\theta$.

3. Find the transformation matrix which will transform the vector $(x', y', z')$ to $(x, y, z)$.

4. A rectangular box of dimension $a \times b \times c$ cm$^3$ is spinning freely with an angular velocity $\omega$. Find the frequency of small oscillation of the axis if the axis of rotation is along the side of length $a$.

5. A uniform charged sphere is spinning freely with angular velocity $\omega$ in a magnetic field **B**. Evaluate the total moment of the magnetic force choosing appropriate axes.

Chapter VIII

# Canonical Transformation

## 8.1 Introduction

In Lagrangian formulation of the equations of motion, the choice of generalised co-ordinates is not subjected to any kind of restriction. For example, we can choose a set

$$(q_1, \cdots, q_n)$$

or any other non-singular set

$$(Q_1, \cdots, Q_n),$$

where

$$q_i = q_i(Q_1, \cdots, Q_n) \tag{8.1}$$

$$\text{or,} \quad Q_i = Q_i(q_1, \cdots, q_n). \tag{8.2}$$

The Lagrangian equations remain form-invariant in either set. Under the transformation (8.1) the Lagrangian becomes

$$L(q_1, \cdots, q_r; \dot{q}_1, \cdots, \dot{q}_r; t) = L'(Q_1, \cdots, Q_n; \dot{Q}_1, \cdots, \dot{Q}_r; t)$$

and

$$\frac{d}{dt}\left(\frac{\partial L'}{\partial \dot{Q}_r}\right) = \frac{d}{dt}\left(\sum_S \frac{\partial L'}{\partial q_s}\frac{\partial q_s}{\partial \dot{Q}_r} + \sum_S \frac{\partial L'}{\partial \dot{q}_s}\frac{\partial \dot{q}_s}{\partial \dot{Q}_r}\right). \tag{8.3}$$

As,

$$q_i = q_i(Q_1, \cdots, Q_n; t),$$

we can easily find

$$\frac{\partial q_i}{\partial Q_r} = \frac{\partial \dot{q}_i}{\partial \dot{Q}_r}.$$

Using this result (8.3) reduces to

$$\frac{d}{dt}\left(\frac{\partial L'}{\partial \dot{Q}_r}\right) = \frac{d}{dt}\left(\sum_S \frac{\partial L}{\partial \dot{q}_s}\cdot \frac{\partial q_s}{\partial Q_r}\right). \tag{8.4}$$

The right-hand side of (8.4) can further be reduced to.

$$\sum_S \frac{d}{dt}\left(\frac{\partial L}{\partial \dot{q}_s}\right)\frac{\partial q_s}{\partial Q_r} + \sum_S \frac{\partial L}{\partial \dot{q}_s}\frac{d}{dt}\left(\frac{\partial q_s}{\partial Q_r}\right)$$

$$= \sum_S \frac{\partial L}{\partial q_s}\cdot\frac{\partial q_s}{\partial Q_r} + \sum \frac{\partial L}{\partial \dot{q}_s}\cdot\frac{\partial \dot{q}_s}{\partial Q_r} = \frac{\partial L}{\partial Q_r} = \frac{\partial L'}{\partial Q_r}.$$

Hence. we find

$$\frac{d}{dt}\left(\frac{\partial L'}{\partial \dot{Q}_r}\right) = \frac{\partial L'}{\partial Q_r}.$$

Thus the Lagrangian equation holds in $Q_r$-variables when it is true in $q_r$-variables. The equation remains form-invariant.

Consider the Hamilton's equation of motion.

$$\dot{p} = -\frac{\partial H(q,p)}{\partial q} \quad \text{and} \quad \dot{q} = \frac{\partial H(q,p)}{\partial p} \tag{8.5}$$

and a transformation (which is not a point transformation) of the following form

$$q \to -P \quad \text{and} \quad p \to Q. \tag{8.6}$$

By (8.6) the canonical conjugate set $(q,p)$ is transformed to another canonical conjugate set $(Q,P)$. We easily see that the $H(q,p)$ becomes under (8.6)

$$H(q,p) \to H'(Q,P) \quad \text{and} \quad \dot{p} = -\frac{\partial H}{\partial q}$$

becomes

$$\dot{Q} = -\frac{\partial H'}{\partial(-P)} \Rightarrow \dot{Q} = \frac{\partial H'(QP)}{\partial P}.$$

Likewise.

$$\dot{q} = \frac{\partial H}{\partial p}$$

transforms to

$$-\dot{P} = \frac{\partial H'(QP)}{\partial Q}, \quad \text{i.e.,} \quad \dot{P} = -\frac{\partial H'(QP)}{\partial Q}.$$

Thus Hamilton's equations of motion retain the same form in new variable as in the original $(q,p)$ co-ordinate. A transformation of the form given in (8.6) is called *canonical.*

We must mention that any arbitrary transformation will not be canonical. There must be some constraint on the choice of such transformation. The detailed discussion follows.

Consider now a general transformation in which

$$q_i \to Q_i \quad \text{and} \quad p_i \to P_i.$$

where $Q_i$ and $P_i$ are in general function of $q_1, \cdots, q_n$ and $p_1, \cdots, p_n$. That is

$$\begin{aligned} q_i &\to Q_i(q_1, \cdots, q_n; p_1, \cdots, p_n; t) \\ \text{and} \quad p_i &\to P_i(q_1, \cdots, q_n; p_1, \cdots, p_n; t). \end{aligned} \tag{8.7}$$

The transformation (8.7) transforms the set of conjugate variables $(q_i, p_i)$ to $(Q_i, P_i)$. Let under (8.7) the Hamiltonian $H(q_i, p_i; t)$ is transformed to $H'(Q_i, P_i; t)$. Suppose the variables $(q_i, p_i)$ satisfy the Hamilton's equation with regard to $H(q_i, p_i; t)$:

$$\dot{q}_i - \frac{\partial H}{\partial p_i} \quad \text{and} \quad \dot{p}_i = \frac{\partial H}{\partial q_i}.$$

Then the transformation (8.7) will be called *canonical* if $H'(Q_i P_i)$ satisfies

$$\dot{Q}_i = \frac{\partial H'}{\partial P_i} \quad \text{and} \quad \dot{P}_i = -\frac{\partial H'}{\partial Q_i}.$$

If

$$\begin{aligned} &L(q_i, \cdots, \dot{q}_i; t) \\ \text{and} \quad &L'(Q_i, \cdots, \dot{Q}_i, \cdots; t) \end{aligned}$$

are the Lagrangian in the original $q_i$ and transformed $Q_i$ variable respectively, then

$$\sum_i p_i \dot{q}_i - H(q_i p_i; t) = L(q_i, \cdots; \dot{q}_i, \cdots; t) \tag{8.8}$$

$$\text{and} \quad \sum_i P_i \dot{Q}_i - H'(Q_i, P_i; t) = L'(Q_i, \cdots, \dot{Q}_i, \cdots; t). \tag{8.9}$$

We know that the Hamilton's equation of motion is derived from

$$\delta \int \left(\sum p\dot{q} - H\right) dt = 0.$$

i.e., from

$$\delta \int L(q, \cdots, \dot{q}; t)\, dt = 0$$

as can be seen from equation (8.8).
Since both $L$ and $L'$ give the same form of equations, the two Lagrangians must be related by a function which is a total time-derivative of a function $F(t)$. Thus,

$$\sum_i p_i \dot{q}_i - H(q_i, p_i; t) - \sum_i P_i \dot{Q}_i + H'(Q_i, P_i; t) = \frac{dF}{dt}. \tag{8.10}$$

where the function $F$ must be quite generally a function of $q_i$, $p_i$ and $Q_i$ and $P_i$. But as $q_i$ and $p_i$ are already related by equation (8.7) therefore $F$ can only be a function of $(q_i, Q_i)$, $(q_i, P_i)$, $(p_i, Q_i)$ or $(p_i, P_i)$. We thus have for $F$, the following

$$F : F_1(q_i, Q_i; t), \quad F_2(q_i, P_i; t), \quad F_3(p_i, Q_i; t) \text{ and } F_4(p_i, P_i; t). \tag{8.11}$$

The $F$'s may also contain time explicitly.
The functions $F_i$; $(i = 1, 2, 3, 4)$ are called the *generators of the canonical transformation.* In the next section, we discuss the properties of the generator functions.

## 8.2 Properties of Generator Functions

Consider the generating function $F_1(q, Q, t)$. We have

$$\sum_i p_i\dot{q}_i - H - \sum_i P_i\dot{Q}_i + H' = \frac{d}{dt}F_1(q_i, Q_i; t). \tag{8.12}$$

The right-hand side of (8.12) is

$$\frac{\partial F_1}{\partial t} + \sum_i \frac{\partial F_1}{\partial q_i}\dot{q}_i + \sum_i \frac{\partial F_1}{\partial Q_i}\dot{Q}_i. \tag{8.13}$$

Substituting (8.13) in (8.12) we get equating the coefficients of $\dot{q}_i$ and $\dot{Q}_i$,

$$p_i = \frac{\partial F_1}{\partial q_i}; \quad P_i = -\frac{\partial F_1}{\partial Q_i} \tag{8.14}$$

and

$$-H + H' = \frac{\partial F_1}{\partial t}. \tag{8.15}$$

Evidently when $F_1$ is known, the transformation equations for $p_i$ and $P_i$ can be obtained. This justifies the name *generator of transformation* for the functions $F_i$'s. We also note the transformed Hamiltonian $H'$ is known when $F_1$ is given. Also from (8.14), we find

$$\frac{\partial p_i}{\partial Q_j} = -\frac{\partial P_j}{\partial q_i}$$

and from (8.15) we get,

$$H' = H + \frac{\partial F_1}{\partial t}. \tag{8.16}$$

We will come back to equation (8.16) later in connection with Hamilton-Jacobi equation.
To determine other generating functions we use Legendre transformation and the equations given in (8.14).

By Legendre transformation we can write

$$F_1(q_i, Q_i; t) - \sum_i Q_i \frac{\partial F_1}{\partial Q_i} = F_2(q_i, P_i; t). \tag{8.17}$$

where

$$\frac{\partial F_1}{\partial Q_i} = -P_i.$$

Therefore from (8.17),

$$F_1(q_i, Q_i; t) + \sum_i Q_i P_i = F_2(q_i, P_i; t).$$

We may also put $F_1$ in terms of $F_2$ as

$$F_1(q_i, Q_i; t) = F_2(q_i, P_i; t) - \sum_i Q_i P_i.$$

Then using

$$\sum_i p_i \dot{q}_i - H - \sum_i P_i \dot{Q}_i + H' = \frac{d}{dt} F_1 = \frac{d}{dt}\left[F_2(q_i, P_i; t) - \sum_i P_i Q_i\right]. \tag{8.18}$$

The right-hand side of (8.18) is

$$\frac{\partial F_2}{\partial t} + \sum_i \frac{\partial F_2}{\partial q_i}\dot{q}_i + \sum_i \frac{\partial F_2}{\partial P_i}\dot{P}_i - \sum_i P_i \dot{Q}_i - \sum_i \dot{P}_i Q_i. \tag{8.19}$$

Substituting (8.19) in (8.18) and equating coefficients from both sides, we get,

$$H' = H + \frac{\partial F_2}{\partial t}; \quad p_i = +\frac{\partial F_2}{\partial q_i}; \quad Q_i = \frac{\partial F_2}{\partial P_i}. \tag{8.20}$$

Hence,

$$\frac{\partial p_i}{\partial P_j} = +\frac{\partial Q_j}{\partial q_i}. \tag{8.21}$$

The generating function $F_3(p_i, Q_i; t)$ can be obtained from $F_1(q_i, Q_i; t)$. We note that

$$p_i = \frac{\partial F_1}{\partial q_i}.$$

Therefore,

$$F_1(q_i, Q_i; t) - \sum_i q_i \frac{\partial F_1}{\partial q_i} = F_3(p_i, Q_i; t).$$

where

$$p_i = \frac{\partial F_1}{\partial q_i}.$$

Then

$$F_1 = F_3(p_i, Q_i; t) + \sum_i q_i p_i. \tag{8.22}$$

Again, applying Legendre transformation twice we get

$$F_1(q_i, Q_i; t) - \sum_i q_i \frac{\partial F_1}{\partial q_i} - \sum_i Q_i \frac{\partial F_1}{\partial Q_i} = F_4(p_i, P_i; t),$$

where

$$\frac{\partial F_1}{\partial q_i} = p_i \quad \text{and} \quad \frac{\partial F_1}{\partial Q_i} = -P_i.$$

Then

$$F_1(q_i, Q_i; t) = F_4(p_i, P_i; t) + \sum_i q_i p_i - \sum_i P_i Q_i. \tag{8.23}$$

From (8.22), we get

$$\left.\begin{aligned} -P_i &= \frac{\partial F_3}{\partial Q_i} \\ -q_i &= \frac{\partial F_3}{\partial p_i} \end{aligned}\right\}. \tag{8.24}$$

Hence,

$$\frac{\partial P_i}{\partial p_j} = \frac{\partial q_j}{\partial Q_i}. \tag{8.25}$$

Similarly from (8.23)

$$\frac{\partial F_4}{\partial p_i} = -q_i \quad \text{and} \quad \frac{\partial F_4}{\partial P_i} = Q_i.$$

Then,

$$\frac{\partial Q_i}{\partial p_j} = -\frac{\partial q_j}{\partial P_i}. \tag{8.26}$$

We thus have the following four important relations between the four conjugate variables.

$$\left.\begin{aligned} &\frac{\partial Q_i}{\partial p_j} = -\frac{\partial q_j}{\partial P_i}; && \frac{\partial P_i}{\partial p_j} = \frac{\partial q_j}{\partial Q_i}; \\ &\frac{\partial Q_j}{\partial q_i} = +\frac{\partial p_i}{\partial P_j} \quad \text{and} && \frac{\partial p_i}{\partial Q_j} = -\frac{\partial P_j}{\partial q_i} \end{aligned}\right\}. \tag{8.27}$$

## 8.3 Invariance of Poisson Bracket under Canonical Transformation

Let the pair $(q_i, p_i)$ is transformed to $(Q_i, P_i)$ by a canonical transformation. To prove that the fundamental Poisson Bracket $[Q_i, P_i]$ is invariant under the canonical transformation $(qp) \to (QP)$ we must show that

$$[Q_i, P_j]_{QP} = [Q_i, P_j]_{qp}.$$

i.e., P.B. calculated using both the basis must be same.
Obviously,

$$[Q_i, P_j]_{QP} = \sum_r \left( \frac{\partial Q_i}{\partial Q_r}\frac{\partial P_j}{\partial P_r} - \frac{\partial Q_i}{\partial P_r}\frac{\partial P_j}{\partial Q_j} \right) = \delta_{ij}.$$

Hence, we must show that $[Q_i, P_j]_{qp}$ is also equal to $\delta_{ij}$, we have

$$\begin{aligned}
[Q_i, P_j]_{qp} &= \sum_k \left( \frac{\partial Q_i}{\partial q_k}\cdot\frac{\partial P_j}{\partial p_k} - \frac{\partial Q_i}{\partial p_k}\cdot\frac{\partial P_j}{\partial q_k} \right) \\
&= \sum_k \left[ \frac{\partial Q_i}{\partial q_k}\cdot\frac{\partial q_k}{\partial Q_j} - \frac{\partial Q_i}{\partial p_k}\left( -\frac{\partial p_k}{\partial Q_j} \right) \right] \\
&= \sum_k \left( \frac{\partial Q_i}{\partial q_k}\cdot\frac{\partial q_k}{\partial Q_j} + \frac{\partial Q_i}{\partial p_k}\cdot\frac{\partial p_k}{\partial Q_j} \right) = \frac{\partial Q_i}{\partial Q_j} = \delta_{ij}.
\end{aligned} \tag{8.28}$$

In proving (8.28), we used the results given in equation (8.27).
In a similar way we can show that

$$[q_i, p_j]_{qp} = [q_i, p_j]_{QP} = \delta_{ij}.$$

Equation (8.28) proves the invariance of fundamental Poisson Bracket under canonical transformation.
In general, for any two dynamical variable $X$ and $Y$, we show that

$$[X, Y]_{QP} = [X, Y]_{qp}. \tag{8.29}$$

Consider $[X, Y]_{QP}$ which is by definition equal to

$$\begin{aligned}
&\sum_i \left( \frac{\partial X}{\partial Q_i}\cdot\frac{\partial Y}{\partial P_i} - \frac{\partial X}{\partial P_i}\cdot\frac{\partial Y}{\partial Q_i} \right) \\
&= \sum_i\sum_j \left[ \frac{\partial X}{\partial Q_i}\left( \frac{\partial Y}{\partial q_j}\cdot\frac{\partial q_i}{\partial P_i} + \frac{\partial Y}{\partial p_j}\cdot\frac{\partial p_j}{\partial P_i} \right) - \frac{\partial X}{\partial P_i}\left( \frac{\partial Y}{\partial q_j}\cdot\frac{\partial q_j}{\partial Q_i} + \frac{\partial Y}{\partial p_j}\cdot\frac{\partial p_j}{\partial Q_i} \right) \right] \\
&= \sum_i\sum_j \left[ \frac{\partial Y}{\partial q_j}\left( \frac{\partial X}{\partial Q_i}\cdot\frac{\partial q_j}{\partial P_i} - \frac{\partial X}{\partial P_i}\cdot\frac{\partial q_j}{\partial Q_i} \right) + \frac{\partial Y}{\partial p_j}\left( \frac{\partial X}{\partial Q_i}\cdot\frac{\partial p_j}{\partial P_i} - \frac{\partial X}{\partial P}\cdot\frac{\partial p_j}{\partial Q_i} \right) \right].
\end{aligned}$$

Using (8.27) we can write the above expression as

$$\sum_j \sum_i \left[ -\frac{\partial Y}{\partial q_j} \left( \frac{\partial X}{\partial Q_i} \cdot \frac{\partial Q_i}{\partial p_j} + \frac{\partial X}{\partial P_i} \cdot \frac{\partial P_i}{\partial p_j} \right) + \frac{\partial Y}{\partial p_j} \left( \frac{\partial X}{\partial Q_i} \cdot \frac{\partial Q_i}{\partial q_j} + \frac{\partial X}{\partial P_i} \cdot \frac{\partial P_i}{\partial q_j} \right) \right]$$
$$= \sum_j \left( -\frac{\partial Y}{\partial q_j} \cdot \frac{\partial X}{\partial p_j} + \frac{\partial Y}{\partial p_j} \cdot \frac{\partial X}{\partial q_j} \right) = [X, Y]_{qp}.$$

Hence, we have showed

$$[X, Y]_{QP} = [X, Y]_{qp}.$$

In other words, the Poisson Bracket remain invariant under canonical transformation. One may also define canonical transformations are those transformations under which Poisson Brackets are invariant.
As a simple illustration we consider the following example.

**Example 1.** *Under what condition the transformation*

$$P = ap + bq; \quad Q = cp + dq$$

*would be canonical.*

***Solution:*** It would be canonical if

$$[Q, P] = 1$$

when it is given that

$$[q, p]_{PB} = 1.$$

To see this we calculate the Poisson Bracket of $[Q, P]$. Then.

$$[Q, P] = [cp + dq; ap + bq] = cb[p, q] + da[q, p] = ad - bc.$$

For $[Q, P]$ to be unity we must have $ad - bc = 1$.

**Example 2.** *To show that the following is a canonical transformation.*

$$Q_1 = q_1^2 + p_1^2; \quad Q_2 = \frac{1}{2}(q_1^2 + q_2^2 + p_1^2 + p_2^2)$$
$$P_1 = \frac{1}{2}\left( -\tan^{-1}\frac{q_1}{p_1} + \tan^{-1}\frac{q_2}{p_2} \right); \quad P_2 = -\tan^{-1}\frac{q_2}{p_2}.$$

***Solution:*** We find.

$$[Q_1, P_1] = \frac{1}{2}\left[ q_1^2 + p_1^2, \ -\tan^{-1}\frac{q_1}{p_1} + \tan^{-1}\frac{q_2}{p_2} \right]$$
$$= \frac{-1}{2}\left[ q_1^2 + p_1^2, \ \tan^{-1}\frac{q_1}{p_1} \right]$$

as

$$[q_1, q_2] = 0 = [q_1, p_2] = \text{ etc.}$$

Thus,

$$[Q_1, P_1] = -q_1\left[q_1, \tan^{-1}\frac{q_1}{p_1}\right] - p_1\left[p_1, \tan^{-1}\frac{q_1}{p_1}\right].$$

Thus,

$$[Q_1, P_1] = -q_1\frac{\partial}{\partial p_1}\tan^{-1}\frac{q_1}{p_1} + p_1\frac{\partial}{\partial q_1}\tan^{-1}\frac{q_1}{p_1} = 1.$$

We can calculate to see $[Q_1, P_2] = 0$ etc. and $[Q_2, P_2] - 1$. Thus, the transformation is canonical.

**Example 3.** *For what value of $\alpha$ and $\beta$, the transformation $Q = q^\alpha \cos\beta p$ and $P = q^\alpha \sin\beta p$ is canonical?*

***Solution:*** We calculate the P.B. namely

$$[Q, P] = \frac{\partial Q}{\partial q}\frac{\partial P}{\partial p} - \frac{\partial Q}{\partial p}\frac{\partial P}{\partial q}.$$

Then

$$[Q, P] = \alpha\beta\cdot q^{2\alpha-1}(\cos^2\beta p + \sin^2\beta p) = \alpha\beta q^{2\alpha-1}.$$

This must be unity for $Q, P$ to be canonical conjugate. Then

$$2\alpha - 1 = 0 \quad \text{and} \quad \alpha\beta = 1$$

then $\alpha$ must be $\frac{1}{2}$ and $\beta = 2$.

## 8.4 Application to Harmonic Oscillator: Some Other Examples

In this section, we show a simple application of the use of canonical transformation to find solutions for dynamical problems. It will be shown later, how Hamilton-Jacobi equation follows by using canonical transformation. Hamilton-Jacobi equation is an important development in classical mechanics and it is sometimes called the *fore-runner* of Schrödinger equation in Quantum mechanics.

The Hamiltonian for the Harmonic oscillator is $H = T + V$ when

$$T = \frac{p^2}{2m} \quad \text{and} \quad V = \frac{1}{2}\mu q^2$$

in one-dimension; $m$ = mass and $\mu$ is the force constant; $p$ = momentum in one dimension.

Consider a generating function $F_1(q, Q)$, where $t$ does not occur explicitly. $F_1$ is given by,

$$F_1 = kq^2 \cot Q; \ : k \text{ is a constant.} \tag{8.30}$$

From (8.30), we find

$$p = \frac{\partial F_1}{\partial q} = 2kq \cot Q \tag{8.31}$$

$$\text{and} \quad P = -\frac{\partial F_1}{\partial Q} = kq^2 \operatorname{cosec}^2 Q. \tag{8.32}$$

From (8.31) and (8.32), we get

$$q = \sqrt{\frac{P}{k}} \sin Q \tag{8.33}$$

$$\text{and} \quad p = 2kq \cot Q = 2\sqrt{Pk} \cdot \cos Q. \tag{8.34}$$

The Hamiltonian $H(q, p)$ can be written in terms of $Q$ and $P$ as

$$H(q, p) = \frac{p^2}{2m} + \frac{1}{2}\mu q^2 = P\left(\frac{2k}{m}\cos^2 Q + \frac{\mu}{2k}\sin^2 Q\right).$$

where we have used (8.31) and (8.32). Thus the Hamiltonian in terms of new canonical variable $P, Q$ becomes

$$H = \frac{\mu P}{2k},$$

where we have used

$$\frac{2k}{m} = \frac{\mu}{2k}, \tag{8.35}$$

which fixes the constant $k$ of the generator function. From (8.35) we find $H$ is cyclic in $Q$, hence, the conjugate momentum $P$ is a constant.

$$P = \beta \text{ (a constant)}.$$

Again, from (8.35)

$$\dot{Q} = \frac{\partial H}{\partial P} = \frac{2k}{m}.$$

Hence, integrating,

$$Q = \frac{2k}{m}t + \alpha, \quad \alpha = \text{constant}.$$

But

$$q = \sqrt{\frac{P}{k}} \sin Q.$$

Therefore,

$$q = \sqrt{\frac{\beta}{k}} \sin\left(\frac{2k}{m}t + \alpha\right) = A\sin(\omega t + \delta). \tag{8.36}$$

where

$$A = \sqrt{\frac{\beta}{k}}, \quad \omega = \frac{2k}{m}, \quad \delta = \text{ a constant}.$$

Equation (8.36) is the desired solution of Harmonic oscillator problem.

In the last example, we have showed how to solve a dynamical problem by using canonical transformation. Below we give well-known examples of canonical transformation, given the generator.

Let a generator $F$ be given by

$$F = \sum_i q_i P_i.$$

Then we have

$$F_2(qP) = \sum_i q_i P_i.$$

Therefore,

$$p_i = \frac{\partial F_2}{\partial q_i} \quad \text{and} \quad Q_i = \frac{\partial F_2}{\partial P_i} = q_i.$$

i.e., $q_i \to Q_i$ and $p_i \to P_i$. This is an Identity transformation. Consider

$$F = -\sum_i q_i P_i.$$

In this case, we have

$$p_i \to -P_i \quad \text{and} \quad q_i \to -Q_i.$$

This is a reflection transformation. Let $a_{ij}$ be an orthogonal matrix and a generating function $F_2$ is

$$F_2 = \sum_i \sum_k a_{ik} q_k P_i.$$

Hence,

$$p_j = \frac{\partial F_2}{\partial q_j} = \sum_i a_{ij} P_i.$$

Since $a_{jl}$ is orthogonal we have,

$$P_l = \sum_j p_j a_{jl} \quad \text{and} \quad Q_l = \frac{\partial F_2}{\partial P_l} = \sum_k a_{lk} q_k.$$

*Perfect differential as a condition for a transformation to be canonical.*

We have from the generating function $F_1(q, Q)$ (assume $F_1$ does not depend on time explicitly)

$$p = \frac{\partial F_1}{\partial q} \quad \text{and} \quad P = -\frac{\partial F_1}{\partial Q}.$$

Since $F_1 = F_1(q, Q)$,

$$dF_1 = \frac{\partial F_1}{\partial q} dq + \frac{\partial F_1}{\partial Q} dQ = pdq - PdQ.$$

In, other words the quantity $pdq - PdQ$ is a perfect differential. Hence a transformation $(q, p) \to (Q, P)$ would be *canonical* if $pdq - PdQ$ is a perfect differential. Let us, check this by direct verification.

**Example 1.** *We show that the transformation*

$$Q = \log\left(\frac{1}{q}\sin p\right) \quad \textit{and} \quad P = q\cot p$$

*is canonical by calculating $PdQ - pdq$.*

***Solution:*** We have

$$PdQ = (q\cot p)\left(-\frac{dq}{q} + \cot pdp\right).$$

Hence

$$\begin{aligned} PdQ - pdq &= (-\cot pdq + q\cot^2 pdp) - pdq \\ &= -\cot pdq + q\,\text{cosec}^2 pdp - qdp - pdq \\ &= d(-\cot p \cdot q) - d(pq) = d(-q\cot p - pq). \end{aligned}$$

Since,

$$PdQ - pdq = -dF_1,$$

so we get

$$F_1 = +q\cot p + pq.$$

The generating function

$$F_3(p, Q) = F_1 - pq = +q\cot p.$$

From

$$Q = \log\left(\frac{1}{q}\cdot\sin p\right)$$

we get

$$q = \sin pe^{-Q}.$$

Hence

$$F_3 = +q\cot p = +\cos pe^{-Q} = (\cos p)e^{-Q}.$$

**Example 2.** *For the time-dependent canonical transformation*

$$Q = p\tan t,$$
$$P = -q\cot\cdot t.$$

*Find $F_1$ and $H'(Q, P)$.*

***Solution:*** From the generating function $F_1(q, Q; t)$, we get,

$$P = -\frac{\partial F_1}{\partial Q} = -q\cot t, \quad \text{i.e.,} \quad \frac{\partial F_1}{\partial Q} = q\cot t \tag{8.37}$$

and

$$p = \frac{\partial F_1}{\partial q} = Q\cot t,$$

where we used

$$Q = p\tan t. \tag{8.38}$$

We can now solve, (8.36) and (8.37) then

$$F_1(q, Q; t) = qQ\cdot\cot t. \tag{8.39}$$

Hence,

$$\frac{\partial F_1}{\partial t} = -qQ\,\mathrm{cosec}^2 t = PQ\tan t\,\mathrm{cosec}^2 t = \frac{2PQ}{\sin 2t}.$$

Therefore,

$$H'(P, Q; t) = H(p, q) + \frac{\partial F_1}{\partial t} = H(Q\cot t, -P\tan t) + \frac{2PQ}{\sin 2t} \tag{8.40}$$

For the time-independent transformation, given by the following:

$$Q = \log(1 + \sqrt{q}\cos p);$$
$$\text{and} \quad P = 2(1 + \sqrt{q}\cos p)\sqrt{q}\sin p.$$

we find for $PdQ - pdq$ the expression

$$\frac{1}{2}d(q\sin 2p) - d(pq).$$

Therefore, the function $F_3$ is

$$-F_3 = \frac{1}{2}q\sin 2p$$

which reduces to $\tan p(e^Q - 1)^2$ by using

$$Q = \log(1 + \sqrt{q}\cos p).$$

We conclude this section by calculating the generating function for a transformation

$$(q_i, p_i) \to (Q_i, P_i); \quad i = 1, 2.$$

The explicit form of the transformation is given by the example.

$$\begin{aligned} Q_1 &= (q_1^2 + p_1^2); \quad Q_2 = \frac{1}{2}(q_1^2 + q_2^2 + p_1^2 + p_2^2) \\ P_1 &= \frac{1}{2}\left(\tan^{-1}\frac{q_2}{p_2} - \tan^{-1}\frac{q_1}{p_1}\right); \quad P_2 = -\tan^{-1}\frac{q_2}{p_2}. \end{aligned}$$

If we use

$$\begin{aligned} p_i &= r_i \cos\theta_i; \quad i = 1, 2 \\ \text{and} \quad q_i &= r_i \sin\theta_i; \quad i = 1, 2 \end{aligned}$$

then

$$\begin{aligned} Q_1 &= r_1^2, \quad Q_2 = \frac{1}{2}(r_1^2 + r_2^2), \\ P_1 &= \frac{1}{2}(\theta_2 - \theta_1), \quad P_2 = -\theta_2. \end{aligned}$$

Then

$$P_1 dQ_1 + P_2 dQ_2 - p_1 dq_1 - p_2 dq_2$$

becomes

$$\begin{aligned} &\frac{1}{2}(\theta_2 - \theta_1)\cdot 2r_1 dr_1 - \theta_2(r_1 dr_1 + r_2 dr_2) \\ &\quad -(r_1\cos\theta_1) d(r_1 \sin\theta_1) - (r_2\cos\theta_2)\, d(r_2\sin\theta_2). \end{aligned}$$

On simplification this reduces to

$$\begin{aligned} &-(\theta_1 r_1 dr_1 + \theta_2 r_2 dr_2) - \cos\theta_1 \sin\theta_1 r_1 dr_1 - r_1^2\cos^2\theta_1 d\theta_1 \\ &\quad - \cos\theta_2 \sin\theta_2 r_2 dr_2 - r_2^2 \cos^2\theta_2 d\theta_2. \end{aligned}$$

This we can rewrite as

$$\begin{aligned} &-\frac{1}{2}d(r_1^2\theta_1 + r_2^2\theta_2) - \frac{1}{2}d(r_1^2 \sin 2\theta_1 + r_2^2 \sin 2\theta_2) \\ &= -\frac{1}{2}d(r_1^2\theta_1 + r_1^2\sin 2\theta_1 + r_2^2\theta_2 + \sin 2\theta_1). \end{aligned}$$

This is a perfect differential. Hence the transformation is canonical and the generating function can be written as

$$-\frac{1}{2}\left[(p_1^2 + q_1^2)\cdot \tan^{-1}\frac{q_1}{p_1} + (q_2^2 + p_2^2)\tan^{-1}\frac{q_2}{p_2} + 2p_1 q_1 + 2p_2 q_2\right].$$

We can use this expression to obtain any desired generating function.

## 8.5 Infinitesimal Contact Transformation

In this section, we discuss another aspect of Poisson Bracket which is revealed by the concept of infinitesimal canonical transformation, also known as *infinitesimal contact transformation*. By infinitesimal transformation we mean the transformation introduced very near the identity transformation. In the last section we took the example of a generating function $F_2(q, P)$, where $F_2$ is given by

$$F_2(q, P) = \sum_i q_i P_i. \tag{8.41}$$

From this we find

$$p_i = \frac{\partial F_2}{\partial q_i} = P_i \quad \text{and} \quad Q_i = \frac{\partial F_2}{\partial P_i} = q_i. \tag{8.42}$$

That is, we have the identity transformation in which

$$Q_i = q_i \quad \text{and} \quad P_i = p_i. \tag{8.43}$$

To find a transformation near the identity transformation we add to $F_2$ a function $G(qP)$ such that,

$$F_2(q_i, P_i) = \sum_i q_i P_i + \varepsilon G(q_i, P_i). \tag{8.44}$$

where $\varepsilon$ is an infinitesimal quantity. From (8.44) we have

$$p_i = P_i + \varepsilon \frac{\partial G}{\partial q_i} \quad \text{and} \quad Q_i = q_i + \varepsilon \frac{\partial G}{\partial P_i}. \tag{8.45}$$

We thus get from (8.45),

$$P_i - p_i = \delta p_i = -\varepsilon \frac{\partial G}{\partial q_i} \quad \text{and} \quad Q_i - q_i = \delta q_i = \varepsilon \frac{\partial G}{\partial P_i}. \tag{8.46}$$

If we keep term correct to first order in $\varepsilon$ only then

$$\delta p_i = -\varepsilon \frac{\partial G}{\partial q_i} \quad \text{and} \quad \delta q_i = \varepsilon \frac{\partial G}{\partial p_i}. \tag{8.47}$$

The function $G(p, q)$ is called *infinitesimal generator of the canonical transformation*. Let us consider the change in the Hamiltonian

$$H = H(q, p)$$

in one-dimension under the infinitesimal transformation given in equation (8.47), we have,

$$\delta H = \frac{\partial H}{\partial p}\delta p + \frac{\partial H}{\partial q}\delta q.$$

Using (8.46), $\delta H$ becomes

$$\frac{\partial H}{\partial p}\left(-\varepsilon\frac{\partial G}{\partial q}\right)+\frac{\partial H}{\partial q}\varepsilon\frac{\partial G}{\partial p}.$$

That is, we have

$$\delta H=\varepsilon\left(\frac{\partial H}{\partial q}\frac{\partial G}{\partial p}-\frac{\partial H}{\partial p}\frac{\partial G}{\partial q}\right)=\varepsilon[H,G]_{P.B}.$$

*Thus, if $G$ is the infinitesimal generator of the canonical transformation, then the change in Hamiltonian is given by the Poisson Bracket of $H$ with $G$.*

If the Hamiltonian is invariant under the canonical transformation, then the generator $G$ is a constant of motion if $G$ does not depend on time explicitly. This is easily seen that, if $\delta H=0$, then $[H,G]=0$ which implies $G$ is a constant of motion.

We know that under the infinitesimal time translation $t\to t+\delta t$, the Hamiltonian remains invariant then

$$\frac{\partial H}{\partial t}=0.$$

But, since

$$\frac{dH}{dt}=\frac{\partial H}{\partial t},$$

we find the energy is a constant of motion. Thus, *Energy is the generator of the infinitesimal time-translation.* That is Hamiltonian $H$ itself is the generator for time-translation. Let us now consider *space-translation* namely $q\to q+\varepsilon$, where $\varepsilon$ is infinitesimal. Therefore, the change in $q$ given by $\delta q$ is equal to $\varepsilon$. But

$$\delta q=\varepsilon\frac{\partial G}{\partial p}.$$

Then

$$\varepsilon=\varepsilon\frac{\partial G}{\partial p}, \quad \text{or,} \quad \frac{\partial G}{\partial p}=1. \tag{8.48}$$

Under *space-translation* $p$ does not change, so $\delta p=0$, which implies

$$0=\varepsilon\frac{\partial G}{\partial q}, \quad \text{or,} \quad \frac{\partial G}{\partial q}=0. \tag{8.49}$$

From (8.48) and (8.49) we then get $G=p$, the momentum of the system. Thus, for infinitesimal space translation $q\to q+\varepsilon$ the generator of the transformation is $p$, the momentum. If $H$ is invariant under the infinitesimal space translation then the generator is a constant of motion. Thus $p$ is a constant of motion if $H$ is invariant under translation.

We now come to the invariance of Hamiltonian under infinitesimal three-dimensional rotation.

If $\delta\phi$ is an infinitesimal amount of rotation then the vector $\mathbf{r}$ changes to $\mathbf{r} + \delta\mathbf{r}$, where

$$\delta\mathbf{r} = (\delta\bar{\phi} \times \mathbf{r}). \tag{8.50}$$

Let

$$\delta\bar{\phi} = (0, 0, \delta\phi) \quad \text{and} \quad \mathbf{r} = (x, y, z).$$

Then, from (8.50),

$$\delta x = -y\delta\phi; \quad \delta y = x\delta\phi; \quad \delta z = 0.$$

But,

$$\delta x = \varepsilon \frac{\partial G}{\partial p_x} = -y\delta\phi = -y\varepsilon.$$

Then

$$\frac{\partial G}{\partial p_x} = -y. \tag{8.51}$$

Again,

$$\delta y = \varepsilon \frac{\partial G}{\partial p_y} = x\delta\phi = x\varepsilon.$$

Then

$$\frac{\partial G}{\partial p_y} = x. \tag{8.52}$$

Solving (8.51) and (8.52), we get,

$$G = xp_y - yp_x. \tag{8.53}$$

Equation (8.53) shows that $G = L_z$, the 3rd component of angular momentum. So, for infinitesimal three-dimensional rotation the infinitesimal generator is the angular momentum, and hence if Hamiltonian is invariant under infinitesimal rotation, the angular momentum is conserved.

These well-known symmetry properties and the associated conservation laws were also discussed previously in the text. Here we have discussed these in the framework of Poisson Bracket formalism. This shows that these results would be valid in the context of quantum theory also.

## 8.6 Liouville's Theorem: Invariance of Phase-space Volume under Canonical Transformations

In Hamiltonian dynamics, the pair of independent variables $q$, the generalised co-ordinate and its canonical conjugate momentum $p$ completely characterize the state of a system. One introduces the concept of a phase-space by taking $q$ and $p$ as a rectangular axes system and a point in this space would specify the state of the system at a given instant of time. Let $q_1, \cdots, q_n$ and $p_1, \cdots, p_n$ be a set of generalized co-ordinates and their corresponding conjugate momentum respectively characterize a given physical system, then the corresponding phase space is $2n$-dimensional space with $q_1, \cdots, q_n$ and $p_1, \cdots, p_n$ denoting the $2n$-number axes system. The elementary volume element in this space is given by

$$dq_1 \cdots dq_n \cdots dp_1 \cdots dp_n = d^{2n}\mathbf{V}.$$

Then the Liouville's theorem states that under canonical transformation this elementary volume is an invariant. If a canonical transformation carries the set $(q_i p_i)$ to $(Q_i P_i)$: $(i = 1, \cdots, n)$ then

$$dq_1 \cdots dq_n dp_1 \cdots dp_n = dQ_1 \cdots dQ_n dP_1 \cdots dP_n.$$

We know the change in the volume-element in two different bases $(q_i, p_i)$ and $(Q_i P_i)$ is related by means of the Jacobian $D$ which is given by the determinant

$$D = \frac{\partial(Q_1, \cdots, Q_n; P_1, \cdots, P_n)}{\partial(q_1, \cdots, q_n; p_1, \cdots, p_n)}. \tag{8.54}$$

To prove the invariance of phase-space volume we must show that the Jacobian $D$ is unity under the canonical transformation. From the property of the Jacobian we can rewrite (8.54) as

$$D = \frac{\partial(Q_1, \cdots, Q_n; P_1, \cdots, P_n)}{\partial(q_1, \cdots, q_n; P_1, \cdots, P_n)} \Bigg/ \frac{\partial(q_1, \cdots, q_n; p_1, \cdots, p_n)}{\partial(q_1, \cdots, q_n; P_1, \cdots, P_n)}. \tag{8.55}$$

In obtaining (8.55) we have divided the numerator and denominator by the same-quantity. The value of the determinant, $D$ does not change.
We now use that

$$\frac{\partial(Q_1, \cdots, Q_n; P_1, \cdots, P_n)}{\partial(q_1, \cdots, q_n; P_1, \cdots, P_n)} = \frac{\partial(Q_1, \cdots, Q_n)}{\partial(q_1, \cdots, q_n)}. \tag{8.56}$$

Therefore

$$D = \frac{\partial(Q_1, \cdots, Q_n)}{\partial(q_1, \cdots, q_n)} \Bigg/ \frac{\partial(p_1, \cdots, p_n)}{\partial(P_1, \cdots, P_n)} = \frac{D_1}{D_2}. \tag{8.57}$$

$D_1$ contains elements like $\partial Q_i/\partial q_j$ while $D_2$ contains elements like $\partial p_i/\partial P_j$; but by (8.27) we have

$$\frac{\partial p_i}{\partial P_j} = \frac{\partial Q_j}{\partial q_i}. \tag{8.58}$$

Equation (8.58) shows that elements of $D_2$ are the same as those occurring in $D_1$ except that row and column interchanged. But the value of the determinant is unchanged by row-column interchange. Hence $D_1 = D_2$. Therefore, $D = 1$. We then find

$$dQ_1 \cdots dQ_n dP_1 \cdots dP_n = dq_1 \cdots dq_n dp_1 \cdots dp_n.$$

This proves the invariance of phase space volume.
In the next chapter, we will consider Hamilton-Jacobi theory of equation of motion which is another aspect of canonical transformation theory results.

## 8.7 Canonical and Gauge Transformation

In this section, we show that if a Hamiltonian is invariant under a suitable canonical transformation, then we are forced to introduce the existence of electromagnetic field which transforms in a definite manner consistent with gauge transformation of second kind as in classical electrodynamics. This way of generating the electromagnetic field in classical mechanics via the use of canonical transformation is similar to that one uses in quantum theory. In quantum theory, one introduces the electromagnetic field by demanding that the Schrödinger equation

$$i\hbar\frac{\partial\psi}{\partial t} = H\psi$$

should remain invariant under the *gauge* (phase) transformation

$$\psi \rightarrow e^{1\alpha}\cdot\psi,$$

where the phase-function $\alpha$ is a real function of $(\mathbf{x}, t)$.
We will follow the same method in the present classical situation by demanding the invariance of Hamiltonian function.
Let us consider a free particle. The Hamilton's equation in classical mechanics are

$$\dot{p}_i = -\frac{\partial H_0}{\partial q_i} \quad \text{and} \quad \dot{q}_i = \frac{\partial H_0}{\partial p_i} \tag{8.59}$$

$$\text{and} \quad H_0 = \frac{1}{2m}\sum_i p_i^2. \tag{8.60}$$

Let us now calculate the canonical transformation generated by the generating function $F_3(p_i, Q_i; t)$. The canonical transformation we consider is

$$Q_i = Q_i(q_i, p_i; t) \quad \text{and} \quad P_i = P_i(q_i, p_i; t). \tag{8.61}$$

The generating function $F_3$ is

$$F_3(p_i, Q_i; t) = -e\alpha(Q_i; t) - \sum_j Q_j p_j. \tag{8.62}$$

Then using (8.24) we have for the transformation and the Hamiltonian $H_0'$ of the new system the following:

$$P_i = -\frac{\partial F_3}{\partial Q_i} = e\frac{\partial \alpha}{\partial Q_i} + p_i \tag{8.63}$$

$$q_i = -\frac{\partial F_3}{\partial p_i} = Q_i \tag{8.64}$$

$$\text{and} \quad H_0' = H_0 + \frac{\partial F_3}{\partial t} = H_0 - e\frac{\partial \alpha}{\partial t}. \tag{8.65}$$

Hence,

$$H_0' = \frac{1}{2m}\sum_i \left(P_i - e\frac{\partial \alpha}{\partial Q_i}\right)^2 - e\frac{\partial \alpha}{\partial t}. \tag{8.66}$$

To which one associates the equation of motion

$$\dot{P}_i = -\frac{\partial H_0'}{\partial Q_i} \quad \text{and} \quad \dot{Q}_i = \frac{\partial H_0'}{\partial P_i}. \tag{8.67}$$

Comparison between equations (8.60) and (8.66) shows that the free Hamiltonian is not invariant under the canonical formation (8.63) and (8.64) generated by

$$F_3 = -e\alpha(Q_i t) - \sum p_i Q_i.$$

We consider this transformation as a gauge transformation.

To achieve invariance of the Hamiltonian $H_0$ we must introduce a 4-vector field $A_i$ and $A_0(q, t)$ so that the $H_0$ becomes

$$H_0 = \frac{1}{2m}\sum_i (p_i - eA_i)^2 - eA_0.$$

The canonical transformation generated by (8.63) and (8.64) will yield for $H_0'$ the following

$$H_0' = \frac{1}{2m}\sum_i \left(P_i - e\frac{\partial \alpha}{\partial Q} - eA_i\right)^2 - e\frac{\partial \alpha}{\partial t} - eA_0. \tag{8.68}$$

Then writing

$$\frac{\partial \alpha}{\partial Q_i} + A_i = A_i'(Q_i t) \quad \text{and} \quad \frac{\partial \alpha}{\partial t} + A_0 = A_0'(Q_i t). \tag{8.69}$$

we get from (8.68)

$$H_0' = \frac{1}{2m}\sum_i (P_i - eA_i')^2 - eA_0'. \tag{8.70}$$

Hence, we get the formal invariance of $H_0$ on the condition that there must exist four vector potentials $A\mu$ and that $A\mu$ must transform as given by

(8.69). The four vector potential $A\mu$ are called *gauge-potential* and the equations (8.62) and (8.63) are called the *gauge transformation*. Invariance of $H_0$ under these gauge transformation forces us to induce the presence of the $A_\mu$ fields. The transformation

$$A_i' = A_i + \frac{\partial\alpha}{\partial Q_i} \quad \text{and} \quad A_0' = A_0 + \frac{\partial\alpha}{\partial t}$$

are known as *gauge transformation of second kind*.
The Hamiltonian (8.70) describes the motion of a charged classical particle in the presence of electro-magnetic field. Thus the Hamiltonian theory in classical mechanics is very similar to the quantum theory. One simply substitutes canonical transformation for gauge transformation.

## Exercises

1. Is the following a canonical transformation?

$$\begin{aligned} P &= q\tan\theta \\ Q &= p\cot\theta. \end{aligned}$$

2. Prove that the following transformation is canonical.

$$\begin{aligned} Q &= \sqrt{2}\sqrt{q+\alpha}\cos p \\ P &= \sqrt{2}\sqrt{q+\alpha}\sin p. \end{aligned}$$

   Find the corresponding generating function.

3. Determine whether the following is a canonical transformation:

$$\begin{aligned} p_1 &= \frac{1}{2}\sqrt{2Q_1}\sin P_1 + \frac{1}{2}\sqrt{2Q_2}\sin P_2 \\ q_1 &= \sqrt{2Q_1}\cos P_1 + \sqrt{2Q_2}\cos P_2 \\ p_2 &= -\frac{1}{2}\sqrt{2Q_1}\sin P_1 + \frac{1}{2}\sqrt{2Q_2}\sin P_2 \\ q_2 &= -\sqrt{2Q_1}\cos P_1 + \sqrt{2Q_2}\cos P_2. \end{aligned}$$

4. Find the generating functions for the following canonical transformations:

   (a) $Q = \log\left(\frac{1}{q}\sin p\right)$ $\quad$ (b) $q = \sqrt{2P}\sin Q$

   $P = q\cot\cdot p.$ $\quad$ $p = \sqrt{2P}\cos Q$

5. Find the infinitesimal generators of the canonical transformation

$$Q = q\cos\theta + p\sin\theta$$
$$P = -q\sin\theta + p\cos\theta.$$

6. Show that

$$[p_k, X]_{QP} = -\frac{\partial X}{\partial q_k},$$

where $X$ is a dynamical quantity.

7. Calculate $p\,dq - P\,dQ$ for the transformation

$$Q = \log(1 + \sqrt{q}\cos p)$$
$$\text{and} \quad P = 2(1 + \sqrt{q}\cos p)q^{1/2}\sin p.$$

Show that it is canonical and find the generating function.

[*Hints.*

$$p\,dq - P\,dQ = p\,dq - 2\sqrt{q}\sin p(1 + \sqrt{q}\cos p) \times \frac{-\sqrt{q}\sin p\,dp + \frac{1}{2}q^{-1/2}\cos p\,dq}{1 + \sqrt{q}\cos p}$$
$$= (p\,dq + q\,dp) - q\,dp - \left(-q2\sin^2 p\,dp + \frac{1}{2}\sin 2p\,dq\right)$$
$$= d(pq) - d\left[\frac{1}{2}(\sin 2p)q\right].$$

Hence

$$F_1 = pq - \frac{\sin 2p}{2}\cdot q.$$
$$F_3 = F_1 - pq = -\frac{\sin 2p}{2}\cdot q = -\sin p\cos p(e^Q - 1)^2/\cos^2 p$$
$$= -\tan p(e^Q - 1)^2.]$$

8. For the above transformation, show that $[Q, P]_{qp} = 1$.

9. Show that equations of transformation in which time $t$ is considered a canonical variable reduce to the ordinary equation (8.20) if the transformation does not affect time-scale.

Chapter IX

# Hamilton-Jacobi Equation; Action Angle Variables and Hannay Angle

## 9.1 Hamilton-Jacobi Equation

In section 3.11 we have discussed the *abbreviated* action $S$ as a function of generalised co-ordinates $q$'s and time $t$ only and showed that it satisfies the equation

$$H + \frac{\partial S}{\partial t} = 0. \tag{9.1}$$

where $H = H(q, p, t)$ is the Hamiltonian and further that

$$\frac{\partial S}{\partial q} = p. \tag{9.2}$$

Using (9.2) in (9.1) we get the explicit form of the Hamilton-Jacobi equation of the Hamiltonian dynamics

$$H\left(q_1, \cdots, q_s; \frac{\partial S}{\partial q_1}, \cdots, \frac{\partial S}{\partial q_s}; t\right) + \frac{\partial S}{\partial t} = 0. \tag{9.3}$$

Equation (9.3) is a partial differential equation satisfied by the action.

$$S(q_1, \cdots, q_s; t).$$

This differential equation is called the *Hamilton-Jacobi equation.*

Like the Lagrangian and Hamilton's canonical equation, Hamilton-Jacobi equation also provides a general method to find the trajectories of the particles. We first note that equation (9.3) contains only first-order-derivative. Every first-order partial differential equation has a solution, depending on an arbitrary function; such a solution is called *general integral* of the equation. General integral of Hamilton-Jacobi equation is not very useful for mechanical applications. For these we need to know the *complete integral* of Hamilton-Jacobi equation. We describe below, how to obtain the solution. Time $t$ and the $s$-co-ordinates $q_1, \cdots, q_s$ are the independent variables in the Hamilton-Jacobi equation given by equation (9.3). Hence, the complete integral of the equation must contain $(s + 1)$ number of constants. As the action, $S$, occurs in (9.3) through first-order derivatives one of the constants would appear as an additive constant: thus

$$S(q_1, \cdots, q_s; t) = f(q_1, \cdots, q_s; t; \alpha_1, \cdots, \alpha_s) + A, \tag{9.4}$$

where $\alpha_1, \cdots, \alpha_s$ and $A$ are the $s+1$ constants of which $A$ appears as an ($\alpha_s$'s may depend on time) additive one. We now, investigate how the solution $S(q_1, \cdots, q_s; t)$, given by (9.4) is related to the solution of the equation of motion variables $q_s(t)$: $s = 1, \cdots, s$. To find this relation we employ canonical transformations.

Consider a canonical transformation $(q_i, p_i) \to (\beta_i, \alpha_i)$ by taking $f(q_s, \alpha_s; t)$ as the generating function, where $\beta_i$ are the new co-ordinates and $\alpha_i$ the new conjugate momenta of $\beta_i$'s. Since $q_s$ are old co-ordinates and $\alpha_i$'s are new momenta, $f(q_s \alpha_s)$ in our notation, corresponds to $F_2(q, P, t)$. From the last chapter, [equation (8.20)],

$$p_i = \frac{\partial F_2}{\partial q_i} \quad \text{and} \quad Q_i = \frac{\partial F_2}{\partial P_i}.$$

Hence, in the present case we derive

$$p_i = \frac{\partial f}{\partial q_i} \quad \text{and} \quad \beta_i = \frac{\partial f}{\partial \alpha_i}$$

and the new Hamiltonian $H'$ is

$$H' = H + \frac{\partial f}{\partial t}.$$

From (9.4) we see

$$\frac{\partial S}{\partial t} = \frac{\partial f}{\partial t}, \quad \text{so, } H' = H + \frac{\partial S}{\partial t}.$$

Therefore, by using (9.1), we get

$$H' = 0. \tag{9.5}$$

Thus, the canonical transformation generated by the generating function $f(q, \alpha; t)$ reduces the new Hamiltonian $H' = 0$. Therefore, from Hamilton's equation

$$\dot{P}_i = -\frac{\partial H'}{\partial Q_i} \quad \text{and} \quad \dot{Q}_i = \frac{\partial H'}{\partial P_i},$$

we see $\dot{P}_i = 0$ and $\dot{Q}_i = 0$, i.e., $\dot{\alpha}_i = 0$ and $\dot{\beta}_i = 0$, i.e., $\alpha_i$ and $\beta_i$ are constant independent of time. We therefore from

$$Q_i = \frac{\partial F_2}{\partial P_i}.$$

the result

$$\beta_i = \frac{\partial f}{\partial \alpha_i}.$$

The $s$-co-ordinates $q_s$'s can be expressed in terms of time $t$ and $2s$ $\alpha_i, \beta_i$ constants. This gives the solutions of the equation of motion for the $q$'s. Therefore, the solutions of the equations of motion satisfied by the co-ordinates are obtained through the solutions of Hamilton-Jacobi equation. The method proceeds as follows.

For any given problem, we first:

(*a*) obtain the Hamiltonian $H(q, p; t)$ of the system;

(*b*) and write down the Hamilton-Jacobi equation

$$H(q_i, p, t) + \frac{\partial S}{\partial t} = 0;$$

(*c*) in the Hamilton-Jacobi equation we replace each $p$ by

$$\frac{\partial S}{\partial q}.$$

Then (*d*) Hamilton-Jacobi equation becomes a first-order differential equation for $S$ in $q$'s and $t$. The solution of this first-order differential equation with appropriate number of constants provide complete-integral, namely

$$S = f(q_s; \alpha_s, t) + A.$$

To find the general solution of the equation of motion for the $q$'s, we differentiate $S$ with respect to $\alpha$'s then

$$\frac{\partial S}{\partial \alpha_i} = \beta_i \quad (\beta_i = \text{ constant}).$$

gives one equation between $q_1, \cdots, q_s$ and time $t$. The momenta $p_i$ is given by $\partial S/\partial q_i$. This gives the complete solution of the problem.

The solution of the Hamilton Jacobi equation becomes much simpler if the Hamiltonian $H$ is a constant of motion. So that $H = E =$ Energy of the system. In this case we can separate the time variable from $S$.

We put

$$S = f(q, \alpha) - Et. \tag{9.6}$$

Then

$$\frac{\partial S}{\partial t} = -E \quad \text{and} \quad H + \frac{\partial S}{\partial t} = 0$$

becomes $H = E$.

The Hamilton-Jacobi equation

$$H(q, p) = E$$

becomes

$$H\left(\frac{\partial S}{\partial q}, q\right) = E. \tag{9.7}$$

therefore, in terms of $f$, one can write

$$H\left(\frac{\partial f}{\partial q}, q\right) = E. \tag{9.8}$$

Equation (9.8) is the time independent form of the Hamilton-Jacobi equation for a conservative system.

Given below are some examples amplifying the method of solution we have described in this section.

## 9.2 Examples in Hamilton-Jacobi Theory

As a simple illustration consider the example of the motion of a projectile in a 2-dimensional vertical plane. Let $m$ be the mass of the projectile. The kinetic energy,

$$T = \frac{1}{2}m(\dot{x}^2 + \dot{y}^2) \quad \text{and} \quad V = mg\,y.$$

We take $y$-axis vertical and $x$-axis horizontal. The Lagrangian

$$L = \frac{m}{2}(\dot{x}^2 + \dot{y}^2) - mg\,y.$$

The conjugate momenta

$$p_x = \frac{\partial L}{\partial \dot{x}} = m\dot{x}, \quad p_y = m\dot{y}.$$

Therefore,

$$H = \frac{1}{2}m(\dot{x}^2 + \dot{y}^2) + mg\,y.$$

That is,

$$H = \frac{(p_x^2 + p_y^2)}{2m} + mg\,y.$$

Note that $H$ does not contain $x$, so, $p_x$ = constant.

Assume

$$p_x = l \text{ (a constant)}.$$

Therefore,

$$H = \frac{p_y^2}{2n} + \frac{l^2}{2m} + mg\,y.$$

Then

$$H = H(p_y, y).$$

We now put

$$p_y = \frac{\partial S}{\partial y} \quad \text{where } S = \text{action.}$$

The Hamilton-Jacobi equation is

$$H\left(\frac{\partial S}{\partial y}, y\right) + \frac{\partial S}{\partial t} = 0.$$

As $H$ is a constant of motion, we put

$$S = f(y) - \alpha t.$$

Then,

$$\frac{\partial S}{\partial t} = -\alpha.$$

Then Hamilton-Jacobi equation becomes $H - \alpha = 0$, i.e.,

$$\frac{1}{2m}\left(\frac{\partial S}{\partial y}\right)^2 + \frac{l^2}{2m} + mg\,y = \alpha \quad \text{(a constant)}.$$

We can find $\partial S/\partial y$ to be,

$$\begin{aligned}\frac{\partial S}{\partial y} &= \sqrt{\left[\left(\alpha - mg\,y - \frac{l^2}{2m}\right)2m\right]}\\ &= \left[(2m\alpha - 2m^2gy - l^2)\right]^{1/2} = \frac{\partial f}{\partial y}.\end{aligned}$$

From this,

$$f = \int \left(2m\alpha - 2m^2g\,y - l^2\right)^{1/2} dy.$$

This gives the complete integral for $S$. For the general solution of the equation of motion we put,

$$\beta = \frac{\partial S}{\partial \alpha}, \quad \text{where } \beta = \text{a constant.}$$

Hence,

$$\beta = \frac{\partial f}{\partial \alpha} - t = -t + \frac{1}{2}\int \frac{2m\,dy}{\sqrt{(2m\alpha - l^2 - 2m^2g\,y)}}.$$

From $p_x = l$ we get $l = m\,u\cos\theta$, where $u\cos\theta$ = velocity at $t = 0$ along $x$ (we use the value at $t = 0$, because $l$ is constant). The particle is projected with velocity $u$ at an angle $\theta$ with the $x$-axis.
The constant $\alpha$ is evaluated from $H$ at $t = 0$.

$$H = \frac{1}{2}mu^2$$

at $t = 0$. so

$$\alpha = \frac{1}{2}mu^2.$$

Using this value of $\alpha$ and $l$. We have

$$\beta = -t + \int \frac{dy \cdot m}{\sqrt{(m^2u^2\sin^2\theta - 2m^2g\,y)}}.$$

From this one obtains $y$ as a function of $t$. namely.

$$\beta = -t - \frac{1}{g}\sqrt{(u^2\sin^2\theta - 2g\,y)}.$$

At $t = 0$. $y = 0$. then.

$$\beta = -\frac{1}{g}\,u\sin\theta.$$

We then finally obtain

$$\left(t - \frac{1}{g}u\sin\theta\right) = -\frac{1}{g}\sqrt{(u^2\sin^2\theta - 2gy)}.$$

Since.

$$u\cos\theta\cdot t = x.$$

the above equation can be rewriteen as

$$\left(\frac{x}{v\cos\theta} - \frac{1}{g}u\sin\theta\right)^2 = \frac{1}{g^2}(u^2\sin^2\theta - 2g\,y).$$

which is an equation of the parabola.

The second *example* is that of the motion of a particle in the field of a central potents $U = V(r)$.

We use spherical polar co-ordinates. The Lagrangian can be written as

$$L = \frac{1}{2}m\left(\dot{r}^2 + r^2\dot{\theta}^2 + r^2\sin^2\theta\dot{\phi}^2\right) - V(r).$$

The conjugate variables.

$$p_r = \frac{\partial L}{\partial \dot{r}} = m\dot{r}. \quad p_\theta = mr^2\dot{\theta} \quad \text{and} \quad p_\phi = mr^2\sin^2\theta\dot{\phi}.$$

We note that Lagrangian is cyclic in $\phi$ so

$$p_\phi = mr^2\sin^2\theta\dot{\phi} = l = \text{ a constant.}$$

The Hamiltonian $H$ is

$$H = \frac{1}{2m}\left(p_r^2 + \frac{p_\theta^2}{r^2} + \frac{l^2}{r^2\sin^2\theta}\right) + V(r).$$

We write

$$p_r = \left(\frac{\partial S}{\partial r}\right), \quad p_\theta = \left(\frac{\partial S}{\partial \theta}\right).$$

Then, the Hamilton-Jacobi equation is

$$\frac{1}{2m}\cdot\left[\left(\frac{\partial S}{\partial r}\right)^2 + \frac{1}{r^2}\left(\frac{\partial S}{\partial \theta}\right)^2 + \frac{l^2}{r^2\sin^2\theta}\right] + V(r) + \frac{\partial S}{\partial t} = 0.$$

As $H$ does not contain time explicitly energy is a constant of motion and it is possible to write

$$S = f(r, \theta, t) - \alpha t,$$

where $\alpha$ is a constant. Hence, $H = \alpha$. Thus,

$$\frac{1}{2m}\left[\left(\frac{\partial f}{\partial r}\right)^2 + \frac{1}{r^2}\left(\frac{\partial f}{\partial \theta}\right)^2 + \frac{l^2}{r^2\sin^2\theta}\right] + V(r) = \alpha.$$

We now carry out *separation of variables* by assuming

$$f(r, \theta) = f_1(r) + f_2(\theta).$$

Then,

$$\frac{\partial f}{\partial r} = \frac{\partial f_1(r)}{\partial r} \quad \text{and} \quad \frac{\partial f}{\partial \theta} = \frac{\partial f_2(\theta)}{\partial \theta}.$$

Substituting these in Hamilton-Jacobi equation we find,

$$\frac{1}{2m}\left\{\left(\frac{\partial f_1}{\partial r}\right)^2 + \frac{1}{r^2}\left[\left(\frac{\partial f_2}{\partial \theta}\right)^2 + \frac{l^2}{\sin^2\theta}\right]\right\} + V(r) = \alpha.$$

Putting

$$\left(\frac{\partial f_2}{\partial \theta}\right)^2 + \frac{l^2}{\sin^2\theta} - \alpha_2 \text{ (a constants)}$$

one obtains

$$\left[\left(\frac{\partial f_1}{\partial r}\right)^2 + \frac{\alpha_2}{r^2} + V(r)\right] = \alpha.$$

Therefore,

$$\frac{\partial f_1}{\partial r} = \left\{2m\left[\alpha - V(r) - \frac{\alpha_2}{r^2}\right]\right\}^{1/2}$$

$$\text{and} \quad \frac{\partial f_2}{\partial \theta} = \left[2m\left(\alpha_2 - \frac{l^2}{\sin^2\theta}\right)\right]^{1/2}.$$

These two equations can be integrated to give

$$f_1(r) = \int dr\left\{2m\left[\alpha - V(r) - \frac{\alpha_2}{r^2}\right]\right\}^{1/2}$$

$$\text{and} \quad f_2(\theta) = \int d\theta\left[2m\left(\alpha_2 - \frac{l^2}{\sin^2\theta}\right)\right]^{1/2}$$

Hence,

$$S = -\alpha t + p_\phi \cdot \phi + \int dr \left\{2m\left[\alpha - V(r) - \frac{\alpha_2}{r^2}\right]\right\}^{1/2} + \int d\theta \left[2m\left(\alpha_2 - \frac{l^2}{\sin^2\theta}\right)\right]^{1/2}.$$

The arbitrary constant are $p_\phi$, $\alpha_2$ and $\alpha(\alpha = \text{Energy of the system})$. On differentiating $S$ with respect to $p_\phi$, $\alpha$ and $\alpha_2$ and equating them with other constants we get the general solution of the equation of motion.

**Example 1.** Consider the *example* of a charged particle moving under the combined coulomb and a uniform field.

$$U = \frac{\alpha}{r} - Fz.$$

We want to find the complete integral of the Hamilton-Jacobi equation of the problem. We use parabolic cylindrical co-ordinates.

$$x = \rho\cos\phi, \qquad y = \rho\sin\phi$$

$$\rho = \sqrt{\xi\eta}, \qquad z = \frac{1}{2}(\xi - \eta).$$

The kinetic energy $= \frac{1}{2}m\left(\dot\rho^2 + \rho^2\dot\phi^2 + \dot z^2\right)$.
In terms of $\xi$, $\eta$ and $\phi$ variables

$$T = \frac{1}{2}m\left[\frac{1}{4}(\xi + \eta)\left(\frac{\dot\xi^2}{\xi} + \frac{\dot\eta^2}{\eta}\right) + \xi\eta\dot\phi^2\right]$$

$$V = \frac{\alpha}{r} - Fz = \frac{\alpha}{\sqrt{(\rho^2 + z^2)}} - Fz = \frac{2\alpha}{(\xi + \eta)} - \frac{1}{2}F(\xi - \eta).$$

We have $p_\phi = m\xi\eta\dot\phi = l$ (a constant, as $\phi$ is cyclic)

$$p_\xi = \frac{m}{4}(\xi + \eta)\frac{\dot\xi}{\xi}; \quad p_\eta = \frac{m}{4}(\xi + \eta)\frac{\dot\eta}{\eta}.$$

Hence,

$$T = \frac{1}{2m}\left[\frac{\xi^2 p_\xi^2 + \eta^2 p_\eta^2}{(\xi + \eta)/4} + \frac{l^2}{\xi\eta}\right].$$

Therefore, the Hamilton-Jacobi equation (as $H$ is independent of time)

$$H = E \text{ (a constant, } E = \text{Energy)}$$

can be written as

$$\frac{1}{2m}\left[4\cdot\frac{\xi^2 p_\xi^2 + \eta^2 p_\eta^2}{(\xi + \eta)} + \frac{l^2}{\xi\eta}\right] + \frac{2\alpha}{\xi + \eta} - \frac{1}{2}F(\xi - \eta) = E.$$

Simplifying and putting

$$p_\xi = \frac{\partial S}{\partial \xi} \quad \text{and} \quad p_\eta = \left(\frac{\partial S}{\partial \eta}\right).$$

the Hamilton-Jacobi equation becomes

$$\frac{1}{2m}\left[4\xi^2\left(\frac{\partial S}{\partial \xi}\right)^2 + 4\eta^2\left(\frac{\partial S}{\partial \eta}\right)^2\right] + \alpha\left(\frac{2}{\xi} + \frac{2}{\eta}\right) - \frac{F(\xi^2 - \eta^2)}{2}$$
$$+ l^2\left(\frac{1}{\xi} + \frac{1}{\eta}\right) = E(\xi + \eta).$$

We separate the equation in $\xi$ and $\eta$ variables by writing

$$S = S_1(\xi) + S_2(\eta).$$

Then

$$4\xi^2\left(\frac{\partial S_1}{\partial \xi}\right)^2 + (2\alpha + l^2)\frac{1}{\xi} - \frac{F\xi^2}{2} - E\xi = \alpha_1$$

and
$$4\eta^2\left(\frac{\partial S_2}{\partial \eta}\right)^2 + (2\alpha + 4l^2)\frac{1}{\eta} + \frac{F\eta^2}{2} - E\eta = -\alpha_1.$$

Here $E, \alpha_1$ and $l$ are constants. We obtain for the complete integral.

$$S = -Et + p_\phi \cdot \phi + S_1(\xi) + S_2(\eta)$$

when

$$S_1(\xi) = \int \frac{d\xi}{2\xi} \cdot \sqrt{2m[\alpha_1 - f_1(\xi)]}$$

and
$$S_2(\eta) = \int \frac{d\eta}{2\eta} \cdot \sqrt{2m[-\alpha_1 - f_2(\eta)]}$$

$$f_1(\xi) = (2\alpha + l^2)\frac{1}{\xi} - \frac{F\xi^2}{2} - E\xi$$

and
$$f_2(\eta) = (2\alpha + l^2)\frac{1}{\eta} + \frac{F\eta^2}{2} - E\eta.$$

This gives the complete integral of the problem. To find the general solution of the equation of motion, we differentiate $S$ with respect to $E$, $\alpha_1$ and $p_\phi$ (where $l = p_\phi$) and equate the resulting expression to three other constants. The general solution of the equation of motion are obtained by performing the integrals.

Other than the parabolic cylindrical co-ordinate one uses elliptic co-ordinate to separate the variable. To see how this works we consider the motion of a particle in the presence of a potential with two centres. That is the particle is moving under

$$U = \frac{\alpha}{r_1} + \frac{\beta}{r_2}.$$

where $r_1$ and $r_2$ are the distances of the particle from two fixed centres. Let

$$(0, 0, +\sigma)$$
$$\text{and} \quad (0, 0, -\sigma)$$

are the co-ordinates of the two fixed points which are $2\sigma$ apart: $\sigma$ = a constant. Thus, the distance $r_1$ and $r_2$ of a point particle with co-ordinate $(x, y, z)$ are

$$r_1 = \sqrt{[x^2 + y^2 + (z - \sigma)^2]}$$
$$r_2 = \sqrt{[x^2 + y^2 + (z + \sigma)^2]}.$$

We use elliptic co-ordinate system.

$$x = \rho \cos\phi, \quad y = \rho \sin\phi$$
$$\rho = \sigma \sinh\xi \cos\eta$$
$$\text{and} \quad z = \sigma \cosh\xi \sin\eta.$$

Then

$$r_1^2 = \sigma^2(\cosh\xi - \sin\eta)^2$$
$$\text{and} \quad r_2^2 = \sigma^2(\cosh\xi + \sin\eta)^2.$$

Hence,

$$V = \frac{\alpha}{\sigma(\cosh\xi - \sin\eta)} + \frac{\beta}{\sigma(\cosh\xi + \sin\eta)}.$$

The kinetic energy

$$T = \frac{1}{2}m\left[(\dot{\rho}^2 + \rho^2\dot{\phi}^2) + \dot{z}^2\right].$$

In terms of $\xi$ and $\eta$ variables

$$T = \frac{1}{2}m\left[\sigma^2 \sinh^2\xi \cos^2\eta\dot{\phi}^2 + \sigma^2(\sinh^2\xi \sin^2\eta + \cosh^2\xi \cos^2\eta)(\dot{\xi}^2 + \dot{\eta}^2)\right].$$

This simplifies to

$$T = \frac{1}{2}m\sigma^2\left[(\cosh^2\xi - \sin^2\eta)(\dot{\xi}^2 + \dot{\eta}^2) + \sinh^2\xi \cos^2\eta\dot{\phi}^2\right].$$

Therefore,

$$\begin{aligned} p_\phi &= m\sigma^2 \sinh^2\xi \cos^2\eta \cdot \dot{\phi} = l \text{ (a constant)}. \\ p_\xi &= m\sigma^2(\cosh^2\xi - \sin^2\eta)\dot{\xi} \\ p_\eta &= m\sigma^2(\cosh^2\xi - \sin^2\eta)\dot{\eta}. \end{aligned}$$

The Hamiltonian $H = T + V$, can then be written as

$$H = \frac{1}{2m\sigma^2}\left[(p_\xi^2 + p_\eta^2)\frac{1}{\cosh^2\xi - \sin^2\eta} + \frac{l^2}{\sinh^2\xi\cos^2\eta}\right] + V(\xi,\eta).$$

So, the Hamilton-Jacobi equation is

$$\frac{1}{2m\sigma^2}\cdot\frac{1}{\cosh^2\xi - \sin^2\eta}\cdot\left[\left(\frac{\partial S}{\partial\xi}\right)^2 + \left(\frac{\partial S}{\partial\eta}\right)^2\right] + \frac{1}{2m\sigma^2}\cdot\frac{l^2}{\sinh^2\xi\cos^2\eta}$$
$$+\frac{\alpha}{\cosh\xi - \sin\eta} + \frac{\beta}{\cosh\xi + \cos\eta} = E.$$

Since,

$$\cosh^2\xi - \sin^2\eta = \sinh^2\xi + \cos^2\eta$$

we have

$$\frac{l^2}{\sinh^2\xi\cos^2\eta}\cdot(\cosh^2\xi - \sin^2\eta) = \frac{l^2}{\sinh^2\xi\cos^2\eta}(\sinh^2\xi + \cos^2\eta).$$

If we use this result then separation of variables can be easily effected.

$$\frac{1}{2m\sigma^2}\left(\frac{\partial S_1}{\partial\xi}\right)^2 + \frac{l^2}{\sinh^2\xi} + (\alpha+\beta)\cosh\xi - E\cosh^2\xi = \alpha_1$$

and

$$\frac{1}{2m\sigma^2}\left(\frac{\partial S_2}{\partial\eta}\right)^2 + \frac{l^2}{\cos^2\eta} + (\alpha-\beta)\sin\eta + E\sin^2\eta = -\alpha_1.$$

These two equation can be solved and the action is

$$S = -Et + p_\phi\cdot\phi + S_1(\xi) + S_2(\eta).$$

This gives the complete integral.

## 9.3 Action Angle Variables

In the last section, we have showed how the Hamilton-Jacobi equation is solved. In particular, we have pointed out the existences of sets of variables in which the separable solutions of the Hamilton-Jacobi equation can be found.

The separation of variables in the solutions of the Hamilton-Jacobi equation is very important for what we are going to study in this section. In this section, we will investigate physical system in which the motion is *periodic*. Very often we are interested in knowing the *frequency* of the periodicity rather than the orbit itself. We will see below that a variant of Hamilton-Jacobi method can be used to handle such problems. By periodic motion we mean the following. For example, in one-dimensional case which may well be characterised by the conjugate variable $(q, p)$, both $q$ and $p$ are periodic and the orbit in the $(q, p)$-phase space is a *closed one*. In other type of periodic motion, $q$ is not periodic, but $q$ is such that when $q$ is increased by same fixed amount, $q_0$ (say) then the system remains unchanged. Indeed for such type of systems the position co-ordinate $q$ is invariably an angle.

In dealing with systems comprising of more than one-degree of freedom, we assume that the Hamilton-Jacobi equation is separable at least in one set of variables $(q_i, p_i)$ $i = 1, 2, \cdots, n$ where $n$ is the degrees of freedom. The motion of the system is said to be periodic if the projection of the system point orbit on each $(q_i, p_i)$ plane is simply periodic in the sense discussed in the case of the system with one-degree of freedom. Since, the variables are separable, these projection points are independent of each other. So, it is not necessary that all $q_i, p_i$ sets will have same frequency. In a three-dimensional harmonic oscillator the force constants along the three axes may all be different but the system may still be called *periodic*. To solve such problems we will use the Hamilton-Jacobi method which is different from the one we had used in the last section. In section (9.1) we found the complete integral for the action $S(q; t)$ in terms of $n$-integration constants, namely

$$S = f(q_i, \alpha_1, \cdots, \alpha_n; t).$$

where $\alpha_i$'s are arbitrary integration constants. To obtain the general solution of the equation of motion we took the $f(q_i, \alpha_i, t)$ as the generating function $F_2(q_i, P_i; t)$ so that the $\alpha_i$'s were interpreted as new momenta. Then, instead, we use suitably defined constants, $J_i$ which form a set of $n$-independent functions of the $\alpha_i$'s and which are known as *action variables*.

The equations of the canonical transformation state that

$$p_i = \frac{\partial f(q_i; \alpha_1, \cdots, \alpha_n; t)}{\partial q_i}. \tag{9.9}$$

Equation (9.9) furnishes each $p_i$ is a function of the corresponding $q_i$ and the $n$-number of constants $\alpha_1, \cdots, \alpha_n$. That is,

$$p_i = p_i(q_i, \alpha_1, \cdots, \alpha_n). \tag{9.10}$$

This equation (9.10) gives $p$ as a function of $q$, i.e., in the phase-space of $(q, p)$; this is the orbit equation. We will have periodic motion of the system if the orbit of equation (9.10) is a closed path, or, is a periodic function of $q_i$. It is not necessary that all $(q_i, p_i)$ sets exhibit same frequency of periodic motion. Indeed, if the separate frequencies are not rational fraction of each other, the particles will not traverse a closed path in space, but will describe an open figure. Such a motion is called *conditionally periodic*.

Having discussed the types of motion we are interested, we can now introduce the *action variables*, $J_1, \cdots, J_n$ which are to replace the constants $\alpha_1, \cdots, \alpha_n$ as the transformed constant momenta, $P_i$. The *action-variables* $J_i$ corresponding to a pair of separation variables $(q_i, p_i)$ is given by

$$J_i = \oint p_i dq_i. \tag{9.11}$$

where the integration has to be performed over a complete period of oscillation or rotation of $q_i$ as the case may be. The name action-variable for $J_i$ defined by (9.11) stems from its similarity in the defination of action $S$ by

$$S = \int p\,dq.$$

By equation (9.9), $J_i$ can be written as

$$J_i = \oint \frac{\partial f(q_i, \alpha_i, \cdots, \alpha_n; t)}{\partial q_i}\, dq_i. \tag{9.12}$$

As, $q_i$'s are integrated away so $J_i$ is a function of the $\alpha_1, \cdots, \alpha_n$.

Further, since, each pair $(q_i, p_i)$ in the separation of variables for the action $S$ the $J_i$'s are independent set of function of the constant $\alpha_i$'s and therefore, they are suitable for the use as a set of new constant momenta. Expressing $\alpha_i$'s as a functions of the $J_i$'s we have

$$f = f(q_1, \cdots, q_n;\ J_1, \cdots, J_n).$$

For the conserved system $S = f - \alpha_1 t$, where $\alpha_1$ is one of the constants, we have from the Hamilton-Jacobi equation

$$H = \alpha_1 = H(J_1, \cdots, J_n) \tag{9.13}$$

as each of $\alpha_i$'s are function of the $J_i$'s. From the definition of

$$J_i = \oint p_i dq_i.$$

we easily note that the dimension of the $J_i$'s are those of angular momenta. For a cyclic variable $q_i$ the corresponding conjugate $p_i$ = constant, hence

$$J_i = p_i \oint dq_i$$

and for rotational periodic motion

$$\oint dq = 2\pi.$$

Thus for a cyclic variable

$$J_i = 2\pi p_i. \tag{9.14}$$

From the generator $f(q_i, J_i)$ we can find the conjugate $\omega_i$ corresponding to $J_i$, they are given by

$$\omega_i = \frac{\partial f(q_1, \cdots, q_n; J_1, \cdots, J_n)}{\partial J_i}. \tag{9.15}$$

As $J_i$'s are new momenta $f$ is the generating function. $\omega_i$ is the new co-ordinate. $J_i$ has the dimension of angular momenta: $\omega_i$ is an angle.
Since $\omega_i$ is the co-ordinate $Q_i$ therefore from Hamilton's equation

$$\dot{Q}_i = \frac{\partial H}{\partial P_i}.$$

We have here,

$$\dot{\omega}_i = \frac{\partial H(J_1, \cdots, J_n)}{\partial J_i} = \nu_i(J_1, \cdots, J_n). \tag{9.16}$$

where $\nu_i$'s are the function of the constant $J_i$'s.
The equation (9.16) has immediate solution

$$\omega_i = \nu_i t + \beta_i. \tag{9.17}$$

That is $\omega_i$ is linear in time. To find the physical meaning of $\nu_i$ consider the following.
Since

$$\omega_i = \frac{\partial f(q_i, J_i, \cdots)}{\partial J_i}.$$

We consider the change in $\omega_i$ when $q_i$ makes one complete revolution. Let $\delta\omega_i$ be an infinitesimal change when $q_i \to q_i + \delta q_i$. Therefore,

$$\delta\omega_i = \frac{\partial \omega_i}{\partial q_j}\delta q_j. \tag{9.18}$$

Let the total change in $\omega_i$ be $\Delta\omega_i$ for one complete revolution.

$$\begin{aligned}\Delta\omega_i &= \oint \delta\omega_i = \oint \frac{\partial \omega_i}{\partial q_j}\delta q_j = \oint \frac{\partial}{\partial q_j}\left(\frac{\partial f}{\partial J_i}\right)\delta q_j \\ &= \frac{\partial}{\partial J_i}\oint \frac{\partial f}{\partial q_j}\delta q_j = \frac{\partial}{\partial J_i}\oint p_j \delta q_j = \frac{\partial J_j}{\partial J_i} = \delta_{ij}.\end{aligned}$$

Hence the change in $\omega_i$ for one complete revolution of $q_i$ is

$$\Delta\omega_i = \delta_{ij}.$$

i.e., the change = 1 unit only.
But

$$\omega_i = \nu_i t + \beta_i; \quad \beta_i = \text{ constant}.$$

So,

$$\Delta\omega_i = \nu_i \tau_i = 1.$$

Therefore,

$$\nu_i = \frac{1}{\tau_i} \tag{9.19}$$

is the frequency associated with the one complete revolution of the periodic motion of $q_i$.
As an illustration consider one dimensional *Harmonic oscillator*.
The Hamiltonian, $H$ is ($m$ = mass, $k$ = force constant)

$$H = \frac{1}{2m}p^2 + \frac{1}{2}kq^2. \tag{9.20}$$

The Hamilton-Jacobi equation,

$$H\left(\frac{\partial S}{\partial q}, q\right) + \frac{\partial S}{\partial t} = 0.$$

For a conserved system

$$S = -\alpha t + f(q, \alpha).$$

Therefore,

$$\frac{\partial S}{\partial t} = -\alpha \quad \text{and} \quad \frac{\partial S}{\partial \alpha} = -t + \frac{\partial f}{\partial \alpha} \quad \text{and} \quad \frac{\partial S}{\partial q} = \frac{\partial f}{\partial q}.$$

Then Hamilton-Jacobi equation becomes

$$H\left(\frac{\partial f}{\partial q}, q\right) = \alpha.$$

For the Hamiltonian (9.20), we have,

$$\frac{1}{2m}\left(\frac{\partial f}{\partial q}\right)^2 + \frac{k}{2}q^2 = \alpha. \tag{9.21}$$

From (9.21), we have

$$\frac{\partial S}{\partial q} = \frac{\partial f}{\partial q} = \sqrt{(2m\alpha - kmq^2)}. \tag{9.22}$$

Equation (9.22) gives us

$$f = \int dq\sqrt{2m\alpha - kmq^2}.$$

Therefore,

$$\beta = \frac{\partial S}{\partial \alpha}$$

yields

$$\beta = -t + \frac{\partial f}{\partial \alpha} = -t + \int \sqrt{\frac{m}{k}}\frac{dq}{\sqrt{\left(\frac{2\alpha}{k} - q^2\right)}}.$$

On integration, we get

$$q = \sqrt{\frac{2\alpha}{k}}\cos\left(\sqrt{\frac{k}{m}}\cdot t + \beta'\right). \tag{9.23}$$

$$= \sqrt{\frac{2\alpha}{k}}\cos(\omega t + \beta'). \tag{9.24}$$

where $\beta'$ = constant. Equation (9.23) gives the general solution of the Hamilton-Jacobi equation for the Harmonic oscillator. Let us now consider the action angle variable, we have action, $J$,

$$J = \oint pdq \text{ over a period of } q.$$

$$= \oint \frac{\partial f}{\partial q}\, dq = \oint \sqrt{(2m\alpha - kmq^2)}\, dq = \sqrt{(km)}\oint \sqrt{\left(\frac{2\alpha}{k} - q^2\right)}\, dq.$$

Substituting,

$$q = \sqrt{\frac{2\alpha}{k}}\sin\theta,$$

we get for $J$,

$$J = 2\pi\alpha\sqrt{\frac{m}{k}}.$$

But $\alpha = H$, therefore,

$$H = \frac{J}{2\pi}\sqrt{\frac{k}{m}}. \tag{9.25}$$

The frequency

$$\nu = \frac{\partial H}{\partial J} = \frac{1}{2\pi}\sqrt{\frac{k}{m}}.$$

$$\text{i.e.,}\quad 2\pi\nu = \sqrt{\frac{k}{m}} = \omega \text{ (angular frequency)}. \tag{9.26}$$

Equation (9.26) is in agreement with (9.24).

The second example we consider is the well known *Kepler-problem* (or the *hydrogen atom problem*).

The particle of mass $m$ is moving in the field of a central potential

$$V(r) = -\frac{k}{r}; \quad k > 0.$$

In three-dimension the Hamiltonian $H$, in spherical polar co-ordinates is

$$H = \frac{1}{2}m(\dot{r}^2 + r^2\dot{\theta}^2 + r^2\sin^2\theta\dot{\phi}^2) + V(r).$$

We note that $H$ does not contain $\phi$ hence $\phi$ is cyclic variable, therefore, $p_\phi$ = constant.*

$$p_\phi = mr^2\sin^2\theta\dot{\phi} = L_z, \tag{9.27}$$

i.e., $L_z$ is a constant, where $L_z$ is the third component of the angular momentum. This shows the motion is confined to a two-dimensional $xy$-plane. Therefore, we consider the Hamiltonian in two-dimensional polar variables

$$H = \frac{1}{2}m(\dot{r}^2 + r^2\dot{\theta}^2) + V(r) = \frac{1}{2m}\left(p_r^2 + \frac{p_\theta^2}{r^2}\right) + V(r), \tag{9.28}$$

where $p_r = m\dot{r}$, $p_\theta = mr^2\dot{\theta}$.

Further, from (9.28) we find that $H$ does not depend on $\theta$ therefore, $\theta$ is cyclic, that is $p_\theta$ is a constant $= l$, say.

Since the Hamiltonian is equal to energy, $E$, we write (9.28) as

$$\frac{1}{2m}\left(p_r^2 + \frac{l^2}{r^2}\right) + V(r) = E.$$

Therefore, the momentum

$$p_r = \left[2m\left(E - \frac{l^2}{r^2} - V\right)\right]^{1/2},$$

where $V = -k/r$. The action variable $J_r$ is then

$$J_r = \oint p_r dr = \oint \left[2m\left(E - \frac{l^2}{r^2} + \frac{k}{r}\right)\right]^{1/2} dr$$

$$= \oint \sqrt{\left(A + \frac{2B}{r} - \frac{C}{r^2}\right)}\, dr \tag{9.29}$$

with $A = 2mE$, $B = mk$ and $C = l^2 = p_\theta^2$. The integral (9.28) is over a closed orbit; we see that for the finite motion $E < 0$ and further the integrand is $p_r$, where $p_r = m\dot{r}$. Therefore, the limits of integration are obtained from $\dot{r} = 0$, i.e., when the integrand is zero. That is the limits are defined by the roots of

$$A + \frac{B}{r} - \frac{C}{r^2},$$

namely $r_1$, $r_2$. If $r_1$ is the inner bound, then the complete cycle in $r$ will involve going from $r_1$ to $r_2$ and back to $r_1$. On the outward journey $p_r$ is positive while for backward journey $p_r$ is negative. The integration therefore involves both the branches of the double-valued function given by

$$\sqrt{\left(A + \frac{B}{r} - \frac{C}{r^2}\right)}$$

with $r_1$ and $r_2$ as the branch points.

So, the complex plane should have a cut along the line joining $r_1$ to $r_2$. As the path of integration encloses the line between the branch points, the method of contour-integration cannot be applied directly. We therefore, choose the path of integration as enclosing all the rest of the complex plane, the direction of integration, now being the clockwise direction. In this region, the integrand is single-valued and method of residue calculation can be applied.

Only two singular points are present one at $r = 0$ and other at $r = \infty$. The path of integration can be distorted into two clock-wise directed circles enclosing these two points. The sign in front of the square-root in the integrand must be negative for the region along the real axis below $r_1$ as can be seen by considering the behaviour of the function in the neighbourhood of $r_1$.

Therefore the residue at $r = 0$ is $-\sqrt{-C}$. Above $r_2$ the sign is positive. To evaluate the residue at $r = \infty$, we change the integrand by substituting $r^{-1} = z$. The integral becomes

$$-\oint \sqrt{(A + 2Bz - Cz^2)/z^2} \cdot dz.$$

The residue at the origin of this gives $-B/\sqrt{A}$.

The total integral is therefore,

$$-2\pi i \text{ (sum of the residues)} = 2\pi i \left(\sqrt{-C} + \frac{B}{\sqrt{A}}\right).$$

Hence

$$J_r = \oint p_r dr = -2\pi l + \pi \frac{2mk}{\sqrt{2mE}}.$$

Since

$$l = p_0 \quad \text{and} \quad J_0 = 2\pi p_0.$$

We get

$$(J_r + J_\theta) = \pi k\sqrt{\frac{2m}{E}}.$$

Hence,

$$E = -\frac{(2\pi^2 mk^2)}{(J_r + J_\theta)^2}.$$

The frequency

$$\nu = \frac{\partial H}{\partial J_r}.$$

Thus,

$$\nu = \frac{4\pi^2 mk^2}{(J_r + J_\theta)^3}.$$

we have used $H = E$ to obtain the frequency.
Therefore, time period

$$\tau = \pi k\sqrt{\frac{m}{(-2E^3)}}.$$

(See Example 20, Miscellaneous Examples)

## 9.4 Adiabatic Theorem and Hannay Angle in Classical Mechanics

Consider an *integrable Hamiltonian*. By an integrable Hamiltonian we mean a Hamiltonian which can be expressed in terms of action and angle-variables, $(I, \theta)$, where

$$I = \frac{J}{2\pi} = \frac{1}{2\pi}\oint p\,dq \quad \text{and} \quad \theta \text{ is canonical to } I.$$

The adiabatic theorem of classical meachanics states that in an adiabatic excursion of a Hamiltonian with respect to time, the action, $I$

$$I = \frac{1}{2\pi}\oint p\,dq$$

is an invariant quantity and the canonical angle $\theta$ is obtained by integrating the instantaneous frequency $\Omega$, given by

$$\Omega = \frac{\partial H(I, \theta, X)}{\partial I}, \quad \text{i.e.,} \quad \theta = \int_0^T dt\,\Omega(t).$$

Here $X(t)$ is the adiabatic parameter.
Hannay and latter Berry showed that if the adiabatic parameter $X(t)$ is changed slowly along a closed curve $C$, and comes back to its original position

after a period of time angle swept is not only the dynamical one due to integration of $\partial H/\partial I$, but, also there is an additional piece, which depends on the geometry of the space spanned by the parameters, and

$$\theta = \int_0^T \Omega(t)\, dt + \Delta\theta.$$

To find $\Delta\theta$ we make a canonical transformation from the variables

$$(p, q) \to (I, \theta)$$

by employing the generating function of the form $F_2(q, P)$. $I$ is $P$ here. A choice for $F_2$ is the same as we did in the last section, namely the action $S^{(\alpha)}(q, I, X(t))$ and is given by

$$S^{(\alpha)} = \int p dq.$$

Then,

$$p = \frac{\partial S^{(\alpha)}}{\partial q} \quad \text{and} \quad \theta = \frac{\partial S^{(\alpha)}}{\partial I}.$$

The index $(\alpha)$ on $S$ indicates various branches of $S^{(\alpha)}$ which is a multi-valued function and can be seen as follows. During a period of $2\pi$ of $q$, the change in $S^{(\alpha)}$ is

$$\Delta S = 2\pi I.$$

Under the canonical transformation through $S^{(\alpha)}$ as a generating function given by $(p, q) \to (I, \theta)$,

$$\mathbf{H}(I, \theta, X) = \mathcal{H}(I, \theta, X) + \frac{\partial S^{(\alpha)}}{\partial t},$$

where

$$\mathcal{H}(I, \theta, X) \equiv H(q, p, X)$$

written in terms of new variables. To make $\mathbf{H}(I, \theta, X)$ a single-valued function, we choose a branch of $S^{(\alpha)}$ such that

$$S(I, \theta, X) \equiv S^{(\alpha)}(q(I, \theta, X), I, X) \quad \text{for } 0 \le \theta \le 2\pi.$$

Since $I$ is an invariant quantity, we get from this expression

$$\frac{\partial S}{\partial X} = \frac{\partial S^{(\alpha)}}{\partial X} + \frac{\partial S^{(\alpha)}}{\partial q} \cdot \frac{\partial q}{\partial X} \quad \text{and} \quad \mathbf{H} = \mathcal{H} + \frac{\partial S^{(\alpha)}}{\partial X}\frac{dX}{dt}.$$

Using the value of $\partial S^{(\alpha)}/\partial X$ from above we have

$$\mathbf{H} = \mathcal{H} + \frac{dX}{dt}\left(\frac{\partial S}{\partial X} - \frac{\partial S^{(\alpha)}}{\partial q} \cdot \frac{\partial q}{\partial X}\right).$$

As.

$$p = \frac{\partial S^{(\alpha)}}{\partial q}.$$

we have for **H**.

$$\mathbf{H} = \mathcal{H} + \frac{dX}{dt}\left(\frac{\partial S}{\partial X} - p\frac{dq}{dX}\right).$$

Therefore.

$$\begin{aligned}\dot{\theta} = \frac{\partial \mathbf{H}}{\partial I} &= \frac{\partial \mathcal{H}}{\partial I} + \frac{\partial}{\partial I}\left[\frac{\partial X}{\partial t}\left(\frac{\partial S}{\partial X} - p\frac{d}{dX}q\right)\right]\\ &= \frac{\partial \mathcal{H}}{\partial I} + \frac{\partial X}{\partial t}\cdot\frac{\partial}{\partial I}\left(\frac{\partial S}{\partial X} - p\frac{\partial}{\partial X}q\right).\end{aligned}$$

Hence. the angle $\theta$ is

$$\begin{aligned}\theta &= \int_0^T \frac{\partial \mathcal{H}}{\partial I}\,dt + \int_0^T dt\cdot\frac{dX}{dt}\cdot\frac{\partial}{\partial I}\left(\frac{\partial S}{\partial X} - p\cdot\frac{\partial}{\partial X}q\right)\\ &= \int_0^T \Omega(t)\,dt + \Delta\theta\\ \Delta\theta &= \int_0^T dt\frac{dX}{dt}\cdot\frac{\partial}{\partial I}\left(\frac{\partial S}{\partial X} - p\frac{\partial q}{\partial X}\right)\\ &= \oint dX\left[-\frac{\partial}{\partial I}\left(p\frac{\partial q}{\partial X}\right)\right]\end{aligned}$$

when the parameter space has many components we write $\Delta\theta$ as

$$\Delta\theta = \oint d\mathbf{X}\left(-\frac{\partial}{\partial I}\right)(p\cdot\bar{\nabla}_X q).$$

Hence the average Hannay angle $\overline{\Delta\theta}$ is given by

$$\begin{aligned}\overline{\Delta\theta} &= -\frac{\partial}{\partial I}\oint d\mathbf{X}\cdot\oint\frac{d\theta}{2\pi}(p\cdot\bar{\nabla}_X q)\\ \overline{\Delta\theta} &= -\frac{\partial}{\partial I}\oint d\mathbf{X}\cdot\mathbf{A}.\end{aligned}$$

where

$$\mathbf{A} = \oint\frac{d\theta}{2\pi}(\mathbf{p}\cdot\bar{\nabla}_X q).$$

This last result for $\overline{\Delta\theta}$ can also be written in terms of surface-integral, $d\bar{\sigma}$

$$\overline{\Delta\theta} = +\frac{\partial}{\partial I}\oint d\bar{\sigma}\cdot\mathbf{B}.$$

where

$$\mathbf{B} = -\oint\frac{d\theta}{2\pi}\,(\bar{\nabla}_X p\times\bar{\nabla}_X q).$$

## Exercises

1. A particle is thrown vertically upward with an initial velocity $u$, against the attraction due to gravity. Write down the Hamilton-Jacobi equation for the motion. Find the complete integral and the general solution of the equation of motion.

[*Hints:* At a height $x$, say,

$$T = \frac{1}{2}m\dot{x}^2, \quad p_x = m\dot{x}.$$

Hence

$$H = T + V = \frac{p_x^2}{2m} + mgx, \quad V = mgx.$$

Hamilton-Jacobi equation

$$H + \frac{\partial S}{\partial t} = 0.$$

Put $S = f - \alpha t$, where $f$ is time independent. Hence,

$$H = \alpha \quad (\alpha = \text{constant}).$$

Therefore, Hamilton-Jacobi equation becomes

$$\frac{1}{2m}\left(\frac{\partial f}{\partial x}\right)^2 + mgx = \alpha, \quad p_x = \frac{\partial S}{\partial x} = \frac{\partial f}{\partial x}.$$

Thus,

$$\frac{\partial f}{\partial x} = [2(\alpha - mgx)m]^{1/2}.$$

Thus,

$$f(x) = \int dx\,[2m(\alpha - mgx)]^{1/2}.$$

Again,

$$\beta = \frac{\partial S}{\partial \alpha} = -t + \frac{\partial f}{\partial \alpha}.$$

so, we get

$$(t + \beta) = \int m dx\,[2m(\alpha - mgx)]^{-1/2}$$

$$\text{or,} \quad (t + \beta) = -\sqrt{\frac{2}{m}}\cdot\frac{1}{g}\sqrt{\alpha - mgx}.$$

$$H = \alpha = \frac{1}{2}mu^2 \quad (u = \text{initial velocity})$$

$$t + \beta = -\sqrt{\frac{2}{m}}\cdot\frac{1}{g}\sqrt{\frac{1}{2}mu^2 - mgx} = -\frac{\sqrt{2}}{g}\sqrt{\left(\frac{1}{2}u^2 - gx\right)}.$$

$t = 0$, $x = 0$ gives,

$$\beta = -\frac{u}{g}.$$

Hence

$$t - \frac{u}{g} = -\frac{\sqrt{2}}{g}\sqrt{\left(\frac{1}{2}u^2 - gx\right)}.$$

this gives

$$x = ut - \frac{1}{2}gt^2.]$$

2. The Lagrangian $L$ is given by

$$L = \sum_i f_i(q_i)\dot{q}_i^2 - \sum_i V_i(q_i).$$

Find the Hamilton-Jacobi equation and the action variables $J_i$. Express the Hamiltonian as a function of $J_i$'s.

[*Hints:*

$$L = \sum_i f_i(q_i)\dot{q}_i^2 - \sum_i V_i(q_i).$$

Hence

$$p_i = \frac{\partial L}{\partial \dot{q}_i} = 2f_i(q_i)\dot{q}_i; \quad \dot{q}_i = \frac{p_i}{2f_i}$$

$$H = \sum_i p_i\dot{q}_i - L = \sum_i \frac{p_i^2}{2f_i} - L$$

$$H = \frac{1}{4}\sum_i \frac{p_i^2}{f_i} + \sum_i V_i(q_i)$$

$$H = \sum_i \left[\frac{p_i^2}{4f_i} - V_i(q_i)\right] = \sum_i H_i(p_i, q_i).$$

Hamilton-Jacobi equation

$$H + \frac{\partial S}{\partial t} = 0.$$

Writing $S = f(q_i\alpha) - \alpha t$.

We get

$$H = \alpha \quad \text{or.} \quad \sum_i [H_i(p_i, q_i) - \alpha_i] = 0.$$

where we put

$$\alpha = \sum_i \alpha_i;$$

$\alpha_i$'s are constants. Since $(q_i, p_i)$ variable are separable we write

$$H_i(p_i, q_i) = \alpha_i.$$

Assuming

$$S(q_1,\cdots,q_2,\cdots;t) = \sum_i [\mathbf{F}_i(q_i) - \alpha_i t],$$

we get

$$\frac{\partial S}{\partial t} = -\sum_i \alpha_i = -\alpha \quad \text{and} \quad \frac{\partial S}{\partial q_i} = \frac{\partial \mathbf{F}_i}{\partial q_i}.$$

Therefore,

$$p_i = \frac{\partial S}{\partial q_i} = \frac{\partial \mathbf{F}_i}{\partial q_i}.$$

Hence from

$$H_i(q_i, p_i) = \alpha_i$$

we get

$$\frac{1}{4f_i}\left(\frac{\partial \mathbf{F}_i}{\partial q_i}\right)^2 - V_i(q_i) = \alpha_i,$$

$$\text{or,} \quad \frac{\partial \mathbf{F}_i}{\partial q_i} = \{4f_i(q_i)\,[\alpha_i - V_i(q_i)]\}^{1/2}.$$

Hence,

$$\mathbf{F}_i = \int [4f_i(\alpha_j - V_i(q_i))]^{1/2}\, dq_i.$$

Hence

$$S = \sum_i \int \{4f_i(q_i)[\alpha_i - V_i(q_i)]\}^{1/2}\, dq_i - \alpha t.$$

For the general solution of equation of motion

$$\beta_i = -t + \frac{\partial F_i}{\partial \alpha_i} = -t + 2\int F_i(q_i) dq_i\, [4F_i(\alpha_i - V_i)]^{-1/2}.$$

The action-variable $J_i$

$$J_i = \oint p_i dq_i = \oint [4f_i(q_i)\alpha_i - V_i(q_i)]^{1/2} dq_i = G_i(\alpha_i)$$

$$\text{or,} \quad \alpha_i = g_i(J_i), \quad H = \sum_i \alpha_i = \sum_i g_i(J_i), \quad \omega_i = \frac{\partial g_i}{\partial J_i}.$$

3. A particle of mass $m$ is constrained to move on a smooth parabolic wire, $x^2 = by$ ($y$-axis vertical) under the action of gravity. Write down the Hamilton-Jacobi equation for the system and find the general solution of the equation of motion. Find the *action-angle* variables.

[*Hints:*

$$T = \frac{1}{2}m(\dot{x}^2 + \dot{y}^2).$$

Particle is moving on the parabola.

$$x^2 = by. \quad \text{Hence } \dot{y} = \left(\frac{2x}{b}\right)\dot{x}.$$

Therefore,

$$T = \frac{1}{2}m\left[\dot{x}^2 + \left(\frac{4x^2}{b^2}\right)\dot{x}^2\right] = \frac{1}{2}m\left(1 + \frac{4x^2}{b^2}\right)\dot{x}^2$$

$$V = mg\,y = \left(\frac{mg}{b}\right)x^2.$$

$$L = \frac{1}{2}m\left(1 + \frac{4x^2}{b^2}\right)\dot{x}^2 - \frac{mg}{b}\cdot x^2$$

$$p_x = \frac{\partial L}{\partial \dot{x}} = m\left(1 + \frac{4x^2}{b^2}\right)\dot{x}$$

$$H = p_x\cdot\dot{x} - L = \frac{1}{2}m\left(1 + \frac{4x^2}{b^2}\right)\dot{x}^2 + mgx^2$$

$$H = \frac{1}{2m}\frac{p_x^2}{\left(1 + \frac{4x^2}{b^2}\right)} + mgx^2.$$

Hamilton-Jacobi equation

$$\frac{1}{2m}\frac{1}{\left(1 + \frac{4x^2}{b^2}\right)}\left(\frac{\partial f}{\partial x}\right)^2 + mgx^2 = \alpha.$$

where $S = f - \alpha t$ is used. From this

$$f = \int\left[2m\left(1 + \frac{4x^2}{b^2}\right)(\alpha - mgx^2)\right]^{1/2} dx.$$

Hence the complete integral is $S = f - \alpha t$

$$S = \int dx\left[2m\left(1 + \frac{4x^2}{b^2}\right)(\alpha - mgx^2)\right]^{1/2} - \alpha t.$$

For general solution of the equation of motion

$$-t + \beta = \frac{\partial f}{\partial \alpha} = \sqrt{2m}\int\frac{dx}{2}\left[\left(1 + \frac{4x^2}{b^2}\right)^{1/2}(\alpha - mgx^2)^{-1/2}\right]$$

$$= \sqrt{\frac{m}{2}}\int dx\left(1 + \frac{4x^2}{b^2}\right)^{1/2}(\alpha - mgx^2)^{-1/2}.]$$

4. A particle is constrained to move on the surface of an inverted cone whose vertex. is at origin. under the action of gravity. Show that the Hamilton-Jacobi equation is separable and find the complete integral of the Hamilton-Jacobi equation.

[*Hints:*

$$\begin{aligned}
L &= \frac{1}{2}m\left(\dot{r}^2 + r^2\dot{\theta}^2 + r^2\sin^2\theta\cdot\dot{\phi}^2\right) - mgr\cos\theta \\
\theta &= \text{constant for a cone of semi-vertical angle } \theta. \\
L &= \frac{1}{2}m(\dot{r}^2 + r^2a\dot{\phi}^2) - mgb\cdot r; a = \sin^2\theta \text{ and } b = \cos\theta \text{ are constant} \\
p_\phi &= \text{constant} = l = r^2a\dot{\phi} \\
p_r &= \frac{\partial L}{\partial \dot{r}} = m\dot{r} \\
H &= \frac{p_r^2}{2m} - \frac{1}{2}mr^2a\dot{\phi}^2 + mgbr = \frac{p_r^2}{2m} + mgb\cdot r - \frac{1}{2}\frac{l^2}{mar^2}.
\end{aligned}$$

Put,

$$p_r = \frac{\partial S}{\partial q_r} = \frac{\partial S}{\partial r}.$$

Then Hamilton-Jacobi equation is

$$H = \alpha,$$

where $S = f - \alpha t$, gives

$$\frac{1}{2m}\left(\frac{\partial f}{\partial r}\right)^2 + mgb\,r - \frac{l^2}{2ma^2r^2} = \alpha.$$

From this we easily find $f$.]

5. A particle is thrown horizontally from the top of a building of height $h$, with an initial velocity $u$. Write down the Hamiltonian of the problem and hence solve the Hamilton-Jacobi equation of the problem.

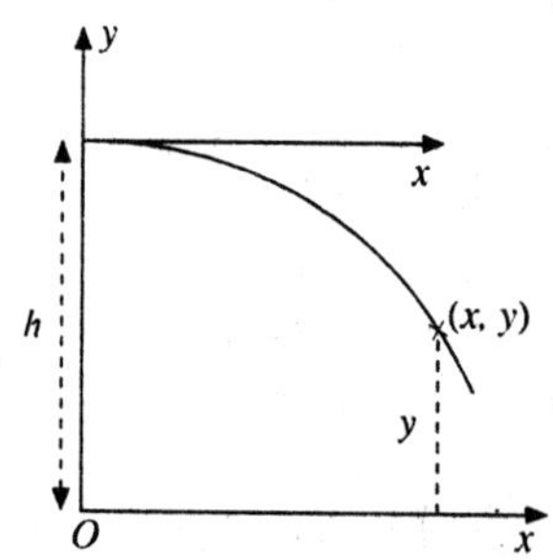


**Fig 9.1**

[*Hints:*

$$\begin{aligned}
\text{K.E.} &= T = \frac{1}{2}m\left(\dot{x}^2 + \dot{y}^2\right), \\
p_x &= m\dot{x}, \\
p_y &= m\dot{y}' \\
V &= -mg(h - y) \\
H &- \frac{1}{2m}(p_x^2 + p_y^2) - mg(h - y),
\end{aligned}$$

as $x$ is cyclic,

$p_x = l$ (constant) $= mu$ (initial value). Hence from $H = \alpha$ (constant)

$$\frac{1}{2m}\left(\frac{\partial f}{\partial y}\right)^2 + \frac{l^2}{2m} - mg(h - y) = \alpha.$$

Then

$$f = \int dy\left[2m\left(\alpha - \frac{l^2}{2m} + mg(h - y)\right)\right]^{1/2}$$

then

$$t+\beta=\frac{\partial f}{\partial \alpha}=\sqrt{2m}\int dy\left[\left(\alpha-\frac{l^2}{2m}+mg(h-y)\right)\right]^{-1/2}$$

$$\alpha=\frac{1}{2}mu^2,\quad \frac{l^2}{2m}=\frac{1}{2}mu^2,$$

therefore,

$$t+\beta=\sqrt{2m}\int\frac{dy}{2}\frac{1}{\sqrt{mg(h-y)}}=-\sqrt{\frac{2}{g}}\sqrt{h-y}$$

at $t=0$, $y=h$, so, $\beta=0$.

$$t^2=\frac{2}{g}(h-y)\ \text{ but } x=ut.$$

Therefore,

$$x^2=\frac{2u^2}{g}(h-y)$$

is the parabola for the path.]

6. For the Harmonic Oscillator assume the motion is given by

$$q=\sqrt{\frac{2E}{\mu}}\sin\left(\sqrt{\frac{\mu}{m}}\cdot t+\delta\right).$$

where $\mu$ = force constant, $E$ = energy. Evaluate action $S$ directly from the Lagrangian and compare the action with that obtained from Hamilton-Jacobi equations.

7. Use Hamilton-Jacobi method to find the equation of the orbit with field of central potential

$$V=-\frac{k}{r};\ \ k>0.$$

[*Hints:* From the Kepler's problem that we discussed in the text, we can write down the Hamiltonian for the orbit of a planet moving round the sun in the field of a central potential

$$V(r)=-\frac{k}{r},\quad H=\frac{1}{2\mu}\left[p_r^2+\frac{p_\theta^2}{r^2}+V(r)\right].$$

where $\mu$ = reduced mass and $p_\theta=l$ = constant. Therefore,

$$H=\frac{1}{2\mu}\left(p_r^2+\frac{l^2}{r^2}-\frac{k}{r}\right).$$

The Hamilton-Jacobi equation is

$$H-\frac{\partial S}{\partial t}=0.$$

reduce to $H=\alpha$, a constant.

Therefore, putting

$$p_r = \frac{\partial S_r}{\partial r},$$

we get

$$\frac{1}{2\mu}\left[\left(\frac{\partial S_r}{\partial r}\right)^2 + \frac{l^2}{r^2} - \frac{k}{r}\right] = \alpha$$

or the action $S_r$ is given by

$$S_r = \int\left[2m\left(\alpha - \frac{l^2}{r^2} + \frac{k}{r}\right)\right]^{1/2} dr.$$

This gives the complete integral for the Hamilton-Jacobi equation. For the general solution of the equation of motion, we have,

$$\beta = -t + \frac{\partial S_r}{\partial \alpha},$$

therefore,

$$\beta = -t + \frac{1}{2}\int dr(2m)^{1/2}\left[\alpha - \frac{l^2}{r^2} + \frac{k}{r}\right]^{-1/2}.$$

Put $\alpha = E$ then the equation becomes

$$t + \beta = \sqrt{\frac{m}{2}}\int dr \frac{1}{\sqrt{\left[E - \frac{l^2}{r^2} + \frac{k}{r}\right]}}, \beta = \text{constant}.$$

This is the equation of the orbit.]

Chapter X

# Relativistic Mechanics

## 10.1 Introduction

So far we have considered non-relativistic mechanics of particles in the sense that the velocities of particles involved were all very less than the velocity of light. In the present section, we want to show how to write a relativistic Lagrangian consistent with special theory of relativity. Newtonian dynamics is valid in an inertial system, i.e., in a frame which is fixed in space or any other reference frame which is moving with a constant velocity with respect to the fixed frame in space. If $\mathbf{r}'$ represents a radius vector from the origin of the moving system to a given point, and $\mathbf{r}$, the corresponding vector in the first fixed frame, then the two vectors are connected by

$$\mathbf{r}' = \mathbf{r} - \mathbf{V}t. \tag{10.1}$$

where $\mathbf{V}$ is the relative constant velocity. Since $\mathbf{V}$ is a constant, then

$$\dot{\mathbf{r}}' = \dot{\mathbf{r}} - \mathbf{V}. \tag{10.2}$$

And we find from (10.2)

$$\ddot{\mathbf{r}}' = \ddot{\mathbf{r}}, \quad \text{i.e.,} \quad \mathbf{a}' = \mathbf{a}.$$

Hence, acceleration is the same in both frames. Therefore Newton's equation $\mathbf{F} = m\mathbf{a}$ remains form-invariant.

The transformation (10.1) is called the *Gallilean transformation*. As Newton's equation remains form-invariant, Newton's equation is thus Gallilean invariant. From (10.2) we notice that velocity in two frames are, however, different. Suppose there is a source of light at the origin of the fixed frame emitting spherical wave-travelling with speed $c$. Let the radius vector $\mathbf{r}$ be the position vector of a point on the surface of the wave. Then the velocity of the point on this wave-surface is $\dot{\mathbf{r}} = c \cdot \mathbf{n}$. Where $\mathbf{n}$ is the unit vector along $\mathbf{r}$. According to (10.2), the corresponding velocity vector in the moving system is $\dot{\mathbf{r}}' = c\mathbf{n} - \mathbf{V}$. In the system moving with respect to the source of light, the magnitude of the wave velocity will in general, no longer be $c$, indeed since it depends on direction the wave will no longer be spherical. From experimental facts we know, velocity of light is the same in all frames. Hence, Gallilean transformation can no longer be valid. We thus postulate that *there is no unique time-scale common to all frames of references*. A point in a ordinary space at a certain instant of time is referred to as an event and may be specified by four co-ordinates $x_1, x_2, x_3$ and $t$. The same event may be identified in any other frame by $x'_1, x'_2, x'_3$ and $t'$. The velocity of light be same in both the frames demands that

$$x_1^2 + x_2^2 + x_3^2 - c^2t^2 = x_1'^2 + x_2'^2 + x_3'^2 - c^2t'^2. \tag{10.3}$$

If the second frame $S'$ is moving with respect to the first frame $S$ with uniform velocity $V$, then the relationship between the two frames is a linear transformation

$$x'_1 = x_1 \qquad x'_2 = x_2$$

$$x'_3 = \frac{x_3 - Vt}{\sqrt{1-\beta^2}} \quad \text{and} \quad t' = \frac{t - Vx_3/c^2}{\sqrt{1-\beta^2}}. \tag{10.4}$$

where $\beta = V/c$, $c$ = speed of light.

We easily check that (10.3) is satisfied by (10.4).

Equation (10.4) is called *Lorentz transformation*. The origins of the two frames of reference coincide initially, the space axes have the same orientation and that the relative motion is along $x_3$.

A more elegant method of description due to Minkowski labels the event with four co-ordinates $(x_1, x_2, x_3, x_4 = ict)$. The four quantities $x_\mu (\mu = 1, 2, 3, 4)$ form the components of a first order four-dimensional cartesian tensor or a four-vector; and the Lorentz transformation relation becomes an orthogonal (i.e., Length-preserving) transformation of such components. It follows that

$$\sum_\mu x_\mu^2 = \sum_\mu x_\mu'^2. \tag{10.5}$$

in consonance with (10.3).

The interval or proper time between two events $x$ and $y$ is defined by

$$\tau = \frac{1}{c}\sqrt{\left[-\sum_\mu (x_\mu - y_\mu)^2\right]}. \tag{10.6}$$

The quantity $\tau$ is an invariant or a scalar, since it represents the length of a four-vector. The motion of a particle can be considered as a continuous succession of events forming a curve in the four-dimension space. The infinitesimal invariant interval between two neighbouring events on this curve is

$$d\tau = \frac{1}{c}\sqrt{\left[-\sum_{\mu}(dx_\mu)^2\right]} = dt\sqrt{\left[1 - \frac{1}{c^2}\sum_{i=1}^{3}(\dot{x}_i^2)\right]}$$
$$= dt\sqrt{1-\beta^2}. \tag{10.7}$$

where

$$\beta c = \sqrt{\sum_i \dot{x}_i^2} = v$$

denotes the ordinary velocity of the particle.

Contrary to the position in Newtonian mechanics the quantities $\dot{x}_i = dx_i/dt$ are no longer vector components. This is because time $t$ itself is a component of a four-vector; rather than a scalar in non-relativistic mechanics. A genuine four-vector must be defined by

$$u_\mu = \frac{dx_\mu}{d\tau}. \tag{10.8}$$

Since,

$$u_\mu = \frac{dx_\mu}{d\tau} = \frac{dx_\mu}{\sqrt{1-\beta^2}dt}$$

it is obvious that the space-like components of this four-vector coincide with the ordinary velocity components in the *non-relativistic limit*, where $\beta \to 0$. This is taken to justify regarding $u_\mu$ the generalisation of the ordinary velocity. The time-like component $u_4$,

$$u_4 = \frac{dx_4}{d\tau} = \frac{icdt}{dt\sqrt{1-\beta^2}} = \frac{ic}{\sqrt{1-\beta^2}}. \tag{10.9}$$

The invariant length of the four-vector $u_\mu$ is

$$\sum_\mu u_\mu^2 = -c^2. \tag{10.10}$$

It is postulated that each particle has associated with it a scalar quantity, $m$ called the *mass*, we then define a quantity $p_\mu$ by

$$p_\mu = mu_\mu \tag{10.11}$$

as the generalisation of the three-vector momentum to four-vector momentum. This is confirmed by the fact three-space 1 time translational invariances of the Lagrangian give rise to four-momentum (three linear momentum + one energy) conservation just as the three-space translational invariance of the Lagrangian in non-relativistic mechanics leads to three-momentum conservation. Thus, of the four-momentum $p_\mu$ the first three-components correspond to ordinary linear momentum, whereas the 4th component correspond to energy. Hence, $p_4$ = energy is given by

$$p_4 = \frac{imc}{\sqrt{1-\beta^2}}. \tag{10.12}$$

## 10.2 Newton's Laws

The use of tensors is primarily to simplify transformation from one frame to another: within any one frame there is no point in changing entirely the familiar notation used so long as to correlate observed facts. In particular, there appears to be no reason to change Newton's laws as the basis for dynamical calculations. Forces have so far not been mentioned in connection with relativity physics. We will still define it by,

$$\text{force} = \text{time-rate of change of momentum}. \tag{10.13}$$

where now momentum is define by (10.11). Thus,

$$F_\mu = \frac{d}{dt}(p_\mu) = \frac{d}{dt}(mu_\mu). \tag{10.14}$$

But although this definition of force, does not present it as a four-vector the first three-components are

$$F_i = \frac{d}{dt}p_i = \frac{d}{dt}\left(\frac{m\dot{x}_i}{\sqrt{1-\beta^2}}\right) \tag{10.15}$$

in the limit $\beta \to 0$ (10.15) gives the familiar equation $\mathbf{F} = m\ddot{\mathbf{X}}$. As in non-relativistic case, we define kinetic energy $T$ by

$$\frac{dT}{dt} = \mathbf{F}\cdot\mathbf{V} = \sum_i F_i\dot{x}_i. \tag{10.16}$$

A development of this relation gives

$$\begin{aligned}\frac{dT}{dt} &= \sum_i u_i\sqrt{1-\beta^2}\frac{d}{dt}(mu_i)\\ &= \sqrt{1-\beta^2}\left[\sum_\mu u_\mu\frac{d}{dt}(mu_\mu) - u_4\frac{d}{dt}p_4\right]\\ &= \sqrt{1-\beta^2}\frac{d}{dt}\left[\sum_\mu \frac{1}{2}mu_\mu^2\right] - ic\frac{dp_4}{dt}\end{aligned}$$

$$\text{or,}\quad \frac{dT}{dt} = -ic\frac{dp_4}{dt}. \quad \left(\text{we used } \sum_\mu u_\mu^2 = -c^2\right). \tag{10.17}$$

Integrating (10.17)

$$p_1 = \frac{i}{c}(T + \text{constant}). \quad (10.18)$$

Referring back to (10.12) and postulating that kinetic energy is zero for a particle at rest, the value of the constant is $mc^2$

$$p_1 = \frac{i}{c}(T + mc^2) = \frac{iE}{c}. \quad (10.19)$$

where $E$ = total energy, $T$ and equivalent mass-energy, $mc^2$.
It now follows that

$$-m^2c^2 = \sum_\mu p_\mu^2 = \mathbf{p}^2 + p_1^2 = \mathbf{p}^2 - \frac{E^2}{c^2}.$$

Therefore,

$$(T + mc^2)^2 = E^2 = m^2c^4 + \mathbf{p}^2 \cdot c^2 = \frac{m^2c^4}{(1 - \beta^2)}. \quad (10.20)$$

This equation gives the general relation between the energy and the three-momentum.

## 10.3 Lagrangian Formulation

If the results obtained in the previous sections are to be consistent with a Lagrangian description, i.e., the new Newton's equation we have obtained in the last section is the same as the Lagrangian equation of motion, then it would be required that the momentum $p_i$ be given by

$$p_j = \frac{\partial L}{\partial \dot{x}_i}, \quad \text{or,} \quad \frac{m\dot{x}_i}{\sqrt{1 - \beta^2}} = \frac{\partial L}{\partial \dot{x}_i}. \quad (10.21)$$

So, integrating this expression we may obtain expression for the Lagrangian, i.e.,

$$L = \int \frac{d\dot{x}_i}{\sqrt{1 - \beta^2}} m\dot{x}_i = \int \frac{d\dot{x}_i m\dot{x}_i}{\sqrt{1 - \frac{1}{c^2}\sum_i \dot{x}_i^2}}.$$

This can be immediately integrated to give,

$$L = -mc^2\sqrt{1 - \beta^2} - V(x_i). \quad (10.22)$$

$V(x_i)$ = a constant of integration. We can interpret $V(x_i)$ as the potential.
From (10.22) we get for

$$\frac{\partial L}{\partial x_i} = \frac{d}{dt}\left(\frac{\partial L}{\partial \dot{x}_i}\right)$$

the following

$$-\frac{\partial V}{\partial x_i} = \frac{d}{dt}\left(\frac{m\dot{x}_i}{\sqrt{(1-\beta^2)}}\right) = \frac{d}{dt}p_i. \tag{10.23}$$

This is a correct equation of motion. For free particle case we put $V(x_i) = 0$. The Lagrangian in non-relativistic limit reduces to $L = \frac{1}{2}m\mathbf{V}^2$.
Let us apply the above consideration to electro-magnetic case. The Lorentz-force of the electro-magnetic case is

$$\mathbf{F} = e\left[\mathbf{E} + \frac{1}{c}(\mathbf{V}\times\mathbf{B})\right]$$

and we recast it in terms of generalised force as

$$F_i = e\left(\frac{d}{dt}\cdot\frac{\partial}{\partial\dot{x}_i} - \frac{\partial}{\partial x_i}\right)\left(\phi - \frac{1}{c}\mathbf{V}\cdot\mathbf{A}\right).$$

Therefore, the relativistic equation of motion can be written as

$$\frac{d}{dt}\left(\frac{m\dot{x}_i}{\sqrt{(1-\beta^2)}}\right) = e\left(\frac{d}{dt}\frac{\partial}{\partial\dot{x}_i} - \frac{\partial}{\partial x_i}\right)\left(\phi - \frac{1}{c}\mathbf{V}\cdot\mathbf{A}\right). \tag{10.24}$$

The equation (10.24) is consistent with experimental results of high energy electrons moving in electro-magnetic field.
We can choose the Lagrangian for the high-energy charged particle in presence of electro-magnetic field as

$$L = -mc^2\sqrt{1-\beta^2} - e\left(\phi - \frac{1}{c}\sum_i \dot{x}_i A_i\right). \tag{10.25}$$

Equation of motion follows from

$$\delta\int L dt = 0.$$

We use (10.25) to find the canonical conjugate momenta $p_i$.
We have

$$p_i = \frac{m\dot{x}_i}{\sqrt{(1-\beta^2)}} + \frac{e}{c}A_i. \tag{10.26}$$

The Hamiltonian $H$ is given by

$$\begin{aligned} H &= \sum_i p_i\dot{x}_i - L \\ &= \sum_i p_i\frac{\sqrt{1-\beta^2}}{m}\left(p_i - \frac{e}{c}A_i\right) + mc^2\sqrt{1-\beta^2} + e\left(\phi - \frac{1}{c}\mathbf{V}\cdot\mathbf{A}\right). \end{aligned}$$

This simplifies to

$$H = c\sqrt{\left[m^2c^2 + \left(\mathbf{p} - \frac{e}{c}\mathbf{A}\right)^2\right]} + e\phi. \tag{10.27}$$

Using (10.26) we write (10.27) as

$$H = \frac{mc^2}{\sqrt{(1-\beta^2)}} + e\phi. \tag{10.28}$$

If $T$ = kinetic energy then

$$\frac{dT}{dt} - \sum_i \dot{x}_i \frac{d}{dt} p_i = \sum_i \dot{x}_i \frac{d}{dt}\left[\frac{m\dot{x}_i}{\sqrt{(1-\beta^2)}}\right] - \frac{d}{dt}\left[\frac{mc^2}{\sqrt{(1-\beta^2)}}\right]. \tag{10.29}$$

Integrating we get

$$T = mc^2\left[\frac{1}{\sqrt{(1-\beta^2)}} - 1\right]. \tag{10.30}$$

where we have used $T = 0$, where $\dot{x}_i = 0$.
Again, from (10.24) and (10.29) we get

$$\frac{dT}{dt} = \sum_i e\dot{x}_i\left(\frac{d}{dt}\cdot\frac{\partial}{\partial \dot{x}_i} - \frac{\partial}{\partial x_i}\right)\left(\phi - \frac{1}{c}\mathbf{V}\cdot\mathbf{A}\right)$$

$$= e\mathbf{V}\cdot\left(E + \frac{1}{c}\mathbf{V}\times\mathbf{B}\right) = e\mathbf{V}\cdot\mathbf{E}. \tag{10.31}$$

Using

$$\mathbf{E} = -\bar{\nabla}\phi - \frac{1}{c}\frac{\partial \mathbf{A}}{\partial t},$$

we get

$$\frac{dT}{dt} = -e\mathbf{V}\cdot\left(\nabla\phi + \frac{1}{c}\frac{\partial}{\partial t}\mathbf{A}\right) = -e\frac{d\phi}{dt} + e\frac{\partial\phi}{\partial t} - \frac{1}{e}\mathbf{V}\cdot\frac{\partial \mathbf{A}}{\partial t}.$$

Hence

$$\frac{d}{dt}(T + e\phi) = e\left(\frac{\partial\phi}{\partial t} - \frac{1}{c}\mathbf{V}\cdot\frac{\partial \mathbf{A}}{\partial t}\right)$$

$$\text{or,}\quad \frac{d}{dt}\left(mc^2 + T + e\phi\right) = e\left(\frac{\partial\phi}{\partial t} - \frac{1}{c}\mathbf{V}\cdot\frac{\partial \mathbf{A}}{\partial t}\right). \tag{10.32}$$

The quantity $mc^2 + T + e\phi = E$ can be interpreted as the total energy. The meaning of the right-hand side of (10.32) is therefore, that they represent in the increase in $E$ of the particle due to the action of charges and currents which are causing the fields to vary with time.
Combining (10.28) with (10.30) we get

$$H = T + e\phi + mc^2. \tag{10.33}$$

Hence, $T + e\phi + mc^2$ is correct to interpret as total energy.
We have thus shown that relativistic formulation of Lagrangian and Hamiltonian description of high energy particle is possible.

## 10.4 Field Equation for Continuous Media

Fields are description of waves. Wave phenomenon occurs when a continuous media is disturbed. So far we have only considered the mechanics of a discrete system with finite degrees of freedom. Continuous system has infinite degrees of freedom. A special case of a continuous system would be a continuous one-dimensional elastic solid and this may be treated as the limiting case of a linear chain of interacting particles. For convenience, the interaction is produced by springs connecting each pair of neighbouring particles.

Let each particle is of mass $m$. The equilibrium separation distance is $a$ and they are connected by springs of force constant, $k$. Let, $\eta_i$ be the displacement of the $i$th particle from its equilibrium state. From the definition of force constant $k$, the force $m\ddot{\eta}_i$ must be

$$m\ddot{\eta}_i = k\left[(\eta_{i+1} - \eta_i) - (\eta_i - \eta_{i-1})\right]. \tag{10.34}$$

This force is derived from a potential $V$, given by

$$V = \frac{1}{2}\sum_i (\eta_{i+1} - \eta_i)^2. \tag{10.35}$$

The kinetic energy

$$T = \frac{1}{2}\sum_i m\dot{\eta}_i^2.$$

Hence, the *Lagrangian of the system* is given by

$$L = T - V = \frac{1}{2}\sum_i \left[m(\dot{\eta}_i^2) - \frac{1}{2}k(\eta_{i+1} - \eta_i)^2\right] \tag{10.36}$$

$$= \sum aL_i. \tag{10.37}$$

From (10.36) we have

$$\frac{d}{dt}\left(\frac{\partial L}{\partial \dot{\eta}_i}\right) = m\ddot{\eta}_i = \frac{\partial L}{\partial \eta_i} = k\left[(\eta_{i+1} - \eta_i) - (\eta_i - \eta_{i-1})\right].$$

The discrete system as given in equation (10.36) and (10.37) may now be made to go over to the continuous system with the following correspondence.

$$a \to dx, \quad \sum_i dx \to \int dx \quad \text{and} \quad L_i \to \mathcal{L}(x).$$

Consider,

$$L = \frac{1}{2}\sum_i a\left[\left(\frac{m}{a}\right)\dot{\eta}_i^2 - ka\left(\frac{\eta_{i+1} - \eta_i}{a}\right)^2\right]$$

$$= \frac{1}{2}\sum_i a\left[\mu\dot{\eta}_i^2 - Y\left(\frac{\eta_{i+1} - \eta_i}{a}\right)^2\right] = \sum_i aL_i.$$

where $Y$ = Young's modulus of elasticity and $\mu$ = mass per unit length.

When $a \to 0$, we have

$$\dot{\eta}_i^2 \to \left[\frac{\partial \eta(x)}{\partial t}\right]^2 \quad \text{and} \quad \left(\frac{\eta_{i+1} - \eta_i}{a}\right)^2 \xrightarrow[a \to 0]{} \left(\frac{\partial \eta}{\partial x}\right)^2 .$$

Thus,

$$L_i \xrightarrow[a \to 0]{} \mathcal{L}(x) = \frac{1}{2}\left[\mu\left(\frac{\partial \eta}{\partial t}\right)^2 - Y\left(\frac{\partial \eta}{\partial x}\right)^2\right].$$

Therefore,

$$L \xrightarrow{a \to 0} \int dx\, \mathcal{L}(x)$$

where

$$\mathcal{L}(x) = \frac{1}{2}\left[\mu\left(\frac{\partial \eta}{\partial t}\right)^2 - Y\left(\frac{\partial \eta}{\partial x}\right)^2\right], \quad \text{i.e.,} \quad \mathcal{L} = \mathcal{L}\left(\frac{\partial \eta}{\partial t}, \frac{\partial \eta}{\partial x}\right). \tag{10.38}$$

Hence, the Lagrangian

$$L = \int dx \mathcal{L}\left(\frac{\partial \eta}{\partial t}, \frac{\partial \eta}{\partial x}\right). \tag{10.39}$$

Quite generally we have for the Lagrangian of a continuous system in one-dimension

$$L = \int dx\; L\left(\eta, \frac{\partial \eta}{\partial x}, \frac{\partial \eta}{\partial t}\right).$$

So, generalising to three-dimension

$$L = \int dx\, dy\, dz\, \mathcal{L}\left[\eta(x, y, z, t), \frac{\partial \eta}{\partial x}, \frac{\partial \eta}{\partial y}, \frac{\partial \eta}{\partial z}, \frac{\partial \eta}{\partial t}\right]. \tag{10.40}$$

The action

$$S = \int L dt = \int dV dt\; \mathcal{L}\left(\eta, \frac{\partial \eta}{\partial x}, \frac{\partial \eta}{\partial y}, \frac{\partial \eta}{\partial z}, \frac{\partial \eta}{\partial t}\right). \tag{10.41}$$

where $dV = dx\, dy\, dz$.

The action (10.41) can be put in four-dimension relativistic form

$$S = \frac{1}{ic}\int d^4x \mathcal{L}\left(\eta, \frac{\partial \eta}{\partial x_\mu}\right) \tag{10.42}$$

where

$$\frac{\partial \eta}{\partial x_\mu} = \left(\frac{\partial \eta}{\partial x_1}, \frac{\partial \eta}{\partial x_2}, \frac{\partial \eta}{\partial x_3}, \frac{\partial \eta}{\partial x_4}\right).$$

where $x_1 = ict$ and $d^4x = dx_1\, dx_2\, dx_3\, dz_4$.

The Lagrangian

$$\mathcal{L}\left(\eta, \frac{\partial \eta}{\partial x_\mu}\right)$$

in (10.42) is called the *Lagrangian density*, for the four-dimensional continuous system. To derive the field equation, i.e., the Euler-Lagrangian equation we use Hamilton's variational principle, namely $\delta S = 0$ which implies

$$\delta \int d^4x\, \mathcal{L}\left(\eta, \frac{\partial \eta}{\partial x_\mu}\right) = 0.$$

To calculate

$$\delta \int \mathcal{L} d^4x = 0$$

we proceed as follows

$$\delta \int \mathcal{L}\left(\eta, \frac{\partial \eta}{\partial x_\mu}\right) d^4x = \int \delta\mathcal{L}\left(\eta, \frac{\partial \eta}{\partial x_\mu}\right) d^4x. \tag{10.43}$$

Consider

$$\int \delta\mathcal{L}\left(\eta, \frac{\partial \eta}{\partial x_\mu}\right) d^4x.$$

this is equal to

$$\int \left[\frac{\partial \mathcal{L}}{\partial \eta}\delta\eta + \frac{\partial \mathcal{L}}{\partial(\partial\eta/\partial x_\mu)}\cdot \delta\left(\frac{\partial \eta}{\partial x_\mu}\right)\right] d^4x. \tag{10.44}$$

Since,

$$\delta\left(\frac{\partial \eta}{\partial x_\mu}\right) = \frac{\partial}{\partial x_\mu}(\delta\eta).$$

we rewrite the second term of (10.44) as

$$\int \frac{\partial \mathcal{L}}{\partial(\partial\eta/\partial x_\mu)}\cdot \frac{\partial}{\partial x_\mu}(\delta\eta)\, d^4x.$$

This can also be written as

$$\int \frac{\partial}{\partial x_\mu}\left[\frac{\partial \mathcal{L}}{\partial(\partial\eta/\partial x_\mu)}\cdot \delta\eta\right] d^4x - \int d^4x \left[\frac{\partial}{\partial x_\mu}\cdot \frac{\partial \mathcal{L}}{\partial(\partial\eta/\partial x_\mu)}\right]\delta\eta. \tag{10.45}$$

The first term of (10.45) is of the form

$$\int d^4x \left[\frac{\partial}{\partial x_\mu}\cdot S_\mu\right],$$

which by four-dimensional Gauss theorem can be put as a surface integral in four-dimension. i.e.,

$$\int_{\text{surface}} d\sigma_\mu S_\mu.$$

and this is zero, if we take the surface at an infinite distance and note that the function $\eta$ should vanish at infinity. Thus, the second term of (10.44) reduce to

$$-\int d^4x \frac{\partial}{\partial x_\mu}\left[\frac{\partial \mathcal{L}}{\partial(\partial\eta/\partial x_\mu)}\right]\cdot\delta\eta. \tag{10.46}$$

Substituting (10.46) in (10.44) we get finally,

$$\int \delta\mathcal{L}\left(\eta, \frac{\partial\eta}{\partial x_\mu}\right) d^4x = \int \left[\frac{\partial\mathcal{L}}{\partial\eta}\delta\eta - \frac{\partial}{\partial x_\mu}\left(\frac{\partial\mathcal{L}}{\partial(\partial\eta/\partial x_\mu)}\right)\delta\eta\right] d^4x$$

$$= \int d^4x \left[\frac{\partial\mathcal{L}}{\partial\eta} - \frac{\partial}{\partial x_\mu}\frac{\partial\mathcal{L}}{\partial(\partial\eta/\partial x_\mu)}\right]\delta\eta. \tag{10.47}$$

Hence,

$$\delta\int \mathcal{L}d^4x = \int \delta\mathcal{L}\cdot d^4x = 0$$

becomes

$$\int d^4x \left\{\frac{\partial\mathcal{L}}{\partial\eta} - \frac{\partial}{\partial x_{mu}}\left[\frac{\partial\mathcal{L}}{\partial(\partial\eta/\partial x_\mu)}\right]\right\}\delta\eta = 0.$$

As variation $\delta\eta$ is arbitrary, the vanishing of the integral implies

$$\frac{\partial\mathcal{L}}{\partial\eta} - \frac{\partial}{\partial x_\mu}\left[\frac{\partial\mathcal{L}}{\partial(\partial\eta/\partial x_\mu)}\right] = 0. \tag{10.48}$$

Equation (10.48) is called the *Eular-Lagrangian equation* for the field $\eta$. The coordinate $\eta_i$ in discrete case goes over to $\eta(x)$ and $\eta(x)$ is called the *field variable*. The Euler-Lagrangian equation (10.48) is called the *Field equation*. For our one dimensional case the Lagrangian density $\mathcal{L}$ is [see equation (10.38)]

$$\mathcal{L} = \frac{1}{2}\mu\left(\frac{\partial\eta}{\partial t}\right)^2 - \frac{1}{2}Y\left(\frac{\partial\eta}{\partial x}\right)^2. \tag{10.49}$$

That is for one-dimension case of equation (10.48) is

$$\frac{\partial\mathcal{L}}{\partial\eta} = \frac{\partial}{\partial x}\left(\frac{\partial\mathcal{L}}{\partial(\partial\eta/\partial x)}\right) + \frac{\partial}{\partial t}\left(\frac{\partial\mathcal{L}}{\partial(\partial\eta/\partial t)}\right).$$

The Euler-Lagrangian equation (10.48) is

$$\frac{\partial\mathcal{L}}{\partial\eta} = \frac{\partial}{\partial x}\left[\frac{\partial\mathcal{L}}{\partial(\partial\eta/\partial x)}\right] + \frac{\partial\mathcal{L}}{\partial y}\left[\frac{\partial\mathcal{L}}{\partial(\partial\eta/\partial y)}\right] + \frac{\partial}{\partial z}\left[\frac{\partial\mathcal{L}}{\partial(\partial\eta/\partial z)}\right] + \frac{\partial}{\partial t}\left[\frac{\partial\mathcal{L}}{\partial(\partial\eta/\partial z)}\right].$$

Hence from (10.49), we get

$$\frac{\partial \mathcal{L}}{\partial \eta} = 0, \qquad \frac{\partial \mathcal{L}}{\partial(\partial \eta/\partial x)} = -\left(\frac{\partial \eta}{\partial x}\right) Y$$

$$\frac{\partial \mathcal{L}}{\partial(\partial \eta/\partial y)} = 0, \qquad \frac{\partial \mathcal{L}}{\partial(\partial \eta/\partial z)} = 0, \quad \text{and} \quad \frac{\partial \mathcal{L}}{\partial(\partial \eta/\partial t)} = +\left(\frac{\partial \eta}{\partial t}\right) \mu.$$

Hence, the field equation for the one-dimensional case is

$$0 = -Y\frac{\partial}{\partial x}\left(\frac{\partial \eta}{\partial x}\right) + \mu\frac{\partial}{\partial t}\left(\frac{\partial \eta}{\partial t}\right), \quad \text{or,} \quad \frac{\partial^2 \eta}{\partial x^2} = \frac{1}{v^2}\left(\frac{\partial^2 \eta}{\partial t^2}\right). \tag{10.50}$$

where $v^2 = Y/\mu$.

Equation (10.50) is a wave-equation. Therefore, the field $\eta(x,t)$ satisfies a wave-equation.

As an example, let us consider a Lagrangian density $\mathcal{L}$ for a scalar field, $\phi$, given by

$$\mathcal{L} = -\frac{1}{2}\left(\frac{\partial \phi}{\partial x_\mu}\right)^2. \tag{10.51}$$

Then,

$$\frac{\partial \mathcal{L}}{\partial \phi} = 0, \qquad \frac{\partial \mathcal{L}}{\partial(\partial \phi/\partial x_\mu)} = -\left(\frac{\partial \phi}{\partial x_\mu}\right).$$

Hence, the *Euler-Lagrangian equation*

$$\frac{\partial \mathcal{L}}{\partial \phi} = \frac{\partial}{\partial x_\mu}\left[\frac{\partial \mathcal{L}}{\partial(\partial \phi/\partial x_\mu)}\right]$$

becomes for the present Lagrangian

$$0 = \frac{\partial}{\partial x_\mu}\left(\frac{\partial \phi}{\partial x_\mu}\right). \tag{10.52}$$

$$\text{i.e.,} \quad \frac{\partial^2}{\partial x_\mu^2}\phi = 0. \tag{10.53}$$

Since

$$\frac{\partial^2}{\partial x_\mu^2} = \frac{\partial^2}{\partial x^2} + \frac{\partial^2}{\partial y^2} + \frac{\partial^2}{\partial z^2} + \frac{\partial^2}{\partial (ict)^2},$$

we get

$$\left(\nabla^2 - \frac{1}{c^2}\frac{\partial^2}{\partial t^2}\right)\phi = 0. \tag{10.54}$$

Equation (10.54) is the wave-equation for the scalar field $\phi$.

As a final example consider

$$\mathcal{L} = -\frac{1}{2}\left(\frac{\partial\phi}{\partial x}\right)^2 + \frac{1}{2}\left(\frac{\partial\phi}{\partial t}\right)^2 - \sin\phi.$$

Then it is easy to see the Euler-Lagrangian equation becomes

$$\frac{\partial^2\phi}{\partial t^2} - \frac{\partial^2\phi}{\partial x^2} + \cos\phi = 0.$$

The wave equation is non-linear. This however has interesting consequences in physics.

## 10.5 Lagrangian Formulation of a String

In section (10.3) we have outlined the Lagrangian formalism for a relativistic particles. We follow the same principle to set up a Lagrangian formalism for a *string*.

For a free particle the Lagrangian $L$ is

$$L = -mc^2\sqrt{1-\beta^2}.$$

The action

$$\begin{aligned} S = \int L dt &= -mc^2 \int \sqrt{1 - \frac{\dot{x}_i^2}{c^2}}dt \\ &= imc \int \sqrt{dx_1^2 + dx_1^2 + dx_1^2 + dx_3^2}. \end{aligned} \tag{10.55}$$

In deriving this we have used $x_4 = ict$ and

$$\dot{x}_i^2 = \left(\frac{dx_1}{dt}\right)^2 + \left(\frac{dx_2}{dt}\right)^2 + \left(\frac{dx_3}{dt}\right)^2.$$

Writing $ds$ as elementary arc length in four-dimension, we have

$$ds^2 = dx_1^2 + dx_2^2 + dx_3^2 + dx_4^2.$$

Therefore (10.54) becomes

$$S = imc\int ds. \tag{10.56}$$

The principle of least action, i.e., $\delta S = 0$ leads to the result that the particle traces out a geodesic ($\delta\int ds = 0$) in space-time.

Following this procedure for the mechanics of a relativistic string we take the Lagrangian as the *area* traced out in space-time by the string, so that the principle of least action leads to a determination of a minimal surface in space-time.

Consider an elementary arc length of the string. We take the $x$-axis as tangent to the string, $y$-axis is the direction of the transverse motion of the string.

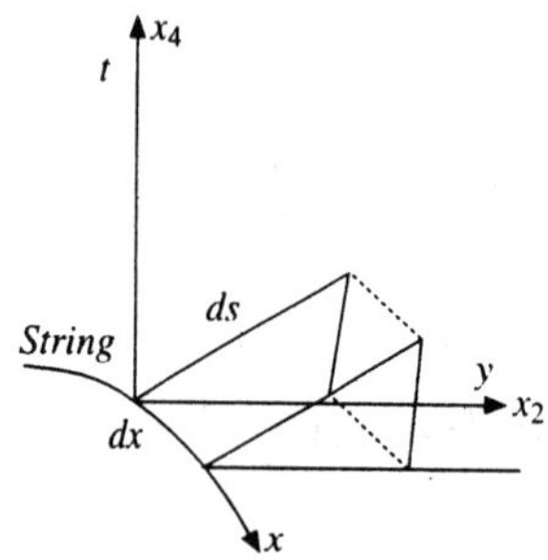


Fig 10.1

From the fig. 10.1,

$$dS^2 = dx_4^2 + dx_2^2,$$

therefore,

$$ds = dx_4 \left[1 + \left(\frac{dy}{dx_4}\right)^2\right]^{1/2}$$
$$= icdt \left[1 - \frac{1}{c}\left(\frac{dy}{dt}\right)^2\right]^{1/2}.$$

We put

$$V_\perp = \frac{dy}{dt},$$

the transverse velocity of the string. The element of area swept out in $dt$ is

$$dx\, ds = ic\, dx\, dt \left(1 - \frac{v_\perp^2}{c^2}\right)^{1/2}.$$

The total area swept out by the string is then

$$ic \int_{t_0}^{t_1} dt \int_{l_0}^{l_1} dl \left(1 - \frac{V_\perp^2}{c^2}\right)^{1/2}$$

We now define the action $S$ as

$$S = -\frac{1}{2\pi\alpha} \int dx_4 \int_{l_0}^{l_1} dl \left(1 - \frac{V_\perp^2}{c^2}\right)^{1/2}$$

and we put

$$L = -\frac{1}{2\pi\alpha} \int_{l_0}^{l_1} dl \left(1 - \frac{V_\perp^2}{c^2}\right)^{1/2} = \int_{l_0}^{l_1} dl\mathcal{L},$$

where $L$ is the Lagrangian and $\mathcal{L}$ is the Lagrangian density. The constant $(2\pi\alpha)^{-1}$ has the dimension so as to make $S$ of the dimension of action. We have for conjugate momentum $p_\perp$

$$p_\perp = \frac{\partial \mathcal{L}}{\partial V_\perp} = \frac{1}{2\pi\alpha} \cdot \frac{V_\perp/c^2}{\left[1 - (V_\perp^2/c^2)\right]^{1/2}}.$$

The energy density $\varepsilon$

$$\varepsilon = p_\perp V_\perp - \mathcal{L} = \frac{1}{2\pi\alpha}\frac{1}{\left[1-(V_\perp^2/c^2)\right]^{1/2}}$$

and the total energy

$$E = \int \varepsilon\, dl = \int_{l_0}^{l_1} dl \frac{1}{2\pi\alpha}\left(1-\frac{V_\perp^2}{c^2}\right)^{-1/2}.$$

As an example, consider a string of constant length rotating with constant angular velocity $\omega$, the ends of the string is moving with velocity $c$. Let $2a$ be the length of the string $(-a < l < a)$ and $\omega =$ angular velocity.
The angular momentum of the string will be given by

$$J = \int_{l_0}^{l_1} dl \cdot l p_\perp = \frac{1}{\pi\alpha}\int_0^a dl\left(\frac{lV_\perp}{c^2}\right)\left(1-\frac{V_\perp^2}{c^2}\right)^{-1/2}$$

and the energy by

$$E = \frac{1}{\pi\alpha}\int_0^a dl\left(1-\frac{V_\perp^2}{c^2}\right)^{-1/2}.$$

Put $l = V_\perp/\omega$.

$$E = \frac{1}{\pi\alpha}\int_0^1\left(\frac{c}{\omega}\right) d\left(\frac{V_\perp}{c}\right)\left[1-\left(\frac{V_\perp}{c}\right)^2\right]^{-1/2} = \frac{c}{2\alpha\omega}$$

$$\text{and} \quad J = \frac{1}{\pi\alpha}\left(\frac{c}{\omega^2}\right)\int_0^1\left(\frac{V_\perp}{c}\right)^2 d\left(\frac{V_\perp}{c}\right)\cdot\left(1-\frac{V_\perp^2}{c^2}\right)^{-1/2} = \frac{c}{4\alpha\omega^2}.$$

Thus,

$$J = \frac{\alpha}{c}E^2.$$

This relation is of importance in physics.

## Exercises

1. A pi-meson of mass $m_\pi$ comes to rest and decays into a $\mu$-meson of mass $m_\mu$ and a neutrino of zero mass. Show that the kinetic energy of $\mu$-meson is

$$T = \frac{(m_\pi - m_\mu)^2 c^2}{2m_\pi}.$$

[*Hints:* $\pi$-meson decays at rest $\mathbf{p} = 0$, and let $\mathbf{p}_\mu$ and $\mathbf{p}_\nu$ are the three-momentum of $\mu$ and $\nu$ particles. Hence by conservation

$$\mathbf{p}_\mu + \mathbf{p}_\nu = 0.$$

$E_\pi$ = energy of $\pi$. $E_\mu$, $E_\nu$ are energies of $\mu$ and $\nu$ particles.

Hence by conservation

$$E_\pi = E_\mu + E_\nu.$$

Put $c = 1$, then

$$E_\pi^2 = m_\pi^2 = (E_\mu + E_\nu)^2.$$

Again,

$$E_\mu^2 = p_\mu^2 + m_\mu^2, \quad E_\nu^2 = p_\nu^2 + m_\nu^2; \quad m_\nu = 0, \quad \text{and} \quad |p_\mu| = |p_\nu|.$$

Hence

$$E_\mu^2 - E_\nu^2 = m_\mu^2 \quad \text{and} \quad (E_\mu + E_\nu) = m_\pi.$$

Therefore,

$$E_\mu - E_\nu = \frac{m_\mu^2}{m_\pi}. \quad \therefore \quad E_\mu = \frac{m_\pi^2 + m_\mu^2}{2m_\pi}.$$

$$T = \text{K.E. of } \mu = E_\mu - m_\mu = \frac{(m_\pi^2 + m_\mu^2)}{2m_\pi} - m_\mu = \frac{(m_\pi - m_\mu)^2}{2m_\pi}$$

with correct dimension

$$T = \frac{(m_\pi - m_\mu)^2 c^2}{2m_\pi}.]$$

2. A photon of mass zero and of momentum $h\nu/c$ hits an electron of mass $m$ at rest and gets scattered off at some angle $\theta$ with new energy $h\nu'$. Show that the scattering angle $\theta$ is given by

$$2\frac{h}{mc}\sin^2\frac{\theta}{2} = \lambda' - \lambda,$$

where $\lambda = c/\nu$ and $\lambda' = c/\nu'$.

3. For the Lagrangian density

$$L = -e^{|\phi|} - \left(\frac{\partial \phi^*}{\partial x_\mu}\right)\left(\frac{\partial \phi}{\partial x_\mu}\right),$$

find the Lagrangian equation of motion for $\phi$ and $\phi^*$.

[*Hints:*

$$\mathcal{L} = -\left(\frac{\partial\phi}{\partial x_\mu}\right)\left(\frac{\partial\phi^*}{\partial x_\mu}\right) - e^{-|\phi|} = -\left(\frac{\partial\phi}{\partial x_\mu}\right)\left(\frac{\partial\phi^*}{\partial x_\mu}\right) - e^{-(\phi^*\phi)^{1/2}}$$

$$\frac{\partial L}{\partial(\partial\phi/\partial x_\mu)} = -\frac{\partial\phi^*}{\partial x_\mu}.$$

$$\frac{\partial L}{\partial\phi} = +\frac{1}{2}e^{-(\phi^*\phi)^{1/2}}(\phi^*\phi)^{-1/2}\phi^* = \frac{1}{2}e^{-|\phi|}\cdot\left(\frac{\phi^*}{\phi}\right)^{1/2}.$$

Hence the Lagrangian equation is

$$\frac{\partial^2}{\partial x_\mu^2}\phi^* + \frac{1}{2}e^{-|\phi|}\left(\frac{\phi^*}{\phi}\right)^{1/2} = 0.$$

4. Obtain the electromagnetic field equation from the Lagrangian density

$$\mathcal{L} = \alpha\sum_{\mu,\nu}\left(\frac{\partial A_\mu}{\partial x_\nu}\right)^2.$$

what is wrong with this Lagrangian?

[*Hints:*

$$\mathcal{L} = \alpha\left(\frac{\partial A_\mu}{\partial x_\nu}\right)^2$$

repeated indices are summed.

Action

$$S = \int \mathcal{L}d^4x = \alpha\int\left(\frac{\partial A_\mu}{\partial x_\nu}\right)^2 d^4x.$$

Therefore,

$$\begin{aligned}\delta S &= \alpha\int\delta\left(\frac{\partial A_\mu}{\partial x_\nu}\right)^2 d^4x = \alpha\int 2\left(\frac{\partial A_\mu}{\partial x_\nu}\right)\cdot\delta\left(\frac{\partial A_\mu}{\partial x_\nu}\right)d^4x\\ &= 2\alpha\int\left(\frac{\partial A_\mu}{\partial x_\nu}\right)\frac{\partial}{\partial x_\nu}(\delta A_\mu)\,d^4x.\end{aligned}$$

By integration by path this can be put as

$$\delta S = 2\alpha\int\left(\frac{\partial A_\mu}{\partial x_\nu}\right)\frac{\partial}{\partial x_\nu}(\delta A_\mu) = -2\alpha\int\left[\frac{\partial}{\partial x_\nu}\left(\frac{\partial A_\mu}{\partial x_\nu}\right)\right]\delta A_\mu d^4x.$$

Hence $\delta S = 0$ implies

$$\frac{\partial}{\partial x_\nu}\left(\frac{\partial A_\mu}{\partial x_\nu}\right) = \Box A_\mu = 0.$$

5. A gas at NTP consists of particles of same mass as that of electron but it behaves mechanically like hard-sphere of radius 10 Å. Consider whether or not the situation can be described adequately by classical non-relativistic mechanics.

6. For a classical rotating string calculate the length of the string, where $J = 1$. [Take $\alpha/c = 1h$ (Gev)$^{-2}$.] [*Ans.* $8 \times 10^{-11}$ cm.]

7. A mass-less particle can be modelled as a non-rotating string of length $b$ moving with velocity $c$ and aligned along with its direction of motion. Calculate $b$ for such a particle having energy 1 Gev [Take $\alpha/c = 1h$ (Gev)$^{-2}$.] [*Ans.* $1.2 \times 10^{-13}$ cm.]

# Miscellaneous Solved Examples

**Example 1** *A light string passes over a fixed smooth pulley and carries at one end a mass 6m, and the other a smooth pulley of mass 3m, over which passes a smooth light string carrying masses 2m and m at the ends. Assuming the system starts from rest, obtain the accelerations of the masses and the movable pulley.*

***Solution:*** $P_1$ and $P_2$ are fixed and movable pulleys. The vertical distance of $3m$ (movable pulley) is $x_1$ from the top of $P_1$. $6m$ has co-ordinate $l_1 - x_1$; $l_1$ = fixed string length over $P_1$. $x_2$ is the distance of $2m$ from $P_2$ so it has coordinate $(x_1 + x_2)$ from $P_1$ and the co-ordinate of $m$ is thus $l_2 - x_2 + x_1$; $l_2$ = fixed string length over $P_2$.
Kinetic energy of $6m$ is

$$\frac{1}{2} \times 6m(\dot{l}_1 - \dot{x}_1)^2 \text{ etc.}$$

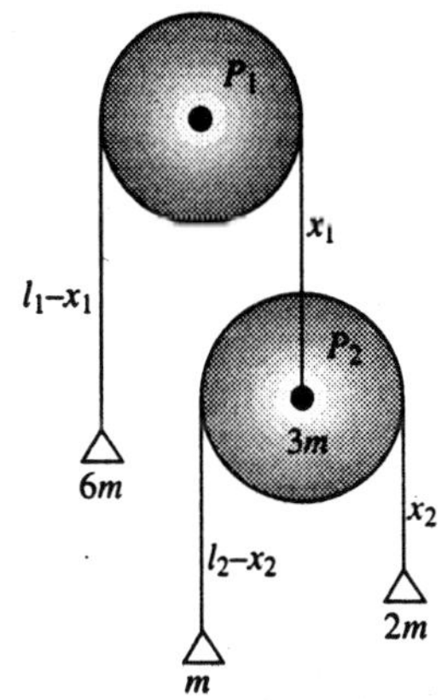


Fig E.1

Thus

$$\begin{aligned} T &= \text{sum of the K.E. of the system} \\ &= \frac{1}{2} \cdot 6m\dot{x}_1^2 + \frac{1}{2} \cdot 3m\dot{x}_1^2 + \frac{1}{2} 2m(\dot{x}_1 + \dot{x}_2)^2 + \frac{1}{2} m(\dot{x}_1 - \dot{x}_2)^2. \end{aligned}$$

The potential energy $= V$ is

$$V = -6mg(l_1 - x_1) - 3mgx_1 - 2mg(x_1 + x_2) - mg(l_2 - x_2 + x_1).$$

The Lagrangian

$$L = T - V.$$

Then from

$$\frac{\partial L}{\partial x_1} = \frac{d}{dt}\left(\frac{\partial L}{\partial \dot{x}_1}\right) \quad \text{and} \quad \frac{\partial L}{\partial x_2} = \frac{d}{dt}\left(\frac{\partial L}{\partial \dot{x}_2}\right)$$

we get

$$12\ddot{x}_1 + \ddot{x}_2 = 0 \quad \text{and} \quad 3\ddot{x}_2 + \ddot{x}_1 = g.$$

Hence

$$\ddot{x}_1 = -\frac{g}{35}; \quad \ddot{x}_2 = \frac{12g}{35}.$$

Acceleration**s**

$$6m \text{ and } 3m : \frac{g}{35}; \quad 2m : \frac{11g}{35}; \quad m : \frac{13g}{35}.$$

**Example 2** *If $(X, Y)$ are the component of forces acting in two dimension on a particles of unit mass, then the Newton's equation can be written as*

$$\frac{d}{dx}\left[\left(Y - X\frac{dy}{dx}\right) \Big/ \frac{d^2y}{dx^2}\right] - 2X = 0.$$

***Solution:*** $m = 1$, hence

$$\frac{d^2x}{dt^2} = v_x\frac{dv_x}{dx} = X$$

$$\text{and} \quad \frac{d^2y}{dt^2} = v_y\frac{dv_y}{dy} = Y.$$

Again,

$$\frac{dy}{dx} = \frac{v_y}{v_x}.$$

Here

$$\frac{d^2y}{dx^2} = \frac{d}{dx}\left(\frac{dy}{dx}\right)$$

$$\frac{d^2y}{dx^2} = \frac{d}{dx}\left(\frac{v_y}{v_x}\right) = \frac{1}{v_x}\cdot\frac{d}{dx}(v_y) - v_y\cdot\frac{1}{v_x^2}\frac{dv_x}{dx} = \frac{1}{v_x}\cdot\frac{dv_y}{dy}\cdot\frac{dy}{dx} - v_y\frac{1}{v_x^2}\frac{dv_x}{dx}$$

$$= \frac{v_y}{v_x^2}\frac{dv_y}{dy} - \frac{v_y}{v_x}\cdot v_x\frac{dv_x}{dx}\cdot\frac{1}{v_x^2} = \frac{1}{v_x^2}\left(v_y\frac{dv_y}{dy} - \frac{dy}{dx}\cdot v_x\frac{dv_x}{dx}\right)$$

$$= \frac{1}{v_x^2}\left(Y - \frac{dy}{dx}\cdot X\right), \quad \text{or,} \quad v_x^2 = \frac{\left(Y - X\frac{dy}{dx}\right)}{d^2y/dx^2}$$

$$\text{or,} \quad \frac{d}{dx}v_x^2 = \frac{d}{dx}\left[\left(Y - X\frac{dy}{dx}\right) \Big/ \frac{d^2y}{dx^2}\right]$$

$$\text{or,} \quad 2v_x\frac{dv_x}{dx} = 2X = \frac{d}{dx}\left[\left(Y - X\frac{dy}{dx}\right) \Big/ \frac{d^2y}{dx^2}\right].$$

**Example 3** *A bead of mass, $m$ is constrained to move on a rod $OA$ in the xy-plane. The rod is made to rotate with constant angular velocity $\omega$ in the xy-plane. The reaction of the rod on the bead is $R$, the force of constraint. Show that the virtual work done by $R$ is zero.*

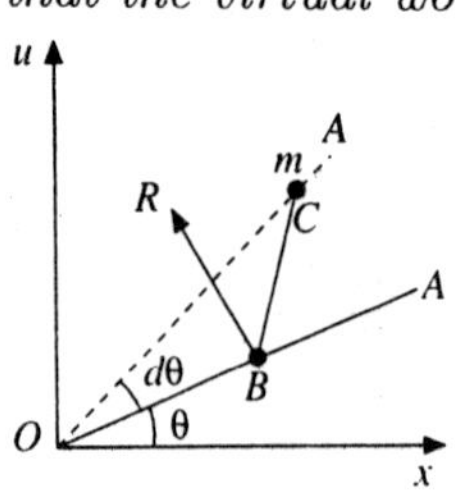


Fig E.2

***Solution:*** $OA$ = rod, $B$ = position of the bead at time $t$. The co-ordinate of $B$

$$x = r\cos\theta, \quad y = r\sin\theta$$

and $\theta = \omega t$; $\omega$ = constant angular velocity of the rod around $O$.

Consider $m$ is confined with the rod, we have therefore,

$$x = r\cos\omega t \quad \text{and} \quad y = r\sin\omega t.$$

Hence for an arbitrary displacement, $\delta s$, in which bead stays with the rod and time varies (account is taken of rod's rotation) i.e., $\delta s$ is a real displacement, then $\delta s = (\delta x, \delta y)$, where

$$\begin{aligned}\delta x &= \delta r\cos\omega t - r\omega\delta t\sin\omega t\\ \delta y &= \delta y\sin\omega t + r\omega\cos\omega t\cdot\delta t.\end{aligned}$$

In this real displacement,

$$\delta s = ac.$$

Therefore we easily see $R\cdot\delta s \neq 0$ as $R$ has component along $ac$, the dotted line in the figure. Hence for real displacement force of constraint does do work.

Consider now the virtual displacement, in which time-variation is not taking place, i.e., $\delta t = 0$. In this case

$$\delta x = \delta r\cos\omega t, \quad \delta y = \delta r\sin\omega t.$$

In this virtual displacement case $\delta s$ is along $\delta r$ therefore,

$$R\cdot\delta s = 0$$

as $R$ is orthogonal to rod, and $r$ = radius vector along the rod.

Hence for virtual displacement, the virtual work done by forces of constraint is zero.

**Example 4** *Set up the Lagrangian equation of motion of a heavy particle placed on the top of a smooth circle in a vertical-plane and allowed to slide. Find the point when the particle flies off the circle.*

***Solution:*** Mass of particle $= m$, radius of circle $= a$. Initially the particle is at $P$. $OP$ makes an angle $\alpha$. At any later time $t$ the position makes an angle $\theta$.

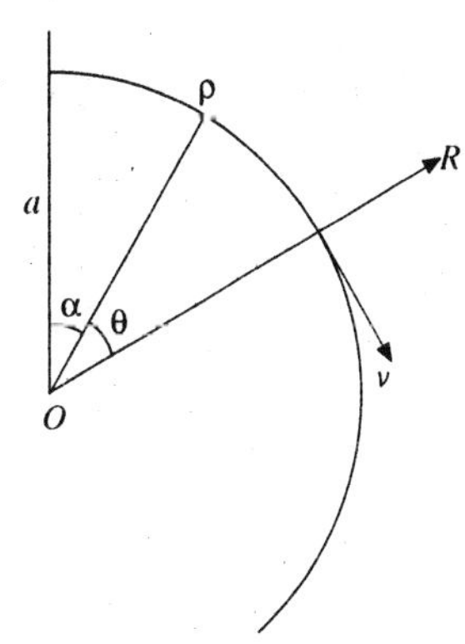

Fig E.3

$$\dot{\theta} = \frac{v}{a}.$$

The K.E. $= \frac{1}{2}mv^2 = \frac{1}{2}ma^2\dot{\theta}^2$.

The P.E. $= -mga(\cos\alpha - \cos\theta)$. $L = T - V$.

Lagrangian equation

$$\frac{d}{dt}\left(\frac{\partial L}{\partial\dot{\theta}}\right) = \frac{\partial L}{\partial\theta}, \quad \text{i.e.,} \quad ma\ddot{\theta} = mg\sin\theta.$$

Hence

$$2ma\dot{\theta}\ddot{\theta} = 2mg\dot{\theta}\sin\theta, \text{ or. } \frac{d}{dt}(ma\dot{\theta}^2) = 2\frac{d}{dt}(-mg\cos\theta).$$

$$\text{or. } \frac{1}{2}ma\dot{\theta}^2 = -mg\cos\theta + \text{ constant}$$

when $\dot{\theta} = 0$ at $t = 0$, $\theta = \alpha$, so

$$\frac{1}{2}ma\dot{\theta}^2 = mga(\cos\alpha - \cos\theta), \quad \text{or,} \quad v^2 = 2g(\cos\alpha - \cos\theta).$$

Again if $R =$ reaction at time $t$ which is along the radius $OP'$. Then

$$\frac{mv^2}{a} = mg\cos\theta - R.$$

Therefore,

$$R = mg(3\cos\theta - 2\cos\alpha).$$

The particle will fly off when $R = 0$ i.e.,

$$\cos\theta = \frac{2}{3}\cos\alpha.$$

**Example 5** *The length of a plane simple pendulum changes with time such that $l = a + bt$, where $a$ and $b$ are constants.*
*Find the Lagrangian equation of motion. Obtain the Hamiltonian and show that it is not invariant under time-translation.*

***Solution:*** If $l$ is the length of the pendulum, $m$ the mass and as $l$ is a function of time, the K.E. is

$$T = \frac{1}{2}m(\dot{l}^2 + l^2\dot{\theta}^2).$$

where $\theta$ is the angle made by the pendulum with vertical. The potential $V$ is

$$V = mgl(1 - \cos\theta) = mg(a + bt)(-\cos\theta).$$

The Lagrangian $L = T - V$. The equation

$$\frac{d}{dt}\left(\frac{\partial L}{\partial\dot{\theta}}\right) = \frac{\partial L}{\partial\theta}$$

yields

$$\frac{d}{dt}\left[m(a + bt)^2\dot{\theta}\right] = -mg(a + bt)\sin\theta,$$

$$\text{i.e.,} \quad (a + bt)\ddot{\theta} + 2b\dot{\theta} + g\theta = 0.$$

we use $\sin\theta \simeq \theta$ for small amplitude of vibration.

$$p_\theta = \frac{\partial L}{\partial \dot\theta} = m(a+bt)^2\dot\theta \quad \text{and} \quad H(p_\theta,\theta,t) = p_\theta \cdot \dot\theta - L$$
$$= \frac{p_\theta^2}{2m(a+bt)^2} + mg(a+bt)(-\cos\theta),$$

we find

$$\frac{\partial H}{\partial t} \neq 0,$$

so Hamiltonian is not invariant under time-translation. Hence energy is not conserved.

**Example 6** *From Stokes law the viscous force $F$ which opposes the motion of a sphere of radius, $a$ in a liquid of coefficient of viscocity, $\eta$ is given by $F = -6\pi\eta a v$, where $v$ is the terminal velocity. Find the Rayleigh dissipation function and obtain the Lagrangian equation for a sphere falling along $z$-axis under gravity in the viscous liquid.*

***Solution:*** If $F_z = -k_z v_z$ is the dissipative force, then the Rayleigh function

$$R = -\frac{1}{2}k_z v_z^2.$$

Here $k_z = 6\pi a\eta$ the Lagrange's equation is

$$\frac{d}{dt}\left(\frac{\partial L}{\partial \dot z}\right) - \frac{\partial L}{\partial z} + \frac{\partial R}{\partial \dot z} = 0$$

and $\dot z = v_z$ and

$$L = \frac{1}{2}Mv_z^2 - Mgz \quad (z \text{ is measured from the ground})$$
$$M = \frac{4}{3}\pi a^3\psi \quad (\psi = \text{density of sphere}).$$

**Example 7** *A particle of mass, $m$ is constrained to move on a circle under the action of gravity. The circle is placed in a vertical-plane and placed on the ground. Find the Lagrangian equation.*

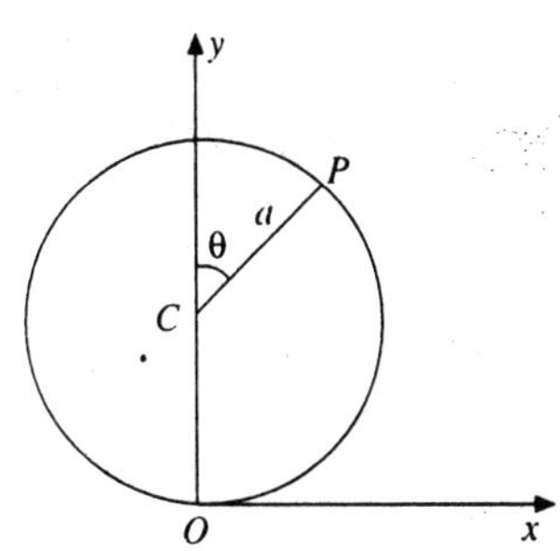

Fig E.4

***Solution:*** $a$ = radius; $CP = OC = a$ and $\angle PCY = \theta$. The co-ordinates of

$$P = (a\sin\theta,\ a + a\cos\theta)$$
$$\text{K.E.} = T = \frac{1}{2}m(\dot x^2 + \dot y^2) = \frac{1}{2}ma^2\dot\theta^2$$
$$V = \text{potential energy } = mga(1+\cos\theta)$$
$$L = T - V, \text{ the Lagrangian equation}$$

$$\frac{d}{dt}\left(\frac{\partial L}{\partial \dot\theta}\right) = \frac{\partial L}{\partial \theta}.$$

$$\ddot\theta = \frac{g}{a}\sin\theta, \quad \text{or,} \quad \frac{d}{dt}(\dot\theta^2) = -\frac{g}{a}2\cdot\frac{d}{dt}(\cos\theta),$$

$$\text{or,} \quad \dot\theta^2 + \frac{2g}{a\cos\theta} = \text{constant} = K.$$

Hence

$$\frac{d\theta}{dt} = \sqrt{\left(K - \frac{2g}{a}\cos\theta\right)}$$

yielding

$$t = \int \frac{d\theta}{\sqrt{\left(K - \frac{2g}{a}\cos\theta\right)}} + \text{constant}.$$

**Example 8** *A body of mass $m_1$ is thrown up an inclined-plane which is moving horizontally with a constant velocity $V$. Use Lagrangian equation to find the locus of the position of the body at any time $t$, after the motion sets in.*

***Solution:*** $OC = Vt$, distance moved by plane in time $t$.

$CP = r$ (position of body at $P$ at time $t$).

Here $(x, y)$ co-ordinate of $P$;

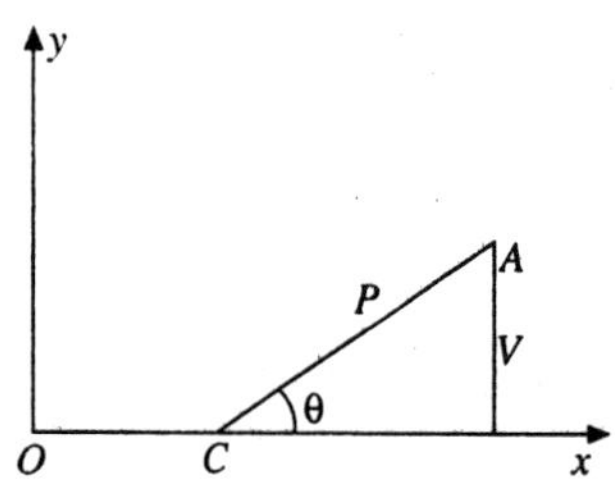


Fig E.5

$$x = Vt + r\cos\theta$$

$$\text{and} \quad y = r\sin\theta \quad (\theta = \text{fixed angle})$$

$$\text{K.E.} = T = \frac{1}{2}m(\dot x^2 + \dot y^2)$$

$$= \frac{m}{2}\left(V^2 + \dot r^2 + 2V\cos\theta\dot r\right);$$

potential energy $U$ is

$$U = mgr\sin\theta. \quad L = \frac{1}{2}m\left(V^2 + \dot r^2 + 2V\cos\theta\dot r\right) - mgr\sin\theta$$

$$2V\cos\theta\dot r = \frac{d}{dt}(2Vr\cos\theta) = \text{total time-derivation, so}$$

$$L = \frac{1}{2}m(V^2 + \dot r^2) - mgr\sin\theta.$$

Lagrange's equation

$$\ddot r = -g\sin\theta.$$

Hence

$$r = ut - \frac{1}{2}g\sin\theta t^2,$$

where $u$ = velocity with which the body was thrown up. Hence the locus is

$$(x - Vt)^2 + y^2 = r^2 = \left(ut - \frac{1}{2}g\cos\theta\cdot t^2\right)^2.$$

**Example 9** *Given*

$$\frac{\partial H}{\partial t} = 0 \quad \textit{and} \quad \frac{dF}{dt} = 0,$$

*where $F$ is an observable. Show that $\frac{\partial F}{\partial t}$ is conserved.*

***Solution:*** We have

$$\frac{dH}{dt} = \frac{\partial H}{\partial t} = 0,$$

so $H$ is conserved. Also $F$ is conserved as $dF/dt = 0$. Therefore, $[F, H]$ is a constant of motion as $H$ and $F$ are constant of motion. But

$$\frac{dF}{dt} = \frac{\partial F}{\partial t} + [F, H]$$

hence

$$\frac{\partial F}{\partial t} = -[F, H] \quad \text{as} \quad \frac{dF}{dt} = 0.$$

As $[F, H]$ is conserved, so $\partial F/\partial t$ is a constant of motion.

**Example 10** *For a 2-degrees of freedom, the kinetic energy:*

$$T = \frac{1}{2}u(\dot{q}_1^2 + \dot{q}_2^2); \quad u = q_1^2 + q_2^2$$

*and the potential $V$ is*

$$V = \frac{1}{u}[\omega_1(q_1) + \omega_2(q_2)]; \quad u = q_1^2 + q_2^2.$$

*Show that the system is soluble.*

***Solution:*** Lagrangian, $L = T - V$ and Hamiltonian $H = T + V$, does not depend on time explicitly, so energy is conserved.

$$T + V = h \text{ (a constant)}.$$

Therefore

$$T = h - V.$$

The Lagrangian equation is

$$\frac{d}{dt}(u\dot{q}_1) = \frac{1}{2}\frac{\partial u}{\partial q_1}(\dot{q}_1^2 + \dot{q}_2^2) - \frac{\partial V}{\partial q_1}$$

and similar equation for $q_2$. Multiply both sides by $2(u\dot{q}_1)$ then

$$\begin{aligned}\frac{d}{dt}(u\dot{q}_1)^2 &= u\frac{\partial u}{\partial q_1}\dot{q}_1(\dot{q}_1^2+\dot{q}_2^2) - 2\frac{\partial V}{\partial q_1}u\dot{q}_1 \\ &= u(\dot{q}_1^2+\dot{q}_2^2)\frac{\partial u}{\partial q_1}\dot{q}_1 + 2u\dot{q}_1\frac{\partial}{\partial q_1}(h-V) \\ &= 2(h-V)\dot{q}_1\frac{\partial u}{\partial q_1} + 2\dot{q}_1 u\frac{\partial}{\partial q_1}(h-V) \\ &= 2\dot{q}_1\frac{\partial}{\partial q_1}\left[u(h-V)\right].\end{aligned}$$

Since

$$u = q_1^2 + q_2^2 = u_1(q_1) + u_2(q_2).$$

therefore,

$$\frac{\partial u}{\partial q_1} = \frac{\partial u_1}{\partial q_1}.$$

Then

$$\frac{d}{dt}(u\dot{q}_1)^2 = 2\dot{q}_1\frac{\partial}{\partial q_1}[u(q_1)h - \omega_1(q_1)]$$

where we used

$$uV = \omega_1(q_1) + \omega_2(q_2) \quad \text{and} \quad \frac{\partial}{\partial q_1}uV = \frac{\partial\omega_1(q_1)}{\partial q_1}.$$

Therefore,

$$\frac{d}{dt}(u\dot{q}_1)^2 = 2\frac{d}{dt}[u_1(q_1)h - \omega_1(q_1)].$$

$$\text{or,}\quad u\dot{q}_1^2 = 2[u_1h - \omega_1] + r_1, \quad r_1 = \text{constant}.$$

Similar equation for $q_2$ with constant $r_2$. Adding these two equations we find $r_1 + r_2 = 0$. And

$$\frac{u\dot{q}_1}{\sqrt{hu_1 - \omega_1 + r_1}} = \frac{u\dot{q}_2}{\sqrt{hu_2 - \omega_2 + r_2}}.$$

$$\text{or,}\quad \int\frac{dq_1}{\sqrt{hu_1 - \omega_1 + r_1}} = \int\frac{dq_2}{\sqrt{hu_2 - \omega_2 + r_2}}.$$

Hence the problem is soluble.

**Example 11** *(a) Is the transformation $Q = 1/p$; $P = qp^2$ canonical? (b) Given the P.B. of $q$ and $f(p)$ as $[q, f(p)] = f(p)$, what is $f(p)$. (c) Is $x\dot{x} + y\dot{y} + x\dot{y} + \dot{x}y =$ constant, a holonomic constraint?*

***Solution:*** (*a*) $Q = \frac{1}{p}$, $P = qp^2$, so,

$$\frac{\partial Q}{\partial q} = 0, \quad \frac{\partial Q}{\partial p} = -\frac{1}{p^2}$$

$$\text{and} \quad \frac{\partial P}{\partial q} = p^2, \quad \text{and} \quad \frac{\partial P}{\partial p} = 2pq.$$

$$[Q, P]_{qp} = \frac{\partial Q}{\partial q}\frac{\partial P}{\partial p} - \frac{\partial Q}{\partial p}\frac{\partial P}{\partial q} = 0 + \left(\frac{1}{p^2}\right)p^2 = 1.$$

Hence the transformation canonical.

(b) $$[q, f(p)] = \frac{\partial q}{\partial q}\frac{\partial f}{\partial p} - \frac{\partial q}{\partial p}\frac{\partial f}{\partial q} = \frac{\partial f}{\partial p}.$$

Now,

$$\frac{\partial f}{\partial p} = f \quad \text{so } f = e^p.$$

(*c*) $x\dot{x} + y\dot{y} + x\dot{y} + \dot{x}y = k$ means

$$\frac{1}{2}\frac{d}{dt}(x^2 + y^2) + \frac{d}{dt}xy = k, \quad \text{i.e.,} \ (x + y)^2 = 2kt + \text{constant}.$$

Hence constraint is holonomic.

**Example 12** *A circular hoop is rolling without slipping down an inclined-plane. Use Lagrange's equation to find the acceleration of the rolling hoop.*

***Solution:*** Let the point of contact $P$ has come down a distance $x$. The C.G. has translational energy $\frac{1}{2}M\dot{x}^2$.

The K.E. of rotation $= \frac{1}{2}I\omega^2$, where $\omega = \dot{\alpha}$, $\alpha =$ angle of rotation.

Since rolling $x = a\alpha$, $a =$ radius of hoop.

Total K.E. $= T = \frac{1}{2}M\dot{x}^2 + \frac{1}{2}(I\dot{x}^2/a^2)$.

Potential energy dropped for rolling a distance $x$ by the centre

$$V = -Mg(x \sin\theta + a\cos\theta).$$

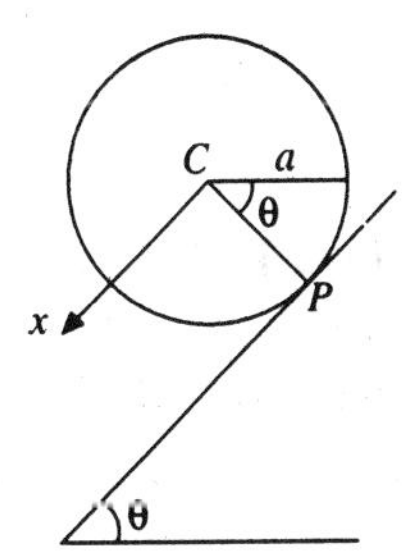


Fig E.6

Hence the

$$L = \frac{1}{2}\left(M + \frac{I}{a^2}\right)\dot{x}^2 + Mg(x\sin\theta + a\cos\theta)$$

$$\frac{d}{dt}\left(\frac{\partial L}{\partial \dot{x}}\right) = \frac{\partial L}{\partial x} \quad \text{yields}$$

$$\left(M + \frac{I}{a^2}\right)\ddot{x} = Mg\sin\theta$$

$I = MK^2$; $K^2 =$ radius of gyration.

Therefore,

$$\left(1+\frac{K^2}{a^2}\right)\ddot{x} = g\sin\theta; \quad K^2 = a^2, \quad \text{so, } \ddot{x} = \frac{1}{2}g\sin\theta.$$

**Example 13** *Find the law of force if the orbit of a planet is given in polar co-ordinate, $r\cdot\theta = k$, where $k$ is a constant.*

***Solution:*** In two dimension, in term of polar co-ordinate acceleration along radius vector $a_r$ and along the tangential $a_\theta$ are given by

$$\begin{aligned} a_r &= \ddot{r} - r\dot{\theta}^2, \\ a_\theta &= r\ddot{\theta} + 2\dot{r}\dot{\theta}. \end{aligned}$$

Hence

$$\mu\ddot{r} - \mu r\dot{\theta}^2 = F(r),$$

where $F$ = force. As $r\theta = k$, so

$$\theta = \frac{k}{r} \quad \text{and} \quad \dot{\theta} = -\frac{k}{r^2}\dot{r}.$$

Since $\mu r^2\dot{\theta}$ = angular momentum = constant = $C$, then

$$\mu r^2\left(-\frac{k}{r^2}\right)\dot{r} = C, \quad \text{or,} \quad \dot{r} = \text{constant}, \ddot{r} = 0.$$

Therefore,

$$F(r) = -\mu r\left(-\frac{k}{r^2}\right)^2 C^2 = -\frac{-\alpha}{r^3}; \quad \alpha = \text{constant}.$$

Hence

$$F \propto \frac{1}{r^3}.$$

**Example 14** *A sphere of mass, m and of radius a is rolling over a fixed rough sphere of radius b, without slipping. Use Lagrange's equation to find the acceleration of the rolling sphere at any instant of time. Determine the frictional force and the coefficient of friction, $\mu$.*

***Solution:*** Let the centre of gravity of small sphere of radius, $a$ be at $C$ which is the centre of the sphere. This has moved a distance, $x$ from vertical with $\omega$ as angular velocity of rotation of the small sphere around its axis. Hence $\dot{x} = a\dot{\alpha}$, $\alpha$ = angle the radius $a$ has rotated around the axis fixed in the body of the small sphere. $\dot{x} = aw$. If the small sphere started rolling at $t = 0$, from the position $\theta = 0$, then

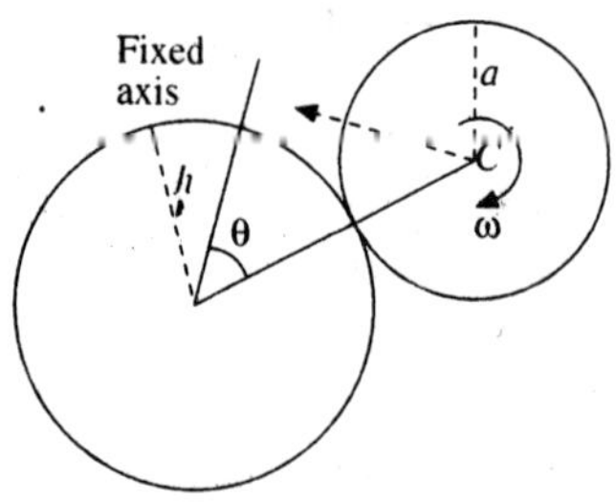

Fig E.7

$$x = a\alpha = (a+b)\theta.$$

The K.E. of translation of rolling sphere $= (1/2)m\dot{x}^2$.
The K.E. of rotation around its body fixed axis $= (1/2)I\omega^2$.
Hence

$$T = \frac{1}{2}m(a+b)^2\dot{\theta}^2 + \frac{1}{2}I\left(\frac{a+b}{a}\right)^2\dot{\theta}^2.$$

The potential energy of $C = V = -mg(a+b)(1-\cos\theta)$.

$$L = \left[\frac{1}{2}m(a+b)^2 + \frac{1}{2}I\left(\frac{a+b}{a}\right)^2\right]\dot{\theta}^2 + mg(a+b)(1-\cos\theta).$$

The Lagrangian equation is

$$m(a+b)^2\ddot{\theta} + I\frac{(a+b)^2}{a^2}\ddot{\theta} = mg(a+b)\sin\theta.$$

Let $K^2 =$ radius of gyration of sphere of radius $a$, then

$$\left(1+\frac{K^2}{a^2}\right)(a+b)^2\ddot{\theta} = g(a+b)\sin\theta.$$

For sphere

$$K^2 = \frac{2}{5}a^2, \quad \text{or.} \quad (a+b)\ddot{\theta} = \frac{5}{7}g\sin\theta. \tag{1}$$

This gives the acceleration at any instant, of the rolling sphere with respect to the fixed sphere measured from the vertical fixed axis. We can solve (1) by multiplying $2\dot{\theta}$ on both sides then

$$\frac{7}{5}(a+b)\frac{d}{dt}\dot{\theta}^2 = -2g\frac{d}{dt}\cos\theta.$$

$$\text{i.e.,} \quad \frac{7}{5}(a+b)\dot{\theta}^2 + 2g\cos\theta = \text{constant}.$$

Let at $t=0$, $\theta=0$ and $\dot{\theta}=0$, so

$$(a+b)\dot{\theta}^2 = 2g(1-\cos\theta)\times\frac{5}{7}.$$

If $F =$ force of friction, then rate of change of angular momentum is the torque of $F$. (Reaction passes along the radius of the small sphere through the point of contact.)
Hence

$$I\dot{\omega} = Fa, \quad m\frac{2}{5}a^2\cdot\frac{a+b}{a}\ddot{\theta} = Fa, \quad \text{or.} \quad F = m\frac{2}{5}(a+b)\ddot{\theta}.$$

For reaction $R$, we have the standard result

$$m(a+b)\dot{\theta}^2 = mg\cos\theta - R \quad \text{and} \quad \mu R = F.$$

From above we have

$$\mu = \frac{(17\cos\theta - 10)}{(2\sin\theta)}.$$

**Example 15** *Discuss the nature of the orbits under a central force. A planet is moving in a parabolic path, what is the speed of the planet at any instant?*

***Solution:*** We have for the orbit, the eccentricity, $e$ is

$$e = \sqrt{1 + \frac{2El^2}{\mu k^2}}.$$

where $E$ is the total energy, $l$ = angular momentum, a constant and potential

$$V = -\frac{k}{r}, \quad k > 0$$

and $\mu$ the reduced mass: the expression for energy

$$E = \frac{\mu}{2}\dot{\mathbf{r}}^2 + V(r) = \frac{1}{2}\mu\dot{r}^2 + \frac{1}{2}\mu r^2\dot{\theta}^2 + V$$

as $l = \mu r^2\dot{\theta}$, so

$$E = \frac{1}{2}\mu\dot{r}^2 + \frac{l^2}{2\mu r^2} - \frac{k}{r} = \frac{1}{2}\mu\dot{r}^2 + V_{\text{eff}}$$

where

$$V_{\text{eff}} = -\frac{k}{r} + \frac{l^2}{2\mu r^2}.$$

Circular orbit, $e = 0$, so

$$E = -\frac{\mu k^2}{2l^2} = V_{\text{eff. min}}.$$

$V_{\text{eff. min}}$ occurs at

$$\frac{d}{dr}V_{\text{eff}} = 0$$

which gives

$$\frac{k}{r^2} - \frac{l^2}{\mu r^3} = 0, \quad \text{i.e., at } r = \frac{l^2}{\mu k}.$$

Then

$$V_{\text{eff. min}} = -\frac{\mu k^2}{2l^2}.$$

The elliptical orbit occurs when $E < 0$ for which $e < 1$, and $V_{\text{eff. min}} < E < 0$ must be satisfied. For parabolic path $E = 0$ as $e = 1$. For hyperbolic path $e > 1$ and $E > 0$.

For parabolic path

$$0 = E = \frac{1}{2}\mu v^2 - \frac{k}{r}.$$

Hence speed

$$v = \frac{\sqrt{2k/\mu}}{\sqrt{r}}, \quad \text{i.e.,} \quad v \propto \frac{1}{r^{1/2}}.$$

Speed is inversely proportional to the square root of the radial distances.

**Example 16** *Two stars each of which has the same mass as that of the sun, form a binary star. The time period of this is the same as that of the earth round the sun. Find the value of the semi-major axis of the ellipse of the binary system.*

***Solution:*** Time period

$$T = \frac{\text{area of the ellipse}}{\text{areal velocity}} = \frac{\pi ab}{(1/2)r^2\dot{\theta}} = \frac{\pi ab}{l/2\mu} = \frac{\pi a^2(1-e^2)^{1/2}}{(l/2\mu)}.$$

From the orbit equation

$$a(1-e^2) = \frac{l^2}{\mu k}.$$

Therefore,

$$T^2 = \frac{4\pi^2\mu a^3}{k}.$$

For binary star, $k = GMM$.

$$\mu = \frac{M^2}{2M} = \frac{M}{2}.$$

Then

$$T^2 = \frac{4\pi^2 a^3}{2GM}.$$

For earth,

$$k = GMm_{\text{earth}}, \quad \mu = m_{\text{earth}}, \quad T^2_{\text{earth}} = \frac{4\pi^2 a'^3}{GM}.$$

$a'$ = distance of earth from sun.
$a$ = distance between the two stars of the binary system. As

$$T = T_{\text{earth}}, \quad \text{so } \frac{a^3}{2} = a'^3, \quad \text{or,} \quad a = 2^{1/3}a'.$$

**Example 17** *Show that the law of addition of two velocities in relativistic mechanics results from the product of two Lorentz transformation applied successively.*

***Solution:*** Lorentz transformation connecting the frame $S_1$ which is moving parallel to $x$-axis w.r.t a fixed frame $S$ with a constant velocity $v_1$ is

$$x_1 = \gamma(x - v_1 t); \quad t_1 = \gamma\left(t - \frac{xv_1}{c^2}\right); \quad \gamma^{-1} = \left(1 - \frac{v_1^2}{c^2}\right)^{1/2}.$$

The L.t. connecting $S_2$ with $S_1$ when $S_2$ is moving with relative velocity $v_2$ w.r.t. $S_1$ is

$$x_2 = \gamma'(x_1 - v_2 t_1); \quad t_2 = \gamma'\left(t_1 - x_1\frac{v_2}{c^2}\right); \quad \gamma' = \left(1 - \frac{v_2^2}{c^2}\right)^{-1/2}.$$

Hence the transformation connection $S_2$ to $S$ is

$$x_2 = \gamma'\left[\gamma(x - v_1 t) - \gamma v_2\left(t - \frac{v_1 x}{c^2}\right)\right] = \gamma'\gamma\left[\left(x + v_1 v_2\frac{x}{c^2}\right) - (v_1 + v_2)t\right]$$

$$= \gamma'\gamma\left(1 + \frac{v_1 v_2}{c^2}\right)\left[x - \frac{(v_1 + v_2)}{1 + \frac{v_1 v_2}{c^2}}\cdot t\right] = \gamma'\gamma\left(1 + \frac{v_1 v_2}{c^2}\right)(x - Vt).$$

But

$$\gamma'\gamma\left(1 + \frac{v_1 v_2}{c^2}\right) = \sqrt{\frac{(1 + v_1 v_2/c^2)^2}{(1 - v_2^2/c^2)(1 - v_1^2/c^2)}} = \sqrt{1 - V^2/c^2}\cdot\left(1 - \frac{V^2}{c^2}\right)^{-1}.$$

where

$$V = \frac{v_1 + v_2}{1 + v_1 v_2/c^2}.$$

Then

$$x_2 = \left(1 - \frac{V^2}{c^2}\right)^{-1/2}(x - \mathbf{V}t),$$

where

$$V = \frac{(v_1 + v_2)}{(1 + v_1 v_2/c^2)}$$

gives the law of addition of two velocities.

We easily check

$$\sqrt{1 - \frac{1}{c^2}\frac{(v_1^2 + v_2^2 + 2v_1 v_2)}{1 + \frac{v_1^2 v_2^2}{c^4} + \frac{2v_1 v_2}{c^2}}} = \frac{\sqrt{1 - \frac{v_1^2}{c^2} - \frac{v_2^2}{c^2} + \frac{v_1^2 v_2^2}{c^4}}}{\sqrt{\left(1 + \frac{v_1 v_2}{c^2}\right)^2}}$$

$$= \sqrt{\left(1 - \frac{v_1^2}{c^2}\right)}\sqrt{\left(1 - \frac{v_2^2}{c^2}\right)}\left(1 + \frac{v_1 v_2}{c^2}\right)^{-1}.$$

$$\text{or,} \quad \left(1 - \frac{V^2}{c^2}\right)^{-1/2} = \gamma'\gamma\left(1 + \frac{v_1 v_2}{c^2}\right).$$

**Example 18** *For the Hamiltonian*

$$H = q_1 p_1 - q_2 p_2 - a q_1^2 + b q_2^2.$$

*Show that* $q_1 q_2 = $ *constant;* $\log q_1 = t +$ *constant and* $(p_2 - b q_2)/q_1 = $ *constant.*

***Solution:***

$$\dot{q}_1 = \frac{\partial H}{\partial p_1} = q_1, \quad \dot{q}_2 = \frac{\partial H}{\partial p_2} = -q_2,$$

$$\dot{p}_1 = -\frac{\partial H}{\partial q_1} = -p_1 + 2aq_1, \quad \dot{p}_2 = -\frac{\partial H}{\partial q_2} = p_2 - 2bq_2;$$

$$\dot{q}_1 q_2 + \dot{q}_2 q_1 = q_1 q_2 - q_2 q_1 = 0.$$

Hence

$$\frac{d}{dt}(q_1 q_2) = 0; \quad q_1 q_2 = \text{constant}; \quad \dot{q}_1 = q_1.$$

Hence

$$\frac{dq_1}{q_1} = dt, \quad \text{or,} \quad \log q_1 = t + \text{ constant}.$$

$$\frac{d}{dt}\left(\frac{p_2 - bq_2}{q_1}\right) = \frac{\dot{p}_2 - b\dot{q}_2}{q_1} - \frac{\dot{q}_1(p_2 - bq_2)}{q_1^2}$$

$$= \frac{1}{q_1}(\dot{p}_2 - b\dot{q}_2 - p_2 + bq_2); \quad [\dot{q}_1 = q_1 \text{ is used}]$$

$$= \frac{1}{q_1}(-2bq_2 + bq_2 + bq_2) = 0.$$

Hence

$$\frac{(p_2 - bq_2)}{q_1} = \text{ constant}.$$

**Example 19** *The Hamiltonian for a two degrees of freedom is given by*

$$H = \frac{1}{2}p_1^2 + \frac{1}{2}p_2^2 + \frac{1}{2}(q_1 - q_2)^2 + \frac{1}{2}(q_1 + q_2)^2$$

*write down the Hamilton-Jacobi equation for the system and obtain the action $S$.*

***Solution:*** Since $H$ is conserved we write

$$S = f - \alpha t,$$

where $\alpha$ = constant. The Hamilton-Jacobi equation

$$H + \frac{\partial S}{\partial t} = 0$$

reduces to $H = \alpha$ (constant = Energy).
Put

$$p_1 = \frac{\partial S}{\partial q_1}, \quad p_2 = \frac{\partial S}{\partial q_1};$$

the Hamilton-Jacobi equation becomes

$$\frac{1}{2}\left(\frac{\partial S}{\partial q_1}\right)^2+\frac{1}{2}\left(\frac{\partial S}{\partial q_2}\right)^2+\frac{1}{2}(q_1-q_2)^2+\frac{1}{2}(q_1+q_2)^2=\alpha.$$

The separation of variable is effected by taking.

$$Q_1=q_1+q_2,\quad Q_2=q_1-q_2.$$

Then

$$\frac{\partial}{\partial q_1}=\frac{\partial}{\partial Q_1}+\frac{\partial}{\partial Q_2};\quad \frac{\partial}{\partial q_2}=\frac{\partial}{\partial Q_1}-\frac{\partial}{\partial Q_2}$$

Therefore,

$$\frac{1}{2}\left\{\left(\frac{\partial S}{\partial q_1}\right)^2+\left(\frac{\partial S}{\partial q_2}\right)^2\right\}=\left(\frac{\partial S}{\partial Q_1}\right)^2+\left(\frac{\partial S}{\partial Q_2}\right)^2.$$

We separate

$$S=S_1(Q_1)+S_2(Q_2),$$

then Hamilton-Jacobi equation reduces to

$$\left(\frac{\partial S_1}{\partial Q_1}\right)^2+\left(\frac{\partial S_2}{\partial Q_2}\right)^2+\frac{1}{2}Q_1^2+\frac{1}{2}Q_2^2=\alpha_1+\alpha_2=\alpha.$$

Thus

$$\left(\frac{\partial S_1}{\partial Q_1}\right)^2+\frac{1}{2}Q_1^2=\alpha_1;\quad \left(\frac{\partial S_2}{\partial Q_2}\right)^2+\frac{1}{2}Q_2^2=\alpha_2$$

Hence

$$S=-\alpha t+\int dQ_1\sqrt{\alpha_1-\frac{1}{2}Q_1^2}+\int dQ_2\sqrt{\alpha_2-\frac{1}{2}Q_2^2}.$$

**Example 20** *Evaluate the action integral*

$$J_r=\oint p_r dr$$

*for the Kepler problem without using contour integral method.*

***Solution:*** The integral for $J_r$ is

$$J_r=\oint\sqrt{2mE+\frac{2mk}{r}-\frac{(J_\theta^2/4\pi^2)}{r^2}}\,dr$$

when $E$ = total energy, for hydrogen atom $E=-|E|$; $k$ = strength of coulomb potential.

$$J_r=2\int_{r_1}^{r_2}\frac{1}{r}\sqrt{2m|E|r^2+2mkr-(J_\theta^2/4\pi^2)}dr$$

where $r_1$, $r_2$ are the roots of $-Ar^2 + Br - C$ where

$$A = 2m|E|, \quad B = 2mk, \quad C = \frac{J_\theta^2}{4\pi^2}.$$

We work out the integral

$$I = \int \frac{dr}{r}\sqrt{-Ar^2 + Br - C}.$$

then

$$I = \frac{1}{2}\int \frac{(-2Ar + B)\,dr}{\sqrt{-Ar^2 + Br - C}} + \frac{B}{2}\int \frac{dr}{\sqrt{-Ar^2 + Br - C}} - C\int \frac{dr}{r}\frac{1}{\sqrt{-Ar^2 + Br + C}}.$$

For the limits $r_1$ and $r_2$ of integration, the first integral does not contribute and the last integral reduces to the second-integral by substitution of $r = \frac{1}{u}$. Then

$$I = \frac{B}{2}\int_{r_1}^{r_2} \frac{dr}{\sqrt{-Ar^2 + Br - C}} - C\int_{u_1}^{u_2} \frac{du}{\sqrt{-Cu^2 + Bu - A}}$$

where $u_1$ and $u_2$ are the smaller $(1/r_2)$ and larger $(1/r_1)$ roots of

$$-Cu^2 + Bu - A = 0.$$

The results of integration is

$$I = \frac{B}{2}\pi\frac{1}{\sqrt{A}} - C\pi\frac{1}{\sqrt{C}}$$

Hence

$$J_r = -J_\theta + \pi k\sqrt{\frac{2m}{|E|}}; \quad |E| = \frac{2m\pi^2k^2}{(J_r + J_\theta)^2}.$$

Hence

$$E = -|E| = \frac{2m\pi^2k^2}{(J_r + J_\theta)^2}.$$

**Example 21** *A flywheel whose moment of inertia about its axis is 200 lb-ft$^2$ makes 10 revolutions per second. What is its kinetic energy (in ft-lb) and what constant couple would reduce it to rest in one revolution?*

***Solution:*** Angular velocity $= 10 \times 2\pi$ radians in one second.
Moment of inertia $= 200$ lb-ft$^2$.

$$\text{K.E.} = \frac{1}{2}I\omega^2.$$

K.E. $= \frac{1}{2} \times 200 \times 100 \times 4\pi^2 \frac{1}{32}$ ft-lb $=$ 12,500 ft-lb.
Angular acceleration $= 10^2 \times 4\pi^2/4\pi = 10^3/\pi$ radian/sec$^2$.
Couple $G = I\alpha$, $\alpha =$ angular acceleration
$= 200 \times 10^3/\pi \cdot 32$ ft-lb $= 6,250/\pi$ ft-lb.

**Example 22** *AB and CD are two rods of lengths 2a and b respectively, the mass of each rod being m per unit length. The rods are rigidly joined together at right angle at C, the mid-point of AB and the system is free to rotate in a vertical-plane about D. If the system is held with AB vertical and let go, calculate the angular velocity, where AB is horizontal.*

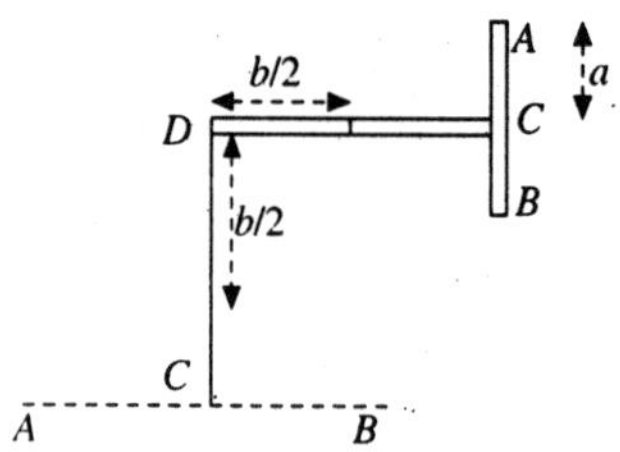

**Fig E.8**

***Solution:*** $AB = 2a$, mass of

$$AB = m_A = 2ma.$$

$CD = b$, mass of

$$CD = m_C = mb.$$

$C =$ mid-point of $AB$. The centre of gravity of $AB$ has fallen the height $CD = b$ when initial $AB$ has been vertical. Hence the potential energy of $AB = 2ma \cdot g \cdot b$.
The centre of gravity of $CD$ has fallen the distance $b/2$. Hence the P.E. of

$$CD \text{ is } mb\frac{b}{2}g.$$

Hence initial P.E. of the system is

$$mg\left(2ab + \frac{b^2}{2}\right).$$

The K.E. due to rotation

$$= \frac{1}{2}I_{AB}\omega^2 + \frac{1}{2}I_{CD}\omega^2,$$

where $I_{AB} =$ moment of inertia of $AB = 2ma\left(\frac{a^2}{3} + b^2\right)$,

$I_{CD} =$ moment of inertia of $CD = mb\left(\frac{b^2}{3}\right)$.

Hence

$$\frac{1}{2}I_{AB}\omega^2 + \frac{1}{2}I_{CD}\omega^2 = \frac{1}{2}m\omega^2\left(\frac{b^3}{3} + \frac{2a^3 + 6ab^2}{3}\right).$$

By energy conservation

$$\frac{1}{2}m\omega^2\left(\frac{b^3}{3} + \frac{2a^3 + 6ab^2}{3}\right) = \frac{mg}{2}(4ab + b^2).$$

Hence

$$\omega = \left[\frac{3gb(4a+b)}{(2a^3+6a^2b+b^3)}\right].$$

**Example 23** *A wheel and axle has moment of inertia, I, about its axis. A mass m is suspended from the axle, which is of radius, r, by a fine thread wrapped round and fastened to the axle. Rotation is opposed by a constant frictional couple G. Find the angular velocity acquired as the mass, m descends a distance of h.*

***Solution:*** Equation of motion of the mass $m\ddot{x} = mg - T$; $T$ = tension in thread.
If $\tau$ = torque, then

$$\tau = G \quad \text{and} \quad Tr = G, \quad \text{or,} \quad T = G/r.$$

Hence the effective force on $m$ is

$$F = m\ddot{x} = \left(mg - \frac{G}{r}\right).$$

If $\omega$ = angular velocity of wheel and axles then K.E. of rotation = $(1/2)I\omega^2$; the K.E. of mass, $m = (1/2)mv^2$, where $v$ is the velocity then $v = r\omega$, $r$ = radius of axle. Total K.E. = $(1/2)(I+mr^2)\omega^2$. Equating this to work done by the force $F$ in descending $h$, we get

$$\frac{1}{2}(I+mr^2)\omega^2 = Fh = \left(mg - \frac{G}{r}\right)h;$$

$$\omega = \left[\frac{2h(mgr - G)}{r(I+mr^2)}\right]^{1/2}.$$

**Example 24** *A uniform sphere is rotating about a diameter, contracts by cooling; show that when its radius is reduced to 1/nth of its original-value the kinetic energy gets multiplied by $n^2$.*

***Solution:*** Angular momentum remains constant, so,

$$I_1\omega_1 = I_2\omega_2.$$

Therefore,

$$\frac{2}{5}mr_1^2\omega_1 = \frac{2}{5}mr_2^2\omega_2, \quad \text{but } r_2 = \frac{r_1}{n}.$$

so,

$$r_1^2\omega_1 = \frac{1}{n^2}r_1^2\omega_2; \quad \omega_2 = n^2\omega_1.$$

$$\text{K.E.} = \frac{1}{2}I_2\omega_2^2 = \frac{1}{2}I_1\omega_1\omega_2 = n^2\frac{1}{2}I_1\omega_1^2 = n^2 \times \text{initial kinetic energy.}$$

**Example 25** *A flywheel of moment of inertia I is rotating with angular velocity, Ω, about a vertical axis. The flywheel contains a pocket at a distance a, from the axis, into which is dropped a sphere of mass, M, and of moment of inertia, i and spin ω, about a vertical axis, without horizontal motion. Find the angular velocity of the system after the sphere has come to relative rest in the pocket.*

***Solution:*** Initial angular momentum of the flywheel $= I\Omega$. At the time of dropping the sphere, the sphere has angular momentum $= i\omega$. Since at the time of dropping there was no horizontal motion of the sphere, therefore, initially at the time of dropping, total initial angular momentum is $(I\Omega+i\omega)$. When the sphere in the pocket has come to relative rust, then both flywheel and sphere has same angular momentum $\Omega'$ and sphere has linear velocity $v = a\Omega'$.
Total rotational + translational energy is

$$\frac{1}{2}(I+i)\Omega'^2 + \frac{1}{2}Mv^2 = \frac{1}{2}(I+i+Ma^2)\Omega'^2.$$

Therefore, the total moment of inertia is $(I + i + Ma^2)$. Hence angular momentum is

$$(I+i+Ma^2)\Omega'.$$

Hence as angular momentum is conserved

$$I\Omega + i\omega = (I+i+Ma^2)\Omega'.$$

Therefore

$$\Omega' = \frac{(I\Omega+i\omega)}{(I+i+Ma^2)}.$$

**Example 26** *A solid rectangular box of dimension 2a, 2b and 2c is suspended from a horizontal axis along the edge 2a. Find the length of the equivalent simple pendulum.*

***Solution:*** $OABC$ denotes a cross-section of the rectangular box. Along the endge $2a$ it is suspended. $O$ is the point of suspension. $OB$ = diagonal. $G$ is the centre of gravity.

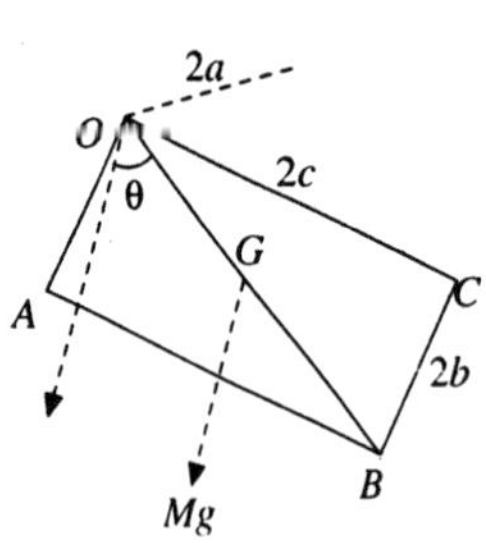

**Fig E.9**

$$OA = 2b, \quad AB = 2c, \quad OG^2 = b^2 + c^2.$$

$OB$ makes an angle $\theta$ with vertical. The rectangle is rotating with angular velocity $\omega = \dot{\theta}$. At any instant from energy conservation.

$$\frac{1}{2}I\omega^2 + Mg\cdot OG(1-\cos\theta) = \text{ constant}.$$

where $I$ is the moment of inertia of the rectangular box around the edge $2a$. $I = I_G + M(OG)^2$, where

$$I_G = \frac{M(b^2+c^2)}{3},$$

(where $M$ = mass) is the moment of inertia about an axes through $G$ and perpendicular to the face $OABC$. Since

$$\omega = \dot{\theta}.$$

We get

$$\frac{1}{2} I \frac{d}{dt}\dot{\theta}^2 + MgOG\frac{d}{dt}(1-\cos\theta) = 0$$

i.e., $$I\ddot{\theta} + Mg \cdot OG \sin\theta = 0;$$

for small amplitude

$$I\ddot{\theta} + MgOG\theta = 0.$$

Hence the time period of the equivalent simple pendulum is

$$\begin{aligned} T &= 2\pi\sqrt{\left(\frac{I}{Mg\cdot OG}\right)} \\ &= 2\pi\sqrt{\left[\frac{M}{3}(b^2+c^2) + M(b^2+c^2)\right] \Big/ \left[Mg(b^2+c^2)^{1/2}\right]} \\ &= 2\pi\sqrt{\left[\frac{4(b^2+c^2)^{1/2}}{3g}\right]}. \end{aligned}$$

**Example 27** *A uniform cube of edge $2a$ is sliding with velocity $v$ on a smooth horizontal table, where its leading edge is suddenly brought to rest by a small ridge on the table. Find the angular velocity immediately after the the impact. Determine the minimum value of $v$ for which the cube topples over rather than falling back.*

***Solution:*** $ABRC$ = cube $AB = 2a$, $G$ = centre of gravity. $R$ = small ridge. $M$ = mass of cube. $v$ = velocity and momentum = $Mv$.
Angular moment before impaot = $Mv\cdot a$. Let $\omega$ = angular velocity with which cube rises at the ridge. Let $I$ = moment of inertia of cube about an axis through $R$ and perpendicular to $ABRC$ face, then angular momentum just after impact = $I\omega$. By angular momentum conservation $I\omega = Mv\cdot a$, so $\omega = \frac{Mv\cdot a}{I}$.

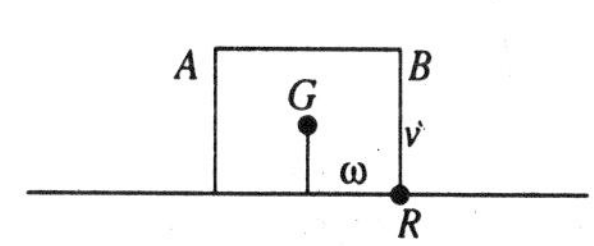

Fig E.10

To topple over the ridge we need K.E. of rotation $= \frac{1}{2}I\omega^2$. This energy must be needed to do the work of lifting the C.G. of cube from the present height $a$ to the position, where $G$ will be over the ridge $R$ i.e., the height will be $\sqrt{2}\,a$. Amount of work to be done $= Mg(\sqrt{2}-1)a$. This must at least be provided by K.E. or rotation.

$$\text{i.e.,}\ \frac{1}{2}I\omega^2 = Mga(\sqrt{2}-1), \quad \text{or,} \quad \omega^2 = \frac{2Mg\cdot a(\sqrt{2}-1)}{I}.$$

$$\text{i.e.,}\ \frac{M^2v^2a^2}{I^2} = \frac{2Mga(\sqrt{2}-1)}{I}, \left(I = \frac{M8a^2}{3}\right).$$

Therefore, $v^2$ should at least be

$$v^2 = \frac{16ga(\sqrt{2}-1)}{3}.$$

**Example 28** *A solid rectangular box of dimension* 10 cm × 6 cm × 2 cm *is spinning freely with angular velocity of* 240 *r.p.m. Find the frequency of small oscillation of the axis of rotation if the axis is the longest side.*

***Solution:*** If $I_1$, $I_2$, $I_3$ are the principal moment of inertia then for the rectangular parallelopiped of side 10 cm, 6 cm and 2 cm along the $x$, $y$ and $z$-axis are

$$M\frac{136}{3},\ M\frac{104}{3}\ \text{and}\ M\frac{40}{3}\ \text{kg-cm}^2.$$

We assumed that the longest axis is along $z$-axis, while the edge with 6 cm lies along $y$-axis. The angular velocity $\omega_3$ about $z$-axis is $\omega_3 = 6$ rev./sec. The frequency of oscillation of the axis of rotation is

$$p^2 = \frac{(I_3 - I_2)(I_1 - I_3)\omega_3^2}{I_1 I_2},$$

where the oscillation is assumed of the form $e^{pt}$. We find from the values of $I_1, I_2$ and $I_3$, $p = \pm i0.6\omega_3$.
Hence oscillation has the form $e^{\pm i0.6\omega_3 t}$ where $\omega_3 = 6$ r.p.s.

**Example 29** *When a small correction $\delta U(r)$ is added to the potential energy*

$$U(r) = -\frac{k}{r},$$

*then the central orbit is no longer a closed path and in each revolution, the perihelian is displaced through a small angle $\delta\phi$. Find $\delta\phi$ when (a) $\delta U = \frac{\beta}{r^2}$, (b) $\delta U = \frac{\gamma}{r^3}$.*

***Solution:*** We have the equation for the parth,

$$\theta = M \int \frac{dr}{r^2} \cdot \left[2m(E-U) - \frac{M^2}{r^2}\right]^{-1/2} \tag{2}$$

From this we get

$$\frac{dr}{d\theta} = \frac{r^2}{M}\left\{2m[E-U(r)] - \frac{M^2}{r^2}\right\}^{1/2} \tag{3}$$

Integrating (1) we get

$$r = \frac{\tilde{l}}{(1+e\cos\theta)}, \quad \tilde{l} = \frac{M^2}{mk}; \quad e = \sqrt{1+\frac{2EM}{mk^2}}.$$

For a closed path when $r$ varies from $r = r_{\min}$ to $r = r_{\max}$ and back the angle charges by $\delta\theta$ given by the following [using (1)].

$$\Delta\theta = 2M\int_{r_{\min}}^{r_{\max}} \frac{dr/r^2}{\sqrt{\left[2m(E-U) - \frac{M^2}{r^2}\right]}}.$$

Rewrite this as

$$\Delta\theta = -2\frac{\partial}{\partial M}\int_{r_{\min}}^{r_{\max}} \sqrt{\left[2m(E-U) - \frac{M^2}{r^2}\right]}\, dr.$$

For potential $U + \delta U$ we find

$$\begin{aligned}\Delta\theta &= 2\frac{\partial}{\partial M}\int_{r_{\min}}^{r_{\max}} \sqrt{2m(E-U) - \frac{M^2}{r^2}}\left[1 - \frac{2m\delta U}{\sqrt{2m(E-U) - \frac{M^2}{r^2}}}\right]^{1/2} dr. \\ &= 2\frac{\partial}{\partial M}\int_{r_{\min}}^{r_{\max}} dr\sqrt{2m(E-U) - \frac{M^2}{r^2}}\left[1 - \frac{1}{2}\frac{\delta U \cdot 2m}{\left[2m(E-U) - \frac{M^2}{r^2}\right]} + \cdots\right]\end{aligned}$$

Hence to order $\delta U$, the change $\delta\theta$ in $\Delta\theta$ is

$$\delta\theta = \frac{\partial}{\partial M}\int dr\, \delta U\left[2m(E-U) - \frac{M^2}{r^2}\right]^{-1/2} 2m.$$

Use now (2) to write

$$\delta\theta = \frac{\partial}{\partial M}2m\int M^{-1}r^2\delta U(r)\, d\theta,$$

where now $\theta$ varies from 0 to $\pi$.

Thus for $(a)$

$$\delta U = \frac{\beta}{r^2}, \quad \text{or,} \quad r^2\delta U = \beta.$$

$$\delta\theta = \frac{\partial}{\partial M}2m\int_0^{\pi} M^{-1}d\theta\beta = -2m\pi\frac{\beta}{M^2}.$$

For $(b)$

$$\begin{aligned}\delta\theta &= \frac{\partial}{\partial M}\int_0^{\pi} d\theta M^{-1}2mr^2\frac{\gamma}{r^3} = \frac{\partial}{\partial M}\gamma\int_0^{\pi} 2mM^{-1}\,d\theta\frac{1}{r}\\ &= \frac{\partial}{\partial M}2m\gamma M^{-1}\int_0^{\pi}(1+e\cos\theta)/\bar{l}\cdot d\theta = \frac{\partial}{\partial M}\cdot\left(2m\gamma\pi\cdot mk/M^3\right)\\ &= -6\pi\gamma m^2\frac{k}{M^4}.\end{aligned}$$

**Example 30** *Show that the equation*

$$\dot{E}_i(t) = \frac{I_i}{C_i} + \sum_j L_{ij}\ddot{I}_j$$

*where $L_{ij}(i \neq j)$ are mutual inductances, $L_{ij}(i = j)$ are self inductances $C_i$'s are capacitances representing the circuital relationship for a net work of coupled reactive impedances in which are flowing a system of currents $I_i$ generated by e.m.f. $E_i$ can be obtained from a Lagrangian. Find the Lagrangian.*

***Solution:*** We easily see that a Lagrangian function given by

$$L = \frac{1}{2}\sum_i\sum_j L_{ij}\dot{I}_i\dot{I}_j - \frac{1}{2}\sum_i\frac{1}{C_i}I_i^2 + \sum_i \dot{E}_i(t)I_i$$

can generate the circuit equations. Because the Euler-Lagrangian equation from $L$ reads

$$\frac{d}{dt}\left(\sum L_{ij}\dot{I}_j\right) = \dot{E}_i - \frac{I_i}{C_i}, \quad \text{i.e.,} \quad \dot{E}_i = \frac{I_i}{C_i} + \sum_j L_{ij}\ddot{I}_j$$

**Example 31** *A uniform bar of length L and mass M is suspended in equilibrium in a horizontal position by two mass-less springs of force constant k for each of them. The springs are attached one at each end of the bar. The motion of the centre of mass of the bar is confined to the vertical X-direction only. Find the normal modes of vibration of the system assuming the motion is only in a vertical plane.*

***Solution:*** Let the vertical displacements be $X_1$ and $X_2$ for the springs from equilibrium. The coordinate of the centre of mass of the bar is

$$\frac{1}{2}(X_1 + X_2);$$

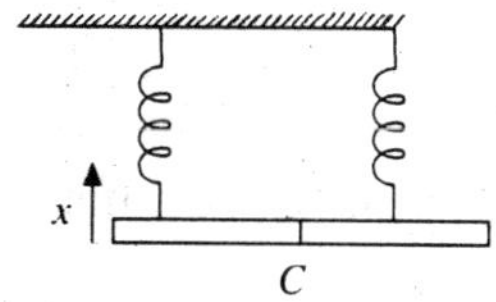

**Fig E.11**

$C$ denotes C.M. Mass of bar $= M$ and length $= L$; the force constant $= k$ for each spring. The equation of motion of the C.M. of the bar is

$$\frac{M}{2}(\ddot{X}_1 + \ddot{X}_2) = -k(X_1 + X_2) - Mg. \tag{4}$$

From the rate of change of angular momentum being equal to the moment of the forces we get

$$\frac{I(\ddot{X}_2 - \ddot{X}_1)}{L} = -\frac{Lk(X_2 - X_1)}{2} \tag{5}$$

$$I = \text{moment of inertia of the bar about its C.M.} = \frac{ML^2}{12}.$$

We can eliminate $Mg$ from (1) using equilibrium condition $Mg = k \cdot 2l$; $l =$ unstretched length of the spring. We can then change the variables $X_1 + l \to X_1$ and $X_2 + l \to X_2$. The resulting equations can be readily solved.

**Example 32** *Two plane simple pendulums are suspended from two points of the same horizontal line. Both the pendulums have same length and mass. The bobs of the two pendulums have been connected by a horizontal spring of force constant k. Find the normal modes of vibration when the system is oscillating in a vertical plane.*

***Solution:*** $AB$ is a horizontal line. Two simple pendulums length of each being $l$ and mass $m$ are suspended from $A$ and $B$. Spring of force constant $- k$ is connecting the two masses. The spring is in a horizontal position. Let $\theta_1$ and the $\theta_2$ are the two small amplitudes of vibrations of the two pendulums. The whole system is lying in a vertical plane.

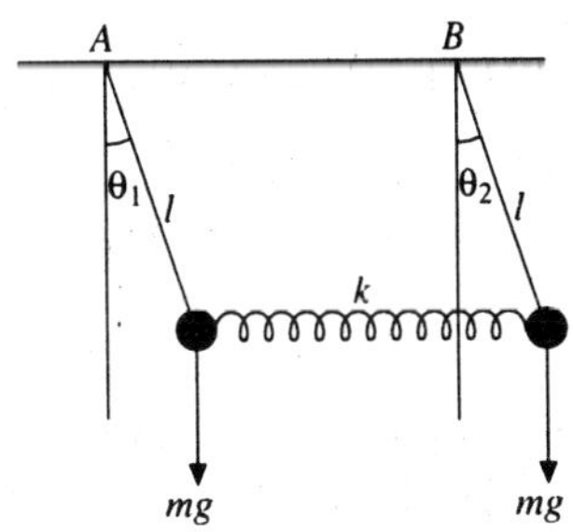

**Fig E.12**

$$\text{K.E.} = T = \frac{1}{2}ml^2(\dot{\theta}_1^2 + \dot{\theta}_2^2).$$

The potential energy

$$V = mgl(1 - \cos\theta_1) + mgl(1 - \cos\theta_2) + \frac{1}{2}kx^2,$$

where $x$ = extension in the spring. As $\theta_1$ and $\theta_2$ are small the extension $x = l\theta_2 - l\theta_1$. For small $\theta_1$ and $\theta_2$ we have for

$$V = \frac{1}{2}mgl(\theta_1^2 + \theta_2^2) + \frac{1}{2}kl^2(\theta_1 - \theta_1)^2.$$

Hence from Lagrangian $L = T - V$ we get

$$\ddot{\theta}_1 + \left(\frac{g}{l} + \frac{k}{m}\right)\theta_1 = \left(\frac{k}{m}\right)\theta_2$$
$$\ddot{\theta}_2 + \left(\frac{g}{l} + \frac{k}{m}\right)\theta_2 = \left(\frac{k}{m}\right)\theta_1.$$

From these we can find the eigenfrequencies.

**Example 33** *For the Lagrangian* $L = \frac{1}{2}\dot{q}^2 - q\dot{q} + q^2$ *find* $p$ *and the Hamiltonian* $H$. *Show that the Hamiltonian* $H$ *reduces to* $\frac{P_2^2}{2} - Q^2$ *under the canonical transformation* $p = P - Q$; $q = Q$.

***Solution:*** From $L$, we get $p = \dot{q} - q$. The Hamiltonian

$$H = p\dot{q} - L = \frac{1}{2}p^2 + pq - \frac{1}{2}q^2.$$

We can write

$$H = \frac{1}{2}(p + q)^2 - q^2.$$

Hence under $p \to P - Q$ and $q \to Q$, we get

$$H \to H' = \frac{1}{2}P^2 - Q^2.$$

**Example 34** *For the Hamiltonian* $H = \frac{(p^2+q^2)}{2}$. *Find* $[\dot{p}, H]$ *and* $[\dot{q}, H]$ *and find the values of* $p$ *and* $q$. *Show that energy is a constant.*

***Solution:*** We easily see that

$$[\dot{p}, H] = \ddot{p} \text{ and } [\dot{q}, H] = \ddot{q}$$
$$\text{and } \dot{p} = -\frac{\partial H}{\partial q} = -q \text{ and } \dot{q} = \frac{\partial H}{\partial p} = p$$

so $\ddot{p} = -[q, H] = -p$ and $\ddot{q} = [\dot{q}, H] = [p, H] = -q$; or $\ddot{p} + p = 0$ and $\ddot{q} + q = 0$. As $\dot{p} = -q$, we write

$$p = \cos t \quad \text{and} \quad q = \sin t.$$

Hence

$$E = \frac{p^2 + q^2}{2} \text{ becomes } \frac{1}{2}, \text{ a constant.}$$

**Example 35** *For a classical rotating string, calculate the length of the string, $2a$, when $J = 1$ in units of $\hbar$. Assume $\alpha' = \hbar c(Gev)^{-2}$.*

***Solution:*** For a rotating string of length $2a$, we have

$$J = \frac{c}{(4\alpha' w^2)},$$

where $w$ = angular velocity. Put

$$w = \frac{c}{a} \text{ then } J = \frac{a^2}{(4\alpha' c)}.$$

Hence for $J = 1 \cdot \hbar$ we get

$$\frac{a^2}{4} = (\hbar c)\alpha', \quad \text{or,} \quad \frac{a^2}{4} = (\hbar c)^2 (Gev)^{-2},$$

$$\text{or, } \frac{a}{2} = \frac{\hbar c}{Gev} = 2 \times 10^{-14} \text{ cm.,}$$

$$\text{or, } 2a = 8 \times 10^{-14} \text{ cm.} = 0.8 \times 10^{-13} \text{ cm.}$$

**Example 36** (*a*) *Show that*

$$[p_k\, X]_{QP} = -\frac{\partial X}{\partial q_k}.$$

(*b*) *$A, B, C$ are dynamical variables and $f(B, C)$ is a function of $B$ and $C$, then show that*

$$[A, f(B, C)]_{P.B} = [A, B]\frac{\partial f}{\partial B} + [A, C]\frac{\partial f}{\partial C}.$$

***Solution:***

(*a*)
$$\begin{aligned}
[p_k . X(p_i . q_i)]_{QP} &= \sum_j \left( \frac{\partial p_k}{\partial Q_j}\frac{\partial X}{\partial P_j} - \frac{\partial p_k}{\partial P_j}\frac{\partial X}{\partial Q_j} \right) \\
&= \sum_j \sum_i \left\{ \frac{\partial p_k}{\partial Q_j} \left( \frac{\partial X}{\partial p_i}\frac{\partial p_i}{\partial P_j} + \frac{\partial X}{\partial q_i}\frac{\partial q_i}{\partial P_j} \right) \right. \\
&\quad \left. - \frac{\partial p_k}{\partial P_j} \left( \frac{\partial X}{\partial p_i}\frac{\partial p_i}{\partial Q_j} + \frac{\partial X}{\partial q_i}\frac{\partial q_i}{\partial Q_j} \right) \right\} \\
&= \sum_i \left\{ \frac{\partial X}{\partial p_i}[p_k . p_i]_{QP} + \frac{\partial X}{\partial q_i}[p_k\, q_i]_{QP} \right\} \\
&= \sum_i [p_k\, q_i]\frac{\partial X}{\partial q_i} = -\sum_i \delta_{ik}\frac{\partial X}{\partial q_i} = -\frac{\partial X}{\partial q_k}.
\end{aligned}$$

(*b*) Use the same procedure to obtain the result.

**Example 37** *Find the generating functions for the canonical transformations:* (*a*) $Q = \log\left(\frac{\sin p}{q}\right)$; $P = q\cot p$; (*b*) $q = \sqrt{2p}\sin Q$; $p = \sqrt{2P}\cos Q$.

***Solution:*** (*a*) We calculate $PdQ - pdq$.

$$\begin{aligned} PdQ - pdq &= q\cot p\left(-\frac{1}{q}dq + \cot p dp\right) - pdq \\ &= -\cot p dq + q\cot^2 p dp + qdp - d(pq) = -d(q\cot p + pq) \end{aligned}$$

Hence

$$F_3 = \cos p e^{-Q}.$$

$$\begin{aligned} (b) \qquad pdq - PdQ &= \sqrt{2P}\cos Q\left(\frac{dP}{\sqrt{2P}}\sin Q + \sqrt{2P}\cos QdQ\right) - PdQ \\ &= \sin Q\cos QdP + P\cos 2QdQ = d\left(\frac{1}{2}P\sin 2Q\right). \end{aligned}$$

Hence

$$F_1 = \frac{1}{2}q^2\cot Q.$$

**Example 38** *If in the relativistic Klien-Gordon equation for* $\phi(\mathbf{X}, t)$, *we put*

$$\phi(\mathbf{X}, t) = \psi(\mathbf{X}, t)\exp\left\{-i\left(\frac{mc^2}{\hbar}\right)t\right\}.$$

*then in the non-relativistic limit, it reduces to Schrödinger equation.*

***Solution:*** Klien-Gorden equation is

$$\left(\bar{\nabla}^2 - \frac{1}{c^2}\frac{\partial^2}{\partial t^2}\right)\phi = \left(\frac{mc}{\hbar}\right)^2\phi.$$

Put

$$\phi = \psi(\mathbf{X}, t)\exp\left\{i\left(\frac{mc^2}{\hbar}\right)t\right\}.$$

We find

$$\left(\bar{\nabla}^2 - \frac{1}{c^2}\frac{\partial'^2}{\partial t^2}\right)\psi + \left(\frac{mc^2}{\hbar}\right)^2\cdot\frac{1}{c^2}\psi + 2i\frac{m}{\hbar}\frac{\partial\psi}{\partial t} = \left(\frac{mc}{\hbar}\right)^2\psi(\mathbf{X}, t)$$

In non-relativistic limit ($c \to \infty$), the second term n the left drops nt and we get after simplification.

$$i\hbar\frac{\partial\psi}{\partial t} = -\frac{\hbar^2}{2m}\bar{\nabla}^2\psi$$

which is a Schrödinger equation for a free particle.

**Example 39** *Consider the Lagrangian density $\mathcal{L}$ for a relativistic massive vector field $\phi_\mu$ interacting with a four-vector current $J_\mu$. namely*

$$\mathcal{L} = -\frac{1}{4}F_{\mu\nu}^2 - \frac{1}{2}m^2\phi_\mu^2 + J_\mu\phi_\mu; \quad F_{\mu\nu} = \frac{\partial\phi_\nu}{\partial x_\mu} - \frac{\partial\phi_\mu}{\partial x_\nu}.$$

*Obtain the field equation. Show that the continuity equation for $J_\mu$ is not gauranteed as it is in Maxwell's case, by the field equation. Show also that the subsidiary condition $\partial_\mu\phi_\mu = 0$, necessary holds if $\partial_\mu J_\mu = 0$.*

***Solution:*** Field equation is obtained from

$$\int \delta\mathcal{L}d^4x = 0, \text{ i.e.,} \quad -\frac{1}{4}\int \delta F_{\mu\nu}^2 dx + \frac{1}{2}m^2\int \delta\phi_\mu^2 d^4x + \int J_\mu\delta\phi_\mu d^4x = 0. \quad (1)$$

Consider the first term

$$-\frac{1}{4}\int \delta F_{\mu\nu}^2 d^4.$$

this is equal to

$$-\frac{1}{4}\cdot 2\int F_{\mu\nu}\delta F_{\mu\nu}d^4x = -\frac{1}{2}\int F_{\mu\nu}\left(\frac{\partial}{\partial x_\mu}\delta\phi_\nu - \frac{\partial}{\partial x_\nu}\delta\phi_\mu\right)d^4x$$

Integration by parts, this becomes equal to

$$\frac{1}{2}\int \frac{\partial F_{\mu\nu}}{\partial x_\mu}\delta\phi_\nu d^4x - \frac{1}{2}\int \frac{\partial F_{\mu\nu}}{\partial x_\nu}\delta\phi_\mu d^4x = \int \frac{\partial F_{\mu\nu}}{\partial x_\mu}\delta\phi_\nu d^4x$$

as $F_{\mu\nu} = -F_{\nu\mu}$. Using this in (1), we get

$$\int\left(\frac{\partial F_{\mu\nu}}{\partial x_\mu} - m^2\phi_\nu + J_\nu\right)\delta\phi_\nu.\, d^4x = 0.$$

Hence the field equation is

$$\frac{\partial F_{\mu\nu}}{\partial x_\mu} - m^2\phi_\nu + J_\nu = 0.$$

Since

$$\frac{\partial^2 F_{\mu\nu}}{\partial x_\mu\partial x_\nu} = 0.$$

we get

$$m^2\frac{\partial\phi_\nu}{\partial x_\nu} = \partial_\nu J_\nu.$$

If $m^2 \neq 0$, then $\partial_\nu\phi_\nu = 0$ holds if $\partial_\nu J_\nu = 0$.
But for Maxwell's case $m = 0$ then $\partial_\nu J_\nu = 0$ holds by field equation itself.

**Example 40** *Show that the energy-momentum tensor density defined by*

$$T_{\mu\nu} = -\frac{\partial\phi}{\partial x_\nu}\cdot\frac{\partial\mathcal{L}}{\partial(\partial\phi/\partial x_\mu)} + \delta_{\mu\nu}\mathcal{L}.$$

*where $\mathcal{L}$ is the Lagrangian-density for a scalar field $\phi$, satisfies*

$$\frac{\partial}{\partial x_\mu}T_{\mu\nu} = 0$$

***Solution:*** We have

$$\frac{\partial\mathcal{L}}{\partial\phi} = \frac{\partial}{\partial x_\mu}\cdot\left(\frac{\partial\mathcal{L}}{\partial(\partial\phi/\partial x_\mu)}\right)$$

from Euler's Lagrangian equation.
Since

$$\mathcal{L} = \mathcal{L}\left(\phi;\frac{\partial\phi}{\partial x_\mu}\right),$$

so we get

$$\frac{\partial\mathcal{L}}{\partial x_\nu}$$

equals to

$$\frac{\partial\mathcal{L}}{\partial x_\nu} = \frac{\partial\mathcal{L}}{\partial\phi}\cdot\frac{\partial\phi}{\partial x_\nu} + \frac{\partial\mathcal{L}}{\partial(\partial\phi/\partial x_\mu)}\frac{\partial}{\partial x_\nu}\left(\frac{\partial\phi}{\partial x_\mu}\right) \tag{1}$$

For $(\partial T_{\mu\nu}/\partial x_\mu)$ we have

$$\begin{aligned}\frac{\partial}{\partial x_\mu}T_{\mu\nu} &= -\frac{\partial}{\partial x_\mu}\left(\frac{\partial\phi}{\partial x_\nu}\right)\cdot\frac{\partial\mathcal{L}}{\partial(\partial\phi/\partial x_\mu)} - \frac{\partial\phi}{\partial x_\nu}\frac{\partial}{\partial x_\mu}\left(\frac{\partial\mathcal{L}}{\partial(\partial\phi/\partial x_\mu)}\right) + \frac{\partial\mathcal{L}}{\partial x_\mu}\delta_{\mu\nu} \\ &= -\left(\frac{\partial\mathcal{L}}{\partial(\partial\phi/\partial x_\mu)}\right)\cdot\frac{\partial}{\partial x_\nu}\left(\frac{\partial\phi}{\partial x_\mu}\right) - \frac{\partial\phi}{\partial x_\nu}\frac{\partial\mathcal{L}}{\partial\phi} + \frac{\partial\mathcal{L}}{\partial x_\nu}\end{aligned}$$

If we use (1) for $(\partial\mathcal{L}/\partial x_\nu)$ then

$$\frac{\partial}{\partial x_\mu}T_{\nu\mu} = 0.$$

# References


1. **Analytical Dynamics** : *E.T. Whittaker* (Cambridge. Univ. Press)
2. **Classical Dynamics** : *E.C.G. Sudarshan* and *N. Mukunda.*
3. **Classical Mechanic** : *E.W. Leech* (Mellenen and Co. Ltd.)
4. **Classical Mechanics** : *H. Goldstein* (Addison-Wesley publishing Company Inc.)
5. **Classical Mechanics** : *L.D. Landau E.M. Lifshitz* (Pargamon Press)
6. **Classical Mechanics** : *T.W.B. Kibble* (Mc Graw-Hill Publications)
7. *B. Boissean* and *G. Barrabes* : Jour Math. Phys. 20, 2058, (1979)
8. *J.A. Hannay* J. Phys : A 18, 221(1985)
9. *M.V. Berry* J. Phys : A-18, 15, (1985)
10. *S.N. Biswas, K.K. Datta* and *S.R. Chandhury* : International J. Phys. A-9, 2603, (1994).

# Index